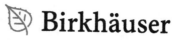

The Theory of Nilpotent Groups

Anthony E. Clement • Stephen Majewicz
Marcos Zyman

The Theory of Nilpotent Groups

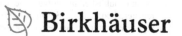 Birkhäuser

Anthony E. Clement
Department of Mathematics
CUNY-Brooklyn College
Brooklyn, NY, USA

Stephen Majewicz
Mathematics and Computer Science
CUNY-Kingsborough Community College
Brooklyn, NY, USA

Marcos Zyman
Department of Mathematics
CUNY-Borough of Manhattan
   Community College
New York, NY, USA

ISBN 978-3-319-88196-6   ISBN 978-3-319-66213-8   (eBook)
DOI 10.1007/978-3-319-66213-8

Printed on acid-free paper

This book is published under the trade name Birkhäuser, www.birkhauser-science.com
The registered company is Springer International Publishing AG
The registered company address is: Gewerbestrasse 11, 6330 Cham, Switzerland

For Gilbert Baumslag, in memoriam.

Preface

Our foremost objective in writing this book was to present a reasonably self-contained treatment of the classical theory of nilpotent groups so that the reader may later be able to study further topics and perhaps undertake research on his or her own. We have also included some recent work by two of the authors.

The theorems and proofs that appear in this work can be found elsewhere in some shape or form, but they are scattered in the literature. We have tried to include some of the omitted details in the original sources and to offer additional computations and explanations whenever we found it appropriate and useful. It is our hope that the examples, constructions, and computations included herein will contribute to a general understanding of both the theory of nilpotent groups and some of the techniques commonly used to study them. With this in mind, we have attempted to produce a single volume that can either be read from cover to cover or used as a reference. This was our main motivation several years ago, when the idea of writing such a book began to materialize.

We expect both working mathematicians and graduate students to benefit from reading this book. We demand from the reader only a solid advanced undergraduate or beginning graduate background in algebra. In particular, we assume that the reader is familiar with groups, rings, fields, modules, and tensor products. We expect the reader to know about direct and semi-direct products. We also assume knowledge about free groups and presentations of groups. Some topology is certainly useful for Chapter 6.

We declare that the choice of topics is based on what we consider to be a coherent discussion of nilpotent groups and mostly responds to our own mathematical interests. We emphasize that some of the more recent developments in nilpotent group theory (especially from the algorithmic, geometric, and model-theoretic perspectives) have been completely excluded and are well-suited topics for future volumes. Furthermore, major results such as the solution of the isomorphism problem for finitely generated nilpotent groups and the Mal'cev correspondence are only mentioned or discussed briefly.

We adopt certain conventions and notations. We sometimes write "1" for the trivial group, the group identity, and the unity of a ring. All functions and morphisms are written on the left unless otherwise told. For example, we write $\varphi(x)$, rather than $x\varphi$ or x^φ. The $n \times n$ identity matrix is always denoted by I_n. Whenever we have a field of characteristic zero, we identity its prime subfield with the rationals. All rings mentioned in this book are associative.

The book is organized in the following manner. In Chapter 1, we discuss the commutator calculus. Chapter 2 is meant to serve as an introduction to nilpotent groups and includes some interesting examples. Chapter 3 deals with the collection process and basic commutators, leading to normal forms in finitely generated free groups and free nilpotent groups. In Chapter 4, we show that finitely generated nilpotent groups are polycyclic, allowing us to obtain another type of normal form in such groups. Chapter 5 is about the theory of isolators, root extraction, and localization. Chapter 6 is a discussion of a classical paper by S. A. Jennings regarding the group ring of finitely generated torsion-free nilpotent groups over a field of characteristic zero. Finally, Chapter 7 contains a selection of additional topics.

We take full responsibility for any errors, mathematical or otherwise, appearing in this work. We have made every effort to accurately cite all pertinent works and do apologize for any omissions.

Brooklyn, NY, USA   Anthony E. Clement
Brooklyn, NY, USA   Stephen Majewicz
New York, NY, USA   Marcos Zyman
October 2017

Acknowledgments

First and foremost, we are deeply indebted to the late Professor Gilbert Baumslag, for the inspiration, encouragement, and support he gave us all.

AEC: I want to express my sincere thanks to the following family members who provided encouragement and emotional support throughout the writing of this book: my wife April Mojica-Clement, my children Mathias and Zayli Clement, and my parents Sir Martin and Lady Margaret Clement, as well as my extended family.

SM: I would like to thank Mrs. Jeanne DeVoy and Mrs. Anne Magliaccio, secretaries in the Department of Mathematics and Computer Science at Kingsborough Community College, CUNY, for assisting me beyond their regular duties at the college.

I thank Ms. Lydia Fischbach for her assistance and in particular for her diligence in proofreading several versions of the manuscript.

Many thanks go to Mr. Igor Melamed, instructor of mathematics at Kingsborough, for translating several articles written in Russian.

Most importantly, I thank my wife Annemarie Majewicz for her patience and support throughout the years.

Lastly, I thank my parents John (in memoriam) and Madeline Majewicz, as well as the rest of my family, for their support.

MZ: I would like to thank Professors Mahmood Sohrabi, Alexei Miasnikov, and Robert Gilman for hosting me at Stevens Institute of Technology during the academic year 2015–2016 and for their input and encouragement. I thank Professor Gretchen Ostheimer for many insightful conversations. I am grateful to Professor Joseph Roitberg (in memoriam), who sparked my interest in nilpotent groups.

Many thanks go to the Department of Mathematical Sciences at Stevens for facilitating me with office space and a pleasant working environment during my sabbatical leave from CUNY in beautiful Hoboken, New Jersey.

Last, but by no means least, I thank my beloved family—my wife Adriana Pérez and my kids María and Clara—for their constant support and patience over the last several years. I also thank my parents Sarita and Salomón and my siblings Jacobo, Sami, and Jayele for always being there.

Contents

Notations

g^h	The conjugate of g by h
g^{-h}	The conjugate of g^{-1} by h
$Z(G)$	The center of G
$Hom(G, H)$	The group of homomorphisms from G to H
$Aut(G)$	The automorphism group of G
$\ker \varphi$	The kernel of φ
$im \varphi$	The image of φ
$G \cong H$	G is isomorphic to H
$G \times H$	The direct product of G and H
$Inn(G)$	The group of inner automorphisms of G
S_n	The symmetric group on the set $\{1, 2, \ldots, n\}$
A_n	The alternating group on the set $\{1, 2, \ldots, n\}$
D_n	The dihedral group of order $2n$
$\mathscr{H}$	The Heisenberg group
$C_G(X)$	The centralizer of a nonempty subset X of a group G
$[G : H]$	The index of H in G
$[g, h]$	The commutator of g and h
$[g_1, \ldots, g_n]$	Simple commutator of weight $n > 1$
$gp(S)$	The subgroup generated by S
$[X_1, X_2]$	The commutator subgroup of X_1 and X_2
$[X_1, \ldots, X_n]$	$[[X_1, \ldots, X_{n-1}], X_n]$
$Ab(G)$	The abelianization of G
$N_G(S)$	The normalizer of S in G
S^T	The subgroup of G generated by all conjugates of elements of S by elements of T
S^H	The normal closure of S in $gp(S, H)$
$\gamma_i G$	The ith lower central subgroup of G
$\zeta_i G$	The ith upper central subgroup of G
Q	The quaternion group
$\langle X \mid Y \rangle$	The group presented by generators X and relators Y

$G \rtimes_{\varphi} H$	The semi-direct product of G and H
$G \wr H$	The wreath product of G and H
$\mathbb{Z}_p$	The cyclic group of order p
$\mathbb{Z}_{p^\infty}$	The Prüfer p-group (or p-quasicyclic group)
$UT_n(R)$	The group of $n \times n$ upper unitriangular matrices over R
I_n	The $n \times n$ identity matrix
$\tau_P(G)$	The set of P-torsion elements of G
$\tau(G)$	The set of torsion elements of G
$|g|$	The order of g; the length of a groupoid element g
$A \otimes B$	The tensor product of A and B
D_∞	The infinite dihedral group
G^n	The subgroup of G generated by the nth powers
A rep a	The rep operation
$lr(X)$	The Lie ring generated by X
$\bigoplus_{j=0}^{\infty} R_j$	The direct sum of R_j
$R[[x_1, \ldots, x_n]]$	The Magnus power series ring in the variables $x_1, \ldots, x_n$ over R.
$\tau_i(\overline{x})$	The ith Hall-Petresco word
G^R	The R-completion of G with respect to a Mal'cev basis
G^*	The Mal'cev completion of G
$H \leq_R G$	H is an R-subgroup of G
$H \trianglelefteq_R G$	H is a normal R-subgroup of G
$gp_R(S)$	The R-subgroup generated by S
$[H_1, H_2]_R$	The commutator R-subgroup of H_1 and H_2
$[H_1, \ldots, H_n]_R$	$[[H_1, \ldots, H_{n-1}]_R, H_n]_R$
$G \cong_R H$	G is R-isomorphic to H
$\tau_\omega(G)$	The set of ω-torsion elements of G
G_π	The π-primary component of G
$ann(g)$	The annihilator of g
G^α	The R-subgroup of G R-generated by αth powers
$I_P(S, G)$	The P-isolator of S in G
$\vartheta(g_1, \ldots, g_n)$	The value of the word ϑ at the r-tuple $(g_1, \ldots, g_n)$
$W(G)$	The verbal subgroup of G
$\vartheta(H_1, \ldots, H_n)$	The generalized verbal subgroup
$H \underset{P}{\sim} K$	H and K are P-equivalent
$\varrho_P(G)$	The maximal P-radicable subgroup of G
$\varrho(G)$	The maximal radicable subgroup of G
G^{p^∞}	$\bigcap_{n=1}^{\infty} G^{p^n}$
$\mathbb{Z}_P$	The ring $\{m/n \in \mathbb{Q} \mid n \neq 0 \text{ is a } P'\text{-number}\}$
RG	The group ring of G over R
$A_R(G), A(G)$	The augmentation ideal of RG
$D_n(R, G), D_n(G)$	The nth dimension subgroup of G over R
$\overline{\gamma}_n G$	The isolator of $\gamma_n G$ in G
$GL(V)$	The group of all invertible R-module endomorphisms

$GL_n(R)$	The group of non-singular $n \times n$ matrices over R
$M_n(\mathbb{Z})$	The ring of $n \times n$ matrices with integral entries
(U, W)	$\mathrm{span}\{(u, w) \mid u \in U,\ w \in W\}$, where U and W are Lie subalgebras
$\overline{FG}$	The completion of FG in the A-adic topology
$\overline{A}$	The completion of the augmentation ideal in the A-adic topology
$1 + \overline{A}$	The group $\{1 + a \mid a \in \overline{A}\}$
$\overline{D}_k$	The group $1 + \overline{A}^{\,k}$
exp	The exponential map
log	The logarithmic map
$\overline{\Lambda} = \mathscr{L}(\overline{A})$	The Lie algebra $\overline{A}$ under commutation
$\log G$	The set $\{\log g \mid g \in G\}$
$\mathscr{L}_F(G)$	The Lie algebra of G over F
$g \sim h\ (g \nsim h)$	g is conjugate (not conjugate) to h
$UT_n^m(R)$	The normal subgroup of $UT_n(R)$ consisting of those matrices whose $m-1$ superdiagonals have 0's in their entries
E_{ij}	The $n \times n$ matrix with 1 in the (i,j) entry and 0's elsewhere
$t_{i,\,j}(r)$	The transvection $I + rE_{ij}$
$Hol(G)$	The holomorph of G
$IA(G)$	The IA-group of G
$\Phi(G)$	The Frattini subgroup of G
$\mathbb{F}_p$	The finite field with p elements
$Fit(G)$	The Fitting subgroup of G

Chapter 1
Commutator Calculus

In this chapter, we introduce the commutator calculus. This is one of the most important tools for studying nilpotent groups. In Section 1.1, the center of a group and other notions surrounding the concept of commutativity are defined. Several results and examples involving central subgroups and central elements are given. Section 1.2 contains the fundamental identities related to commutators of group elements. By definition, the commutator of two elements g and h in a group G is the element $[g, h] = g^{-1}h^{-1}gh$. Clearly, $[g, h] = 1$ whenever g and h commute. This leads to a natural connection between central elements and trivial commutators. The commutator identities allow us to develop properties of commutator subgroups. This is the main focus of Section 1.3.

1.1 The Center of a Group

The commutator calculus is an essential tool which is used for working with nilpotent groups. In this section, we collect various results on commutators which will be used throughout the book. This material can be found in various places in the literature (see [1–6]).

1.1.1 Conjugates and Central Elements

We begin by defining the conjugate of a group element.

Definition 1.1 Let g and h be elements of a group G. The *conjugate* of g by h, denoted by g^h, is the element $h^{-1}gh$ of G.

© Springer International Publishing AG 2017
A.E. Clement et al., *The Theory of Nilpotent Groups*,
DOI 10.1007/978-3-319-66213-8_1

The conjugate of g^{-1} by h is written as g^{-h}. Notice that

$$g^{-h} = \left(g^{-1}\right)^h = h^{-1}g^{-1}h = \left(h^{-1}gh\right)^{-1} = \left(g^h\right)^{-1}.$$

Furthermore, if $k \in G$, then

$$(gh)^k = k^{-1}ghk = (k^{-1}gk)(k^{-1}hk) = g^k h^k$$

and

$$\left(g^h\right)^k = \left(h^{-1}gh\right)^k = k^{-1}h^{-1}ghk = (hk)^{-1}g(hk) = g^{hk}.$$

We summarize these in the next lemma.

Lemma 1.1 *Suppose that g, h, and k are elements of any group. Then* $(gh)^k = g^k h^k$, $\left(g^{-1}\right)^h = \left(g^h\right)^{-1}$, *and* $\left(g^h\right)^k = g^{hk}$.

The notion of conjugacy extends to subgroups in a natural way.

Definition 1.2 Two subgroups H and K of a group G are called *conjugate* if $g^{-1}Hg = K$ for some $g \in G$.

In particular, every normal subgroup of G is conjugate to itself.

Definition 1.3 Let G be a group. An element $g \in G$ is called *central* if it commutes with every element of G. The set of all central elements of G is called the *center* of G and is denoted by $Z(G)$. Thus,

$$Z(G) = \{g \in G \mid gh = hg \text{ for all } h \in G\}$$
$$= \{g \in G \mid g^h = g \text{ for all } h \in G\}.$$

It is easy to verify that $Z(G)$ is a normal abelian subgroup of G, and the conjugate of a central element $g \in G$ by any element of G is just g itself.

If G and H are groups, then the (internal and external) direct product of G and H will be written as $G \times H$.

Lemma 1.2 *If G_1 and G_2 are groups, then* $Z(G_1 \times G_2) = Z(G_1) \times Z(G_2)$.

Proof Suppose that $(g_1, g_2) \in Z(G_1 \times G_2)$. Then $(g_1, g_2)(x, y) = (x, y)(g_1, g_2)$ for all $(x, y) \in G_1 \times G_2$. This implies that $(g_1x, g_2y) = (xg_1, yg_2)$, and thus $g_1x = xg_1$ and $g_2y = yg_2$. Hence, $g_1 \in Z(G_1)$ and $g_2 \in Z(G_2)$. Therefore, (g_1, g_2) is contained in $Z(G_1) \times Z(G_2)$. And so, $Z(G_1 \times G_2) \subseteq Z(G_1) \times Z(G_2)$. In a similar way, one can show that $Z(G_1) \times Z(G_2) \subseteq Z(G_1 \times G_2)$.   □

Lemma 1.3 *If G_1 and G_2 are any two groups, then*

$$\frac{G_1 \times G_2}{Z(G_1 \times G_2)} \cong \frac{G_1}{Z(G_1)} \times \frac{G_2}{Z(G_2)}.$$

Proof The map from $G_1 \times G_2$ to $(G_1/Z(G_1)) \times (G_2/Z(G_2))$ defined by

$$(g_1, g_2) \mapsto (g_1 Z(G), g_2 Z(G))$$

is a surjective homomorphism whose kernel is $Z(G_1 \times G_2)$. The result follows from the First Isomorphism Theorem.   □

Let G and H be any two groups. The set of homomorphisms from G to H will be denoted by $Hom(G, H)$, and the group of automorphisms of G by $Aut(G)$. The kernel and image of $\varphi \in Hom(G, H)$ are abbreviated as $ker\ \varphi$ and $im\ \varphi = \varphi(G)$ respectively. If G and H are isomorphic groups, then we write $G \cong H$.

Let G be a group and $h \in G$. Using Lemma 1.1, it is easy to show that the map

$$\varphi_h : G \to G \text{ defined by } \varphi_h(g) = g^h$$

is contained in $Aut(G)$.

Definition 1.4 The map φ_h is called the *conjugation map* or *inner automorphism* induced by h.

It is easy to see that the set of all inner automorphisms of G forms a group under composition. This group is denoted by $Inn(G)$. There is a natural connection between the center of a group and the inner automorphisms of the group.

Theorem 1.1 *Let G be a group and $h \in G$. The map*

$$\varrho : G \to Aut(G) \text{ defined by } \varrho(h) = \varphi_h, \text{ where } \varphi_h(g) = g^h,$$

is a homomorphism with $ker\ \varrho = Z(G)$ *and* $im\ \varrho = Inn(G)$.

Proof The result follows from Lemma 1.1.   □

By Theorem 1.1 and the First Isomorphism Theorem, we have:

Corollary 1.1 *If G is any group, then* $G/Z(G) \cong Inn(G)$.

1.1.2 Examples Involving the Center

In the next few examples, we give the center of various groups.

Example 1.1 A group G is abelian if and only if $Z(G) = G$.

Example 1.2 Let S_n be the symmetric group on the set $S = \{1, 2, \ldots, n\}$, and let "e" denote the identity element of S_n. Clearly, S_1 has trivial center because $S_1 = \{e\}$. Furthermore, $Z(S_2) = S_2$ since S_2 is abelian.

We show that $Z(S_n) = \{e\}$ for $n > 2$. Suppose, on the contrary, that $Z(S_n)$ is nontrivial. Let $\sigma \in Z(S_n)$ be a nonidentity element. There exist distinct elements $a, b \in S$ such that $\sigma(a) = b$. Choose an element $c \in S$ different from a and b, and let

τ be the transposition $(b\ c)$. A direct calculation shows that $(\sigma \circ \tau)(a) \neq (\tau \circ \sigma)(a)$, contradicting the assumption that σ is in the center of $Z(S_n)$.

Example 1.3 Let A_n be the alternating group on the set $S = \{1,\ 2,\ \ldots,\ n\}$. This is the subgroup of S_n consisting of all even permutations. Note that $A_1 = A_2 = \{e\}$, and A_3 is cyclic since it has order 3. Thus, $Z(A_n) = A_n$ for $n = 1, 2,$ and 3 according to Example 1.1.

The center of A_4 is trivial. The proof is similar to the one used in Example 1.2. Assume that $Z(A_n)$ is nontrivial, and let σ be a nonidentity element of $Z(A_n)$. There exist distinct elements $a,\ b \in S$ such that $\sigma(a) = b$. Choose two elements c and d in S different from a and b, and let $\tau = (b\ c\ d)$. It is easy to see that $(\sigma \circ \tau)(a) \neq (\tau \circ \sigma)(a)$, contradicting the assumption that the center is nontrivial.

Using the same argument as above, one can show that A_n has trivial center whenever $n \geq 5$. We provide an alternative proof which uses the fact that A_n is a simple group whenever $n \geq 5$. Since this is the case, either $Z(A_n) = \{e\}$ or $Z(A_n) = A_n$. If it were true that $Z(A_n) = A_n$, then A_n would be abelian by Example 1.1. However, a quick calculation shows that

$$(1\ 2\ 3)(3\ 4\ 5) \neq (3\ 4\ 5)(1\ 2\ 3).$$

Thus, A_n is non-abelian and $Z(A_n) \neq A_n$. We conclude that $Z(A_n) = \{e\}$ for $n \geq 5$.

Example 1.4 Let D_n be the dihedral group of order $2n$, the group of isometries of the plane which preserve a regular n-gon. If y is a reflection across a line through a vertex and x is the counterclockwise rotation by $2\pi/n$ radians, then the elements of D_n are

$$1,\ x,\ x^2,\ \ldots,\ x^{n-1},\ y,\ xy,\ x^2y,\ \ldots,\ x^{n-1}y,$$

and the equalities

$$x^n = 1,\ y^2 = 1,\ \text{and } xy = yx^{-1}$$

hold in D_n.

Both D_1 and D_2 are abelian, so $Z(D_1) = D_1$ and $Z(D_2) = D_2$. We determine $Z(D_n)$ when $n \geq 3$. Since $xy = yx^{-1}$, we have

$$x^r y = yx^{-r} \quad (r \in \mathbb{Z}). \tag{1.1}$$

We claim that no element of the form $x^t y$ for any $t \in \{0,\ 1,\ \ldots,\ n-1\}$ is central. Assume, on the contrary, that $x^t y \in Z(D_n)$ for some such t. Then $x^t y$ commutes with x. Hence, $x^{-1}(x^t y)x = x^t y$, and thus $x^{t-1}yx = x^t y$. Applying (1.1) to both sides of this equality yields $yx^{1-t}x = yx^{-t}$. After canceling the y's, we get $x^{2-t} = x^{-t}$. This means that $x^2 = 1$, a contradiction. Therefore, $x^t y \notin Z(D_n)$ for any

$t \in \{0, 1, \ldots, n-1\}$. Consequently, an element of $Z(D_n)$ must take the form x^t for some $t \in \{0, 1, \ldots, n-1\}$. Clearly, $x^0 = 1 \in Z(D_n)$.

Suppose $x^t \in Z(D_n)$ for some $t \in \{1, \ldots, n-1\}$. By (1.1), we have

$$yx^t = x^t y = yx^{-t}.$$

Hence, $x^t = x^{-t}$; that is, $x^{2t} = 1$. Since x has order n, it must be that n divides $2t$. Hence, there exists $k \in \mathbb{N}$ such that $2t = nk$. If $k \geq 2$, then $2t \geq 2n$. This cannot happen since $1 \leq t \leq n-1$. This means that $k = 1$, and thus $2t = n$. Now, if n is odd, then no such t exists. We conclude that $Z(D_n)$ is trivial when n is odd. If n is even, then $t = \frac{n}{2}$, and consequently, $x^{n/2} \in Z(D_n)$. Therefore, $Z(D_n)$ is the cyclic group of order 2 generated by $x^{n/2}$ when n is even.

Example 1.5   Let $\mathcal{H}$ be the group of 3×3 upper unitriangular matrices over $\mathbb{Z}$ with the group operation being matrix multiplication. Thus,

$$\mathcal{H} = \left\{ \begin{pmatrix} 1 & a_{12} & a_{13} \\ 0 & 1 & a_{23} \\ 0 & 0 & 1 \end{pmatrix} \;\middle|\; a_{ij} \in \mathbb{Z} \right\}.$$

This group is called the *Heisenberg group*. The identity element in $\mathcal{H}$ is clearly the 3×3 identity matrix and will be denoted by I_3. It is easy to show that

$$Z(\mathcal{H}) = \left\{ \begin{pmatrix} 1 & 0 & c \\ 0 & 1 & 0 \\ 0 & 0 & 1 \end{pmatrix} \;\middle|\; c \in \mathbb{Z} \right\}.$$

1.1.3   Central Subgroups and the Centralizer

Definition 1.5   A subgroup H of a group G is called *central* if $H \leq Z(G)$.

Related to the center of a group is the centralizer of a subset of a group.

Definition 1.6   The *centralizer* of a nonempty subset X of a group G is

$$C_G(X) = \{g \in G \mid g^{-1}xg = x \text{ for all } x \in X\}.$$

It is easy to verify that $C_G(X)$ is a subgroup of G. If $X = \{x\}$, then we write $C_G(x)$ for the centralizer of x. Clearly,

$$C_G(G) = \bigcap_{x \in G} C_G(x) = Z(G).$$

Notice that $C_G(x)$ is just the stabilizer of x under the action of G on itself by conjugation. The orbit of x under this action, called the *conjugacy class* of x, is the set $\{g^{-1}xg \mid g \in G\}$. When G is finite, we get the *class equation* of G :

$$|G| = |Z(G)| + \sum_k [G : C_G(x_k)], \tag{1.2}$$

where one x_k is chosen from each conjugacy class containing at least two elements. Here, $|G|$ stands for the order of G and $[G : H]$ is the index of a subgroup H in G. These notations are standard and will be used throughout the book. We will also write $|g|$ for the order of an element $g \in G$.

1.1.4 The Center of a p-Group

Definition 1.7 Let p be any prime. A group G is called a *p-group* if every element of G has order a power of p.

Finite p-groups are the building blocks of finite groups. The next fact regarding their central structure is important in the study of finite groups.

Theorem 1.2 *If G is a nontrivial finite p-group for some prime p, then $Z(G) \neq 1$.*

Proof Suppose that $|G| = n$. Consider the class equation (1.2) of G. If $x_k \in G$ is not central for some $1 \leq k \leq n$, then $C_G(x_k)$ is a proper subgroup of G. Hence, $[G : C_G(x_k)]$ is a positive power of p. Consequently, each summand in the sum

$$\sum_k [G : C_G(x_k)]$$

is divisible by p. Since p divides $|G|$ by hypothesis, p also divides $|Z(G)|$. Therefore, $Z(G)$ contains nontrivial elements.   □

Remark 1.1 It is important to emphasize that G must be finite in Theorem 1.2. An infinite p-group does not necessarily have nontrivial center. This notion is discussed in Remark 2.8.

1.2 The Commutator of Group Elements

One can determine whether or not two group elements commute by calculating their commutator.

Definition 1.8 Let g and h be elements of a group G. The *commutator* of g and h, written as $[g, h]$, is

$$[g, h] = g^{-1}h^{-1}gh = g^{-1}g^h.$$

Clearly, g and h commute if and only if $[g, h] = 1$. Thus, the center of G can also be characterized as

$$Z(G) = \{g \in G \mid [g, h] = 1 \text{ for all } h \in G\}.$$

Definition 1.9 Let $S = \{g_1, g_2, \ldots, g_n\}$ be a set of elements of a group G. A *simple commutator*, or *left-normed commutator*, of *weight* $n \geq 1$ is defined recursively as follows:

1. The simple commutators of weight 1 are the elements of S, written as $g_j = [g_j]$.
2. The simple commutators of weight $n > 1$ are $[g_1, \ldots, g_n] = [[g_1, \ldots, g_{n-1}], g_n]$.

We collect some commutator identities which are of utmost importance.

Lemma 1.4 *Let* x, y, *and* z *be elements of a group* G.

(i) $xy = yx[x, y]$.
(ii) $x^y = x[x, y]$.
(iii) $[x, y] = [y, x]^{-1}$.
(iv) $[x, y]^z = [x^z, y^z]$.
(v) $[xy, z] = [x, z]^y[y, z] = [x, z][x, z, y][y, z]$.
(vi) $[x, yz] = [x, z][x, y]^z = [x, z][x, y][x, y, z]$.
(vii) $\left[x, y^{-1}\right] = \left([x, y]^{y^{-1}}\right)^{-1}$.
(viii) $\left[x^{-1}, y\right] = \left([x, y]^{x^{-1}}\right)^{-1}$.

Proof

(i) $xy = yx\left(x^{-1}y^{-1}xy\right) = yx[x, y]$.

(ii) $x^y = y^{-1}xy = x\left(x^{-1}y^{-1}xy\right) = x[x, y]$.

(iii) $[x, y] = x^{-1}y^{-1}xy = \left(y^{-1}x^{-1}yx\right)^{-1} = [y, x]^{-1}$.

(iv) We have

$$[x, y]^z = z^{-1}\left(x^{-1}y^{-1}xy\right)z$$

$$= \left(z^{-1}x^{-1}z\right)\left(z^{-1}y^{-1}z\right)\left(z^{-1}xz\right)\left(z^{-1}yz\right)$$

$$= \left(z^{-1}xz\right)^{-1}\left(z^{-1}yz\right)^{-1}\left(z^{-1}xz\right)\left(z^{-1}yz\right)$$

$$= [x^z, y^z].$$

(v) Observe that

$$[xy, z] = (xy)^{-1}z^{-1}xyz$$
$$= y^{-1}x^{-1}z^{-1}xyz$$
$$= y^{-1}\left(x^{-1}z^{-1}xz\right)y\left(y^{-1}z^{-1}yz\right)$$
$$= y^{-1}[x, z]y[y, z]$$
$$= [x, z]^y[y, z]$$
$$= [x, z][x, z, y][y, z] \text{ by (ii)}.$$

A similar computation gives (vi). By (vi), we have

$$1 = \left[x, \, yy^{-1}\right] = \left[x, \, y^{-1}\right][x, \, y]^{y^{-1}}. \tag{1.3}$$

This establishes (vii), and (viii) follows from (v) in a similar way.   □

Lemma 1.5 (The Hall-Witt Identities) *If x, y, and z are elements of a group, then*

$$\left[x, \, y^{-1}, \, z\right]^y\left[y, \, z^{-1}, \, x\right]^z\left[z, \, x^{-1}, \, y\right]^x = 1$$

and

$$\left[x, \, y, \, z^x\right]\left[z, \, x, \, y^z\right]\left[y, \, z, \, x^y\right] = 1.$$

Proof By Lemma 1.4 (iii), we have

$$\left[x, \, y^{-1}, \, z\right]^y = y^{-1}\left[\left[x, \, y^{-1}\right], \, z\right]y$$
$$= y^{-1}\left[x, \, y^{-1}\right]^{-1}z^{-1}\left[x, \, y^{-1}\right]zy$$
$$= y^{-1}\left[y^{-1}, \, x\right]z^{-1}\left[x, \, y^{-1}\right]zy$$
$$= x^{-1}y^{-1}xz^{-1}x^{-1}yxy^{-1}zy$$
$$= \left(xzx^{-1}yx\right)^{-1}yxy^{-1}zy.$$

Similarly,

$$\left[y, \, z^{-1}, \, x\right]^z = \left(yxy^{-1}zy\right)^{-1}zyz^{-1}xz$$

and

$$\left[z, x^{-1}, y\right]^x = \left(zyz^{-1}xz\right)^{-1}xzx^{-1}yx.$$

It follows that $\left[x, y^{-1}, z\right]^y \left[y, z^{-1}, x\right]^z \left[z, x^{-1}, y\right]^x = 1$. One can prove the other identity in a similar way. □

1.3 Commutator Subgroups

The notion of the commutator of elements of a group can be generalized to the commutator of subsets of a group.

Definition 1.10 Let G be a group with subset $S = \{s_1, s_2, \ldots\}$. The subgroup of G *generated* by S, denoted by

$$gp(S) = gp(s_1, s_2, \ldots),$$

is the smallest subgroup of G containing S. We call S a set of *generators* for $gp(S)$.

The subgroup $gp(S)$ of G can be obtained by taking the intersection of all subgroups of G that contain S. A typical element of $gp(S)$ is of the form

$$s_{i_1}^{\varepsilon_1} s_{i_2}^{\varepsilon_2} \cdots s_{i_n}^{\varepsilon_n},$$

where $s_{i_j} \in S$ and $\varepsilon_j \in \{-1, 1\}$ for $1 \leq j \leq n$. If $g \in G$, then $gp(g)$ is just the cyclic subgroup of G generated by g. If $S_1, \ldots, S_n$ are subsets of G, then the subgroup $gp(S_1 \bigcup \cdots \bigcup S_n)$ is written as $gp(S_1, \ldots, S_n)$.

Definition 1.11 Let X_1 and X_2 be nonempty subsets of a group G. The *commutator subgroup* of X_1 and X_2 is defined as

$$[X_1, X_2] = gp\left([x_1, x_2] \mid x_1 \in X_1, x_2 \in X_2\right).$$

Thus, $[X_1, X_2]$ is the subgroup of G generated by *all* commutators $[x_1, x_2]$, where x_1 varies over X_1 and x_2 varies over X_2. In particular, $[G, G] = G'$ is the *commutator subgroup* or *derived subgroup* of G.

Remark 1.2 The *set* of all commutators

$$S = \{[x_1, x_2] \mid x_1 \in X_1, x_2 \in X_2\}$$

does not necessarily form a subgroup of G. For instance, $[x_1, x_2]^{-1}$ may not be in S for some $[x_1, x_2] \in S$.

If $X_1 = X_2 = G$, then the inverse of every element of S is contained in S by
Lemma 1.4 (iii). However, it may be that S is not a subgroup of G because the
product of two or more commutators in S is not necessarily a commutator in S.
Consider, for example, the *special linear group* $SL_2(\mathbb{R})$ whose elements are the
2×2 matrices with real entries and determinant 1 (the group operation is matrix
multiplication). Let I_2 denote the 2×2 identity matrix, and set

$$A = \begin{pmatrix} 1 & 0 \\ -1 & 1 \end{pmatrix} \quad \text{and} \quad B = \begin{pmatrix} 1 & 1 \\ 0 & 1 \end{pmatrix}.$$

A routine check shows that $-I_2 = (ABA)^2$,

$$A = \left[\begin{pmatrix} 1 & 0 \\ \frac{4}{3} & 1 \end{pmatrix}, \begin{pmatrix} \frac{1}{2} & 0 \\ 0 & 2 \end{pmatrix} \right], \quad \text{and} \quad B = \left[\begin{pmatrix} 2 & 0 \\ 0 & \frac{1}{2} \end{pmatrix}, \begin{pmatrix} 1 & \frac{4}{3} \\ 0 & 1 \end{pmatrix} \right].$$

Thus, $-I_2$ is a product of commutators. However, $-I_2$ is not the commutator of two
elements of $SL_2(\mathbb{R})$. To see this, assume, on the contrary, that $-I_2 = [C, D]$ for
some $C, D \in SL_2(\mathbb{R})$. Rewriting this gives $C^{-1}DC = -D$, and thus D and $-D$ are
similar matrices. Since the trace of a square matrix equals the trace of any matrix
similar to it, D and $-D$ have equal trace. Consequently, the trace of D equals 0. Since
the determinant of D equals 1, the characteristic polynomial of D is $f(\lambda) = \lambda^2 + 1$.
And so, D has eigenvalues $\pm i$. This means that D is similar to the matrix $\begin{pmatrix} 0 & 1 \\ -1 & 0 \end{pmatrix}$.

Without loss of generality, we may as well assume that $D = \begin{pmatrix} 0 & 1 \\ -1 & 0 \end{pmatrix}$. Suppose that

$C = \begin{pmatrix} a & b \\ c & d \end{pmatrix}$. Since $CD = -DC$ by assumption, a computation shows that $d = -a$
and $c = b$. Using the fact that C has determinant 1, it follows that $-a^2 - b^2 = 1$.
This contradicts the fact that $a, b \in \mathbb{R}$.

Definition 1.11 can be generalized. If $\{X_1, X_2, \ldots\}$ is a collection of nonempty
subsets of G, then

$$[X_1, \ldots, X_n] = [[X_1, \ldots, X_{n-1}], X_n],$$

where $n \geq 2$. Note that $[X_1, \ldots, X_n]$ contains all simple commutators of the form
$[x_1, \ldots, x_n]$, where $x_1 \in X_1, \ldots, x_n \in X_n$. Thus,

$$[X_1, \ldots, X_n] \geq gp([x_1, \ldots, x_n] \mid x_1 \in X_1, \ldots, x_n \in X_n).$$

However, $[X_1, \ldots, X_n]$ may not equal $gp([x_1, \ldots, x_n] \mid x_1 \in X_1, \ldots, x_n \in X_n)$ if
$n \geq 3$. For example (see [6]), consider the cyclic subgroups

$$H_1 = gp((1 \ 2)), \ H_2 = gp((2 \ 3)), \ \text{and} \ H_3 = gp((3 \ 4))$$

of the symmetric group S_4. A routine check confirms that $[H_1, H_2, H_3]$ equals A_4, while $gp([h_1, h_2, h_3] \mid h_1 \in H_1, h_2 \in H_2, h_3 \in H_3)$ equals $gp((1\ 3\ 4))$. Thus,

$$[H_1, H_2, H_3] \neq gp([h_1, h_2, h_3] \mid h_1 \in H_1, h_2 \in H_2, h_3 \in H_3).$$

Lemma 1.6 *Let G be any group.*

(i) *If $H \leq G$ and $[G, G] \leq H$, then $H \trianglelefteq G$ and G/H is abelian. Thus, $[G, G] \trianglelefteq G$ and $G/[G, G]$ is abelian.*

(ii) *If $N \trianglelefteq G$ and G/N is abelian, then $[G, G] \trianglelefteq N$.*

Thus, the commutator subgroup of a group is the smallest normal subgroup inducing an abelian quotient. The factor group $Ab(G) = G/[G, G]$ is called the *abelianization* of G.

Proof

(i) Let $g \in G$ and $h \in H$. By Lemma 1.4 (ii),

$$h^g = g^{-1}hg = h[h, g] \in H$$

because H contains $[G, G]$. Therefore, $g^{-1}Hg = H$, and thus H is normal in G. If g_1H and g_2H are elements of G/H, then

$$(g_1H)(g_2H) = g_1g_2H = g_2g_1[g_1, g_2]H = g_2g_1H = (g_2H)(g_1H)$$

by Lemma 1.4 (i). Therefore, G/H is abelian.

(ii) If $gN, hN \in G/N$, then $(gN)(hN) = (hN)(gN)$. Hence,

$$(gN)^{-1}(hN)^{-1}(gN)(hN) = N.$$

We thus have $g^{-1}h^{-1}gh = [g, h] \in N$. It follows that $[G, G] \trianglelefteq N$. □

Lemma 1.6 allows one to conveniently calculate the derived subgroup. This is illustrated in the next few examples.

Example 1.6 Any two elements of an abelian group G commute. Thus, $[G, G] = 1$.

Example 1.7 We compute the commutator subgroup of the alternating group A_n on the set $S = \{1, 2, \ldots, n\}$. Clearly, $[A_n, A_n] = \{e\}$ for $n = 1, 2, 3$ by Example 1.6.

We find the commutator subgroup of A_4. It is well known that A_4 contains a unique nontrivial normal subgroup

$$K = \{e, (1\ 2)(3\ 4), (1\ 3)(2\ 4), (1\ 4)(2\ 3)\},$$

which is an isomorphic copy of the Klein 4-group (see [1]). Since $[A_4 : K] = 3$, the quotient A_4/K is abelian. Therefore, $[A_4, A_4] \trianglelefteq K$, and thus $[A_4, A_4] = K$.

Lastly, we consider the case when $n \geq 5$. In this case, A_n is simple. Thus, the only normal subgroups of A_n are $\{e\}$ and A_n. Since A_n is not abelian, $[A_n, A_n] = A_n$.

Example 1.8 We find the commutator subgroup of the symmetric group S_n on the set $S = \{1, 2, \ldots, n\}$. By Example 1.6, $[S_n, S_n] = \{e\}$ for $n = 1, 2$.

In order to find $[S_n, S_n]$ for $n \geq 3$, we use the fact that A_n is a normal subgroup of index 2 in S_n, and thus S_n/A_n is an abelian group. First, we find $[S_3, S_3]$. Since S_3/A_3 is abelian, we know that $[S_3, S_3] \trianglelefteq A_3$. Furthermore, each element of A_3 can be written as a commutator of elements in S_3 (this is obvious for the identity permutation):

$$(1 \ 2 \ 3) = [(2 \ 3), (1 \ 3 \ 2)] \text{ and } (1 \ 3 \ 2) = [(2 \ 3), (1 \ 2 \ 3)].$$

Therefore, A_3 is contained in $[S_3, S_3]$, and consequently, $[S_3, S_3] = A_3$.

Next, we show that $[S_4, S_4] = A_4$. Let $(a \ b \ c)$ be any 3-cycle for some distinct elements $a, b, c \in S$. This 3-cycle can be written as a commutator of elements in S_4 as

$$(a \ b \ c) = [(a \ b), (a \ c \ b)].$$

It follows that $A_4 \leq [S_4, S_4]$ because A_4 is generated by 3-cycles. Since S_4/A_4 is abelian, $[S_4, S_4] \trianglelefteq A_4$. We conclude that $[S_4, S_4] = A_4$.

Finally, consider the case when $n \geq 5$. Once again, $[S_n, S_n] \trianglelefteq A_n$ because S_n/A_n is abelian. Since the only nontrivial normal subgroup of S_n is A_n, it must be that $[S_n, S_n] = A_n$.

Example 1.9 We find the derived subgroup of the dihedral group D_n. Recall from Example 1.6 that

$$D_n = \{1, x, x^2, \ldots, x^{n-1}, y, xy, x^2y, \ldots, x^{n-1}y\},$$

where

$$x^n = 1, \ y^2 = 1, \text{ and } xy = yx^{-1}. \tag{1.4}$$

It follows from the last equality in (1.4) that

$$x^r y = yx^{-r} \text{ and } x^r y = (x^r y)^{-1} \quad (r \in \mathbb{Z}). \tag{1.5}$$

Now, it is clear that $[D_1, D_1] = [D_2, D_2] = 1$ by Example 1.6 because D_1 and D_2 are abelian. We claim that $[D_n, D_n] = gp(x^2)$ for $n \geq 3$.

From this point on, suppose $n \geq 3$ and let r and s denote integers. Choose $x^{2r} \in gp(x^2)$, and observe that this element can be written as a commutator as follows:

$$x^{2r} = x^r y^{-1} yx^r = x^r y^{-1} x^{-r} y = [x^{-r}, y],$$

where the second equality is a consequence of (1.5). Thus, $gp\left(x^2\right) \le [D_n, D_n]$. To prove that $[D_n, D_n] \le gp\left(x^2\right)$, we use (1.4) and (1.5). Suppose that $[a, b] \in [D_n, D_n]$. There are four possible cases for a and b.

- If $a = x^r$ and $b = x^s$, then $[x^r, x^s] = 1 \in gp\left(x^2\right)$.
- If $a = x^r$ and $b = x^s y$, then

$$[x^r, x^s y] = x^{-r}\left(x^s y\right)^{-1} x^r x^s y = x^{-r} y^{-1} x^{-s} x^r x^s y$$
$$= x^{-r} y^{-1} x^r y = x^{-r} y^{-1} y x^{-r} = x^{-2r} \in gp\left(x^2\right).$$

- Suppose $a = x^r y$ and $b = x^s$. Then

$$[x^r y, x^s] = [x^s, x^r y]^{-1} \in gp\left(x^2\right)$$

by the previous case and Lemma 1.4 (iii).
- Suppose $a = x^r y$ and $b = x^s y$. Then

$$[x^r y, x^s y] = \left(x^r y\right)^{-1}\left(x^s y\right)^{-1} x^r y x^s y = x^r y x^s y x^r y x^s y$$
$$= x^r x^{-s} y y x^r x^{-s} y y = x^{2r-2s} \in gp\left(x^2\right).$$

It follows that $[D_n, D_n] \le gp\left(x^2\right)$. And so, $[D_n, D_n] = gp\left(x^2\right)$ for $n \ge 3$ as claimed. In fact, $[D_n, D_n] = gp\left(x^2\right) = gp\left(x\right)$ whenever $n \ge 3$ is odd.

Example 1.10 We show that the derived subgroup of the Heisenberg group $\mathcal{H}$ equals its center. By Example 1.5, the center of $\mathcal{H}$ is

$$Z(\mathcal{H}) = \left\{ \begin{pmatrix} 1 & 0 & c \\ 0 & 1 & 0 \\ 0 & 0 & 1 \end{pmatrix} \,\middle|\, c \in \mathbb{Z} \right\} = gp\left(\begin{pmatrix} 1 & 0 & 1 \\ 0 & 1 & 0 \\ 0 & 0 & 1 \end{pmatrix} \right). \tag{1.6}$$

Let

$$a = \begin{pmatrix} 1 & a_1 & a_2 \\ 0 & 1 & a_3 \\ 0 & 0 & 1 \end{pmatrix} \text{ and } b = \begin{pmatrix} 1 & b_1 & b_2 \\ 0 & 1 & b_3 \\ 0 & 0 & 1 \end{pmatrix}$$

be elements of $\mathcal{H}$. A simple calculation shows that

$$[a, b] = \begin{pmatrix} 1 & 0 & a_1 b_3 - b_1 a_3 \\ 0 & 1 & 0 \\ 0 & 0 & 1 \end{pmatrix}.$$

Hence, each commutator of elements of $\mathcal{H}$ is central, and thus $[\mathcal{H}, \mathcal{H}] \leq Z(\mathcal{H})$. In addition, the generator of $Z(\mathcal{H})$ in (1.6) is a commutator of elements of $\mathcal{H}$:

$$\begin{pmatrix} 1 & 0 & 1 \\ 0 & 1 & 0 \\ 0 & 0 & 1 \end{pmatrix} = \left[\begin{pmatrix} 1 & 1 & 0 \\ 0 & 1 & 0 \\ 0 & 0 & 1 \end{pmatrix}, \begin{pmatrix} 1 & 0 & 0 \\ 0 & 1 & 1 \\ 0 & 0 & 1 \end{pmatrix} \right].$$

It follows that $[\mathcal{H}, \mathcal{H}] = Z(\mathcal{H})$.

1.3.1   Properties of Commutator Subgroups

We collect several properties of commutator subgroups.

Definition 1.12 Let G be any group, and let S be a nonempty subset of G. The *normalizer* of S in G, denoted by $N_G(S)$, is

$$N_G(S) = \{g \in G \mid gS = Sg\}.$$

If H is a subgroup of G, then $N_G(H)$ is the largest subgroup of G in which H is normal. If K is another subgroup of G, then K *normalizes* H if $K \leq N_G(H)$. Clearly, $N_G(H) = G$ if and only if $H \trianglelefteq G$.

Theorem 1.3 *Let G be a group and $H \leq G$. Then $C_G(H) \trianglelefteq N_G(H)$ and the factor group $N_G(H)/C_G(H)$ is isomorphic to a subgroup of Aut(H).*

In particular, we obtain Corollary 1.1 when $H = G$.

Proof By Theorem 1.1, the map

$$\varrho : G \to Aut(H) \text{ defined by } \varrho(h) = \varphi_h, \text{ where } \varphi_h(g) = g^h,$$

is a homomorphism. Thus, $\varrho|_{N_G(H)}$, the restriction of ϱ to $N_G(H)$, is a homomorphism. It is easy to verify that $\varrho|_{N_G(H)}$ has kernel $C_G(H)$. The result follows from the First Isomorphism Theorem.   □

Proposition 1.1 *Let G be any group with subgroups H and K.*

(i) $[H, K] = [K, H]$.
(ii) $[H, K] \leq H$ *if and only if K normalizes H. In particular, $[H, G] < H$ if and only if $H \trianglelefteq G$.*
(iii) *If $H_1 < G$ and $K_1 < G$ such that $H_1 \leq H$ and $K_1 \leq K$, then $[H_1, K_1] \leq [H, K]$.*

We point out that (i) is valid for any two *subsets* H and K of G.

Proof

(i) By Lemma 1.4 (iii),

$$[H,\ K] = gp\ ([h,\ k] \mid h \in H,\ k \in K)$$
$$= gp\ ([k,\ h]^{-1} \mid h \in H,\ k \in K)$$
$$= [K,\ H].$$

(ii) If $[H,\ K] \le H$, then $[h,\ k] \in H$ for any $h \in H$ and $k \in K$. This means that $k^{-1}hk \in H$, and consequently, $k^{-1}Hk < H$. Similarly, we have $kHk^{-1} \le H$. Therefore, $k^{-1}Hk = H$; that is, $k \in N_G(H)$. Conversely, if $k \in N_G(H)$, then $[h,\ k] \in H$ for all $h \in H$. A routine check confirms that $[H,\ K] \le H$.

(iii) The proof is straightforward.   □

Definition 1.13 Let G be a group with $H \le G$, and let $A \subseteq Aut(G)$.

(i) If $\varphi(h) \in H$ for every $\varphi \in A$ and $h \in H$, then H is called *A-invariant*.
(ii) If H is $Aut(G)$-invariant, then H is called *characteristic* in G.
(iii) If every endomorphism of G restricts to an endomorphism of H, then H is *fully invariant*.

Clearly, every fully invariant subgroup must be characteristic. Furthermore, every characteristic subgroup is normal. We record this as a lemma.

Lemma 1.7 *Let G be any group. If H is a characteristic subgroup of G, then $H \trianglelefteq G$.*

Proof If H is a characteristic subgroup of G, then $\varphi(H) = H$ for every $\varphi \in Aut(G)$. In particular, $\varphi_g(H) = H$, where $g \in G$ and φ_g is the inner automorphism induced by g. Thus, $g^{-1}Hg = H$ for every $g \in G$; that is, $H \trianglelefteq G$.   □

The next property of characteristic subgroups will be useful later.

Lemma 1.8 *Let G be a group with subgroups H and K. If H is characteristic in K and $K \triangleleft G$, then $H \triangleleft G$.*

Proof Choose any element $g \in G$. Since $K \triangleleft G$, there is an endomorphism

$$\varphi_g : K \to K \text{ defined by } \varphi_g(x) = x^g.$$

It is easy to verify that $\varphi_g \in Aut(K)$. Since H is characteristic in K, $\varphi_g(H) = H$. And so, $g^{-1}Hg = H$. This is true for all $g \in G$ since g was arbitrarily chosen.   □

Proposition 1.2 *Let G and H be groups, and let G_1 and G_2 be subgroups of G.*

(i) *If $\theta \in Hom(G,\ H)$, then $\theta([G_1,\ G_2]) = [\theta(G_1),\ \theta(G_2)]$.*
(ii) *Let $A \subseteq Aut(G)$. If G_1 and G_2 are A-invariant, then $[G_1,\ G_2]$ is also A-invariant.*

Proof

(i) If $g_{1_j} \in G_1$, $g_{2_j} \in G_2$, and $\varepsilon_j \in \{-1, 1\}$ for $1 \leq j \leq k$, then

$$\theta \left(\prod_{j=1}^{k} \left[g_{1_j}, g_{2_j} \right]^{\varepsilon_j} \right) = \prod_{j=1}^{k} \theta \left(\left[g_{1_j}, g_{2_j} \right] \right)^{\varepsilon_j}$$

$$= \prod_{j=1}^{k} \theta \left(g_{1_j}^{-1} g_{2_j}^{-1} g_{1_j} g_{2_j} \right)^{\varepsilon_j}$$

$$= \prod_{j=1}^{k} \left[\theta \left(g_{1_j} \right)^{-1} \theta \left(g_{2_j} \right)^{-1} \theta \left(g_{1_j} \right) \theta \left(g_{2_j} \right) \right]^{\varepsilon_j}$$

$$= \prod_{j=1}^{k} \left[\theta \left(g_{1_j} \right), \ \theta \left(g_{2_j} \right) \right]^{\varepsilon_j}.$$

(ii) We show that $\varphi \left([G_1, G_2] \right) \leq [G_1, G_2]$ for any $\varphi \in A$. Let

$$\prod_{j=1}^{k} \left[g_{1_j}, g_{2_j} \right]^{\varepsilon_j} \in [G_1, G_2]$$

as above. Since G_1 and G_2 are A-invariant subgroups, a computation similar to (i) gives

$$\varphi \left(\prod_{j=1}^{k} \left[g_{1_j}, g_{2_j} \right]^{\varepsilon_j} \right) = \prod_{j=1}^{k} \left[\varphi \left(g_{1_j} \right), \ \varphi \left(g_{2_j} \right) \right]^{\varepsilon_j} \in [G_1, G_2].$$

This completes the proof.   □

It follows from Proposition 1.2 (i) that the derived subgroup of any group is always fully invariant.

Corollary 1.2 *Let G be a group and $N \trianglelefteq G$. If $H \leq G$ and $K \leq G$, then*

$$[HN/N, KN/N] = [H, K]N/N.$$

Proof If $\pi : G \to G/N$ is the natural homomorphism, then $\pi(H) = HN/N$ and $\pi(K) = KN/N$. Apply Proposition 1.2 (i).   □

Lemma 1.9 *Let G be a group, and suppose that $N \trianglelefteq G$ and $H < G$. Then $[H, G] \leq N$ if and only if $HN/N \leq Z(G/N)$.*

Proof Suppose that $HN/N \leq Z(G/N)$. If $h \in H$, then $(hN)(gN) = (gN)(hN)$ for any $g \in G$. This means that $[hN, gN] = N$. Since $[hN, gN] = [h, g]N$, it follows that $[h, g] \in N$. Consequently, every commutator of an element of H and an element of G is contained in N. It follows that $[H, G]$ is a subgroup of N.

Conversely, suppose that $[H, G] \leq N$ and let $hN \in HN/N$ and $gN \in G/N$. Since $[hN, gN] = [h, g]N$ and $[h, g] \in N$ by hypothesis, we have $[hN, gN] = N$. Thus, $hN \in Z(G/N)$. $\qquad\square$

Theorem 1.4 *Let G be a group. If H and K are normal subgroups of G, then $[H, K] \trianglelefteq G$ and $[H, K] \leq H \cap K$. In particular, every element of H commutes with every element of K whenever $H \cap K = 1$.*

Proof Suppose that $g \in G$ and $\prod_{i=1}^{n}[h_i, k_i]^{\varepsilon_i} \in [H, K]$, where $h_i \in H$, $k_i \in K$, and $\varepsilon_i \in \{-1, 1\}$. By Lemmas 1.1 and 1.4 (iv),

$$\left([h_1, k_1]^{\varepsilon_1}[h_2, k_2]^{\varepsilon_2}\cdots[h_n, k_n]^{\varepsilon_n}\right)^g = \left([h_1, k_1]^{\varepsilon_1}\right)^g\left([h_2, k_2]^{\varepsilon_2}\right)^g\cdots\left([h_n, k_n]^{\varepsilon_n}\right)^g$$

$$= \left(\left[h_1^g, k_1^g\right]^{\varepsilon_1}\right)\left(\left[h_2^g, k_2^g\right]^{\varepsilon_2}\right)\cdots\left(\left[h_n^g, k_n^g\right]^{\varepsilon_n}\right)$$

is contained in $[H, K]$ since H and K are normal in G. Thus, $[H, K] \trianglelefteq G$. Furthermore,

$$\prod_{i=1}^{n}[h_i, k_i]^{\varepsilon_i} = \prod_{i=1}^{n}\left(h_i^{-1}\left(k_i^{-1}h_ik_i\right)\right)^{\varepsilon_i} \in H$$

and

$$\prod_{i=1}^{n}[h_i, k_i]^{\varepsilon_i} = \prod_{i=1}^{n}\left(\left(h_i^{-1}k_i^{-1}h_i\right)k_i\right)^{\varepsilon_i} \in K.$$

Thus, $\prod_{i=1}^{n}[h_i, k_i]^{\varepsilon_i} \in H \cap K$, and therefore, $[H, K] \leq H \cap K$. $\qquad\square$

An easy induction argument gives:

Corollary 1.3 *If $G_1, \ldots, G_n$ are normal subgroups of a group G, then the subgroup $[G_1, \ldots, G_n]$ is normal in G.*

Lemma 1.10 *If H, K, and L are normal subgroups of a group G, then*

$$[HK, L] = [H, L][K, L] \text{ and } [H, KL] = [H, K][H, L].$$

Proof The result follows from Lemma 1.4 (v) and (vi), together with Theorem 1.4.   □

More generally, we have:

Lemma 1.11 *If* $\{G_1, \ldots, G_n, H_1, H_2\}$ *is a set of normal subgroups of a group* G, *then*

(i) $[G_1, \ldots, G_n, H_1 H_2] = \prod_{i=1}^{2} [G_1, \ldots, G_n, H_i]$;

(ii) $[H_1 H_2, G_1, \ldots, G_n] = \prod_{i=1}^{2} [H_i, G_1, \ldots, G_n]$;

(iii) $[G_1, \ldots, G_{m-1}, H_1 H_2, G_{m+1}, \ldots, G_n] =$
$\prod_{i=1}^{2} [G_1, \ldots, G_{m-1}, H_i, G_{m+1}, \ldots, G_n]$ *for* $1 < m < n$.

Proof

(i) Note that $[G_1, \ldots, G_n] \trianglelefteq G$ by Corollary 1.3. The result follows from Lemma 1.10.

(ii) Set $n = 2$. By Lemma 1.10,

$$[H_1 H_2, G_1, G_2] = [[H_1 H_2, G_1], G_2]$$
$$= [[H_1, G_1][H_2, G_1], G_2].$$

Now, $[H_1, G_1]$ and $[H_2, G_1]$ are normal in G by Theorem 1.4. Another application of Lemma 1.10 gives

$$[[H_1, G_1][H_2, G_1], G_2] = [H_1, G_1, G_2][H_2, G_1, G_2].$$

We iterate this procedure for any n to obtain the desired result.

(iii) Let $C = [G_1, \ldots, G_{m-1}]$. By Corollary 1.3, Lemma 1.10, and (ii) above, we have

$$[C, H_1 H_2, G_{m+1}, \ldots, G_n] = [[C, H_1 H_2], G_{m+1}, \ldots, G_n]$$
$$= [[C, H_1][C, H_2], G_{m+1}, \ldots, G_n]$$
$$= [[C, H_1], G_{m+1}, \ldots, G_n][[C, H_2], G_{m+1},$$
$$\ldots, G_n]$$
$$= [C, H_1, G_{m+1}, \ldots, G_n][C, H_2, G_{m+1}, \ldots,$$
$$G_n].$$

This completes the proof.   □

The next two lemmas pertain to central commutator subgroups.

Lemma 1.12 (P. Hall) *Let G be a group with subgroups H and K, and suppose that $[H, K] \leq Z(G)$. For any $a \in H$ and $b \in K$, the maps*

$$\varphi_a : K \to Z(G) \ \ defined\ by\ \ \varphi_a(k) = [a,\ k]$$

and

$$\varphi_b : H \to Z(G) \ \ defined\ by\ \ \varphi_b(h) = [h,\ b]$$

are homomorphisms.

Proof Suppose that k_1, $k_2 \in K$. By Lemma 1.4 (vi),

$$[a,\ k_1 k_2] = [a,\ k_2][a,\ k_1]^{k_2} = [a,\ k_2][a,\ k_1].$$

Therefore, φ_a is a homomorphism. In a similar way, one can show that φ_b is a homomorphism. $\qquad\qquad\square$

Lemma 1.13 *Let G be any group. If $[g, h] \in Z(G)$ for some g, $h \in G$ and $n \in \mathbb{Z}$, then*

$$[g^n,\ h] = [g,\ h]^n = [g,\ h^n].$$

Proof The result is obvious for $n = 0$ and $n = 1$, and Lemma 1.12 gives the result when $n \geq 2$. Suppose that $n < 0$. Since $[g, h]$ is central, so is $[g, h]^{-n}$. This, together with Lemma 1.4 (viii), implies

$$[g^n,\ h] = \left[(g^{-n})^{-1},\ h\right] = \left([g^{-n},\ h]^{g^n}\right)^{-1}$$

$$= \left(([g,\ h]^{-n})^{g^n}\right)^{-1} = ([g,\ h]^{-n})^{-1}$$

$$= [g,\ h]^n .$$

In a similar way, one can show that $[g, h^n] = [g, h]^n$. $\qquad\qquad\square$

1.3.2 The Normal Closure

Let S and T be nonempty subsets of a group G. Denote by S^T, the subgroup of G generated by all conjugates of elements of S by elements of T :

$$S^T = gp\left(t^{-1}st \ \middle|\ s \in S,\ t \in T\right).$$

It is easy to see that if $H \leq G$, then S^H is the smallest normal subgroup of $gp(S, H)$ containing S. We call S^H the *normal closure* of S in $gp(S, H)$.

We record some fundamental properties on normal closures and commutator subgroups.

Proposition 1.3 *Let G be a group with $H \leq G$ and $\emptyset \neq S \subseteq G$.*

(i) $S^H = gp\,(S, [S, H])$.

(ii) $[S, H]^H = [S, H]$.

(iii) *If $H = gp(T)$ for some $\emptyset \neq T \subseteq G$, then $[S, H] = [S, T]^H$ and $[H, S] = [T, S]^H$.*

Proof

(i) Note first that $S \subseteq S^H$ because $H \leq G$. Moreover, any generator of $[S, H]$ can be written as $[s, h] = s^{-1}s^h$ with $s \in S$ and $h \in H$. Thus, $gp\,(S, [S, H]) \leq S^H$. It follows from Lemma 1.4 (ii) that $S^H \leq gp\,(S, [S, H])$.

(ii) Since $H \leq G$, $[S, H] \subseteq [S, H]^H$. We establish the reverse inclusion. By definition and Lemma 1.1,

$$[S, H]^H = gp\left(x^h \,\middle|\, x \in [S, H],\ h \in H\right)$$

$$= gp\left([s, h_1]^{h_2} \,\middle|\, s \in S,\ h_1 \in H,\ h_2 \in H\right).$$

By Lemma 1.4 (vi),

$$[s, h_1]^{h_2} = [s, h_2]^{-1}[s, h_1 h_2].$$

Consequently, $[s, h_1]^{h_2} \in [S, H]$, and thus $[S, H]^H = [S, H]$.

(iii) It is enough to prove that $[S, H] = [S, T]^H$. First, observe that $[S, T] \leq [S, H]$ because $T \subseteq H$. This implies that $[S, T]^H \leq [S, H]^H$. By (ii), $[S, T]^H \leq [S, H]$. It suffices to show that $[s, h] \in [S, T]^H$ for any $s \in S$ and $h \in H$. Since $H = gp(T)$, we can write

$$h = t_1^{\varepsilon_1} t_2^{\varepsilon_2} \cdots t_m^{\varepsilon_m}$$

for $t_i \in T$ and $\varepsilon_i \in \{-1, 1\}$. The proof is done by induction on m. If $m = 1$ and $\varepsilon_1 = 1$, then $[s, t_1] \in [S, T]^H$. If $m = 1$ and $\varepsilon_1 = -1$, then $\left[s, t_1^{-1}\right] = \left([s, t_1]^{t_1^{-1}}\right)^{-1}$ is also contained in $[S, T]^H$.

Assume that the result is true for $m - 1$. If $m > 1$, then Lemma 1.4 (vi), together with induction, implies that

$$[s, h] = \left[s, t_1^{\varepsilon_1} t_2^{\varepsilon_2} \cdots t_{m-1}^{\varepsilon_{m-1}} t_m^{\varepsilon_m}\right] = \left[s, t_m^{\varepsilon_m}\right]\left[s, t_1^{\varepsilon_1} t_2^{\varepsilon_2} \cdots t_{m-1}^{\varepsilon_{m-1}}\right]^{t_m^{\varepsilon_m}}$$

is contained in $[S, T]^H$. □

Corollary 1.4 *If G is a group with $H \leq G$ and $K \leq G$, then $[H, K] \trianglelefteq gp(H, K)$.*

Proof By Proposition 1.3 (ii), $[H, K]^K = [H, K]$ and $[K, H]^H = [K, H]$. Hence,

$$[H, K]^K = [H, K] = [H, K]^H$$

by Proposition 1.1 (i). Consequently, both H and K normalize $[H, K]$. □

The next two corollaries follow from Proposition 1.3 (iii).

Corollary 1.5 *Let H and K be subgroups of a group G, and let S and T be nonempty subsets of G. If $H = gp(S)$ and $K = gp(T)$, then $[H, K] = \left([S, T]^H\right)^K$.*

Corollary 1.6 *If G is a group and $H_1, \ldots, H_n$ are normal subgroups of G, then*

$$[H_1, \ldots, H_n] = gp\left([h_1, \ldots, h_n] \mid h_i \in H_i \text{ for } i = 1, \ldots, n\right).$$

References

1. B. Baumslag, B. Chandler, *Schaum's Outline of Theory and Problems of Group Theory* (McGraw-Hill, New York, 1968)
2. P. Hall, *The Edmonton Notes on Nilpotent Groups*. Queen Mary College Mathematics Notes. Mathematics Department (Queen Mary College, London 1969). MR0283083
3. D.J.S. Robinson, *Finiteness Conditions and Generalized Soluble Groups. Part 1* (Springer, New York, 1972). MR0332989
4. D.J.S. Robinson, *A Course in the Theory of Groups*. Graduate Texts in Mathematics, vol. 80, 2nd edn (Springer, New York, 1996). MR1357169
5. J.S. Rose, *A Course on Group Theory* (Dover Publications, New York, 1994). MR1298629. Reprint of the 1978 original [Dover, New York; MR0498810 (58 #16847)]
6. M. Suzuki, *Group Theory. II.* Grundlehren der Mathematischen Wissenschaften, vol. 248 (Springer, New York, 1986). MR0815926. Translated from the Japanese

Chapter 2
Introduction to Nilpotent Groups

The aim of this chapter is to introduce the reader to the study of nilpotent groups. In Section 2.1, we define a nilpotent group, as well as the lower and upper central series of a group. Section 2.2 contains some classical examples of nilpotent groups. In particular, we prove that every finite p-group is nilpotent for a prime p. In Section 2.3, numerous properties of nilpotent groups are derived. For example, we prove that every subgroup of a nilpotent group is subnormal, and thus satisfies the so-called normalizer condition. Section 2.4 is devoted to the characterization of finite nilpotent groups. In Section 2.5, we use tensor products to show that certain properties of a nilpotent group are inherited from its abelianization. We focus on torsion nilpotent groups in Section 2.6. We prove that every finitely generated torsion nilpotent group must be finite, and that the set of torsion elements of a nilpotent group form a subgroup. Section 2.7 deals with the upper central series and its factors. Among other things, we illustrate how the center of a group influences the structure of the group.

2.1 The Lower and Upper Central Series

In this section, we define a nilpotent group and discuss the lower and upper central series of a group. First, we provide some standard terminology.

2.1.1 Series of Subgroups

Definition 2.1 Let G be a group. A *series* for G is a finite chain of subgroups

$$1 = G_0 \leq G_1 \leq \cdots \leq G_n = G.$$

© Springer International Publishing AG 2017
A.E. Clement et al., *The Theory of Nilpotent Groups*,
DOI 10.1007/978-3-319-66213-8_2

If the subgroups G_0, ..., G_n are distinct, then n is called the *length* of the series. The series is called *normal* if $G_i \trianglelefteq G$ for $0 \leq i \leq n$, and *subnormal* if $G_i \trianglelefteq G_{i+1}$ for $0 \leq i \leq n - 1$. The *factors* of a subnormal series are the quotients G_{i+1}/G_i for $0 \leq i \leq n - 1$.

Clearly, every normal series of a group is subnormal. On the other hand, not every subnormal series is normal. Consider, for example, the symmetric group S_4. Let $G_1 = \{e, (1\ 2)(3\ 4)\}$ and $G_2 = \{e, (1\ 2)(3\ 4), (1\ 3)(2\ 4), (1\ 4)(2\ 3)\}$. It can be shown that the series

$$\{e\} \trianglelefteq G_1 \trianglelefteq G_2 \trianglelefteq A_4 \trianglelefteq S_4$$

is subnormal. It is not normal, however, because G_1 is not a normal subgroup of S_4. Notice, for instance, that

$$(1\ 2\ 3\ 4)^{-1}(1\ 2)(3\ 4)(1\ 2\ 3\ 4) \notin G_1.$$

Definition 2.2 Let G_1, G_2, G_3, ... be a sequence of subgroups of a group G.

(i) If $G_i \leq G_j$ for $1 \leq i \leq j$, then

$$G_1 \leq G_2 \leq G_3 \leq \cdots \tag{2.1}$$

is an *ascending series* (or an *ascending chain of subgroups*).
(ii) If $G_i \geq G_j$ for $1 \leq i \leq j$, then

$$G_1 \geq G_2 \geq G_3 \geq \cdots \tag{2.2}$$

is a *descending series* (or a *descending chain of subgroups*).

An ascending series may not reach G. If it does, then we say that the series *terminates in G*. Similarly, a descending series which reaches the identity is said to *terminate in the identity*. If there exists an integer $m > 1$ such that $G_{m-1} \neq G_m$ and $G_m = G_{m+1} = G_{m+2} = \cdots$ in either (2.1) or (2.2), then the series is said to *stabilize in G_m*.

2.1.2 Definition of a Nilpotent Group

Definition 2.3 A group G is called *nilpotent* if it has a normal series

$$1 = G_0 \leq G_1 \leq \cdots \leq G_n = G \tag{2.3}$$

such that

$$G_{i+1}/G_i \leq Z(G/G_i)$$

for $i = 0, 1, \ldots, n - 1$. Such a series (2.3) is called a *central series* for G. The shortest length of all central series for G is called the *nilpotency class*, or simply the *class*, of G.

An equivalent definition of a central series which involves commutators is given in the next lemma.

Lemma 2.1 *Let G be a group with a series*

$$1 = G_0 \leq G_1 \leq \cdots \leq G_n = G. \tag{2.4}$$

The series (2.4) is central if and only if $[G_{i+1}, G] \leq G_i$ for $0 \leq i \leq n - 1$.

Proof If the series (2.4) is central, then setting $H = G_{i+1}$ and $N = G_i$ in Lemma 1.9 yields the desired result.

Conversely, suppose that $[G_{i+1}, G] \leq G_i$ for $1 \leq i \leq n - 1$. We claim that (2.4) is a normal series. Let $g \in G$ and $g_i \in G_i$ for some $i = 1, 2, \ldots, n$. By Lemma 1.4 (ii), we have

$$g_i^g = g_i[g_i, g] \in G_i G_{i-1} = G_i.$$

Thus $G_i \trianglelefteq G$. The rest follows from Lemma 1.9. □

The trivial group is regarded as a nilpotent group of class 0, and nontrivial abelian groups are nilpotent of class 1. To see why this is the case, suppose that G is a nontrivial abelian group. Since $Z(G) = G$, the series $1 < G$ is a central series for G of shortest length (simply take $G_1 = G$ in Definition 2.3). More examples of nilpotent groups are given in the next section.

The following lemma shows that the only nilpotent group with trivial center is the trivial group.

Lemma 2.2 *If G is a nontrivial nilpotent group, then $Z(G) \neq 1$.*

Proof Suppose that $1 = G_0 \leq G_1 \leq \cdots \leq G_n = G$ is a central series for G. There exists an integer $i \geq 0$ such that $G_i = 1$ and $G_{i+1} \neq 1$. Thus, $G_{i+1}/G_i \leq Z(G/G_i)$ becomes $G_{i+1} \leq Z(G)$. And so, $Z(G) \neq 1$. □

Remark 2.1 An important collection of groups which arises in many areas of research (Galois theory, for example) are solvable groups. A group G is *solvable* if it has a subnormal series

$$1 = G_0 \trianglelefteq G_1 \trianglelefteq \cdots \trianglelefteq G_n = G$$

such that G_{i+1}/G_i is abelian for $0 \leq i \leq n - 1$. Every nilpotent group is solvable since the series (2.3) is subnormal and each factor is abelian. On the other hand, not every solvable group is nilpotent. For example, S_3 is a solvable group because it has a series $1 \triangleleft A_3 \triangleleft S_3$ which is subnormal and has abelian factors. However, Lemma 2.2 shows that S_3 is not nilpotent because it has trivial center (refer to Example 1.2).

2.1.3 The Lower Central Series

One series which is fundamental in the study of nilpotent groups is the lower central series.

Definition 2.4 Let G be a group. The descending series

$$G = \gamma_1 G \geq \gamma_2 G \geq \cdots \tag{2.5}$$

recursively defined by $\gamma_{i+1} G = [\gamma_i G, G]$ for $i \in \mathbb{N}$ is called the *lower central series* of G. Its terms are called the *lower central subgroups* of G.

In particular, $\gamma_2 G = [G, G] = G'$ is the commutator (or derived) subgroup of G. By definition,

$$\gamma_i G = \underbrace{[G, \cdots, G]}_{i}$$

for $i \geq 2$. Thus, $\gamma_i G \trianglelefteq G$ for $i \geq 1$ by Corollary 1.3.

Remark 2.2 Let G be any group.

(i) If $\gamma_i G = 1$ for some $i \geq 1$, then $\gamma_{i+1} G = [\gamma_i G, G] = [1, G] = 1$. It follows by induction on j that $\gamma_j G = 1$ for all $j = i, i+1, \ldots$. In this case, the lower central series of G is a central series in the sense of Definition 2.3 (see Lemma 2.1).

(ii) If $\gamma_2 G = G$, then $\gamma_3 G = [\gamma_2 G, G] = [G, G] = \gamma_2 G = G$. Continuing this argument shows that $\gamma_j G = G$ for all $j \geq 1$.

In the next examples, we give the lower central subgroups of some groups.

Example 2.1 If G is an abelian group, then $[G, G] = 1$ (see Example 1.6). Thus, $\gamma_i G = 1$ for all $i \geq 2$ by Remark 2.2 (i).

Example 2.2 We find the lower central subgroups of A_n, the alternating group on $S = \{1, 2, \ldots, n\}$. By Example 2.1, $\gamma_i A_n = \{e\}$ for $n = 1, 2, 3$ and $i \geq 2$ since A_1, A_2, and A_3 are abelian.

It was shown in Example 1.7 that $[A_4, A_4] = K$. We claim that $\gamma_i A_4 = K$ for $i \geq 3$. It suffices to consider the case $i = 3$. We begin by noting that any nonidentity element of K can be written as

$$(a \ \ d)(b \ \ c) = [(a \ \ c \ \ b), (a \ \ b)(c \ \ d)],$$

where a, b, c, d are distinct elements of $S = \{1, 2, 3, 4\}$. Since A_4 is generated by 3-cycles, $K \leq [K, A_4]$. Consequently, $K = [K, A_4] = \gamma_3 A_4$ as claimed. We conclude that $\gamma_i A_4 = K$ for $i \geq 2$.

If $n \geq 5$, then each lower central subgroup of A_n equals A_n. This is a consequence of Example 1.7 and Remark 2.2 (ii).

This example illustrates that the lower central series (2.5) may not descend to the identity, and consequently, may not be a central series in the sense of Definition 2.3.

Example 2.3 As before, let S_n be the symmetric group on $S = \{1, 2, \ldots, n\}$. Clearly, $\gamma_i S_1 = \gamma_i S_2 = \{e\}$ for $i \geq 2$. We claim that $\gamma_i S_n = A_n$ for $i \geq 2$ and $n \geq 3$. Consider the case $n = 3$. By Example 1.8, $[S_3, S_3] = A_3$. It is easy to verify that

$$[(a \ c \ b), (a \ b)] = (a \ c \ b) \text{ for distinct } a, b, c \in S.$$

It follows that $\gamma_3 S_3 = [\gamma_2 S_3, S_3] = [A_3, S_3] = A_3$. And so, $\gamma_i S_3 = A_3$ for $i \geq 2$ as claimed.

Next, consider the case $n = 4$. We found that $[S_4, S_4] = A_4$ in Example 1.8. The computation given in the same example also shows that $A_4 \leq [S_4, A_4]$. Thus,

$$[S_4, A_4] \leq [S_4, S_4] = A_4 \leq [S_4, A_4].$$

We conclude that $A_4 = [S_4, A_4]$. Hence, $\gamma_3 S_4 = A_4$, and thus $\gamma_i S_4 = A_4$ for $i \geq 2$.

Finally, suppose that $n \geq 5$. We know that $[S_n, S_n] = A_n$ whenever $n \geq 5$ from Example 1.8. Furthermore, $A_n = [A_n, A_n]$ whenever $n \geq 5$ from Example 2.2. Thus,

$$A_n = [A_n, A_n] \leq [S_n, A_n] \leq [S_n, S_n] = A_n.$$

This implies that $\gamma_3 S_n = [S_n, A_n] = A_n$, and in general, $\gamma_i S_n = A_n$ for $i \geq 2$.

Example 2.4 We give the lower central subgroups of the dihedral group D_n. See Examples 1.4 and 1.9 for notations and [4] for details.

- If $n \geq 3$ is odd, then $\gamma_2 D_n = gp\left(x^2\right) = gp(x)$ and

$$\gamma_3 D_n = [\gamma_2 D_n, D_n] = [gp(x), D_n] = gp(x).$$

  Thus, $\gamma_i D_n = gp(x)$ for $i \geq 2$.
- If $n = 2^k m$, where $m \geq 3$ is odd and $k \geq 1$, then

$$\gamma_2 D_{2^k m} = gp\left(x^2\right), \ \gamma_3 D_{2^k m} = gp\left(x^4\right), \ \ldots, \ \gamma_i D_{2^k m} = gp\left(x^{2^{i-1}}\right)$$

  for $2 \leq i \leq k+1$. Since x^{2^k} has odd order m, $\gamma_i D_{2^k m} = gp\left(x^{2^k}\right)$ when $i \geq k+1$.
- If $n = 2^k$ for some $k > 1$, then

$$\gamma_2 D_{2^k} = gp\left(x^2\right), \ \gamma_3 D_{2^k} = gp\left(x^4\right), \ \ldots, \ \gamma_i D_{2^k} = gp\left(x^{2^{i-1}}\right)$$

  for $2 \leq i \leq k+1$. In particular, $\gamma_{k+1} D_{2^k} = gp\left(x^{2^k}\right) = 1$, and thus $\gamma_i D_{2^k} = 1$ for $i \geq k+1$ by Remark 2.2. This shows that the lower central series of D_{2^k} is central in the sense of Definition 2.3. Therefore, D_{2^k} is nilpotent.

Example 2.5 Consider the Heisenberg group. It was shown in Example 1.10 that $[\mathscr{H}, \mathscr{H}] = Z(\mathscr{H})$ or, equivalently, $\gamma_2\mathscr{H} = Z(\mathscr{H})$. Clearly,

$$\gamma_3\mathscr{H} = [Z(\mathscr{H}), \mathscr{H}] = I_3.$$

By Remark 2.2 (i), $\gamma_i\mathscr{H} = I_3$ for $i \geq 3$. Hence, the lower central series of $\mathscr{H}$ is central in the sense of Definition 2.3. Therefore, $\mathscr{H}$ is nilpotent.

We give some useful properties enjoyed by the lower central subgroups.

Lemma 2.3 *The lower central subgroups of a group are fully invariant (hence, characteristic).*

Proof Apply Proposition 1.2 (i) repeatedly.   □

Lemma 2.4 *If G is any group and $H \leq G$, then $\gamma_i H \leq \gamma_i G$ for each $i \in \mathbb{N}$.*

Proof The proof is done by induction on i. If $i = 1$, then the result is obvious. Assume that $\gamma_i H \leq \gamma_i G$ holds for $i > 1$. Then $\gamma_{i+1}H = [\gamma_i H, H] \leq [\gamma_i G, G] = \gamma_{i+1}G$.   □

Lemma 2.5 *Let G and K be groups. If $\varphi : G \to K$ is a homomorphism, then $\varphi(\gamma_i G) = \gamma_i(\varphi(G))$ for each $i \in \mathbb{N}$. Thus, $\varphi(\gamma_i G) \leq \gamma_i K$ with equality when φ is surjective.*

Proof The proof is done by induction on i. If $i = 1$, then

$$\varphi(\gamma_1 G) = \varphi(G) = \gamma_1(\varphi(G)).$$

Assume that $\varphi(\gamma_i G) = \gamma_i(\varphi(G))$ holds for $i > 1$. By Proposition 1.2 (i), we obtain

$$\varphi(\gamma_{i+1}G) = \varphi([\gamma_i G, G]) = [\varphi(\gamma_i G), \varphi(G)]$$
$$= [\gamma_i(\varphi(G)), \varphi(G)] = \gamma_{i+1}(\varphi(G)).$$

This completes the proof.   □

Corollary 2.1 *If G is a group and $N \trianglelefteq G$, then $\gamma_i(G/N) = (\gamma_i G)N/N$ for each $i \in \mathbb{N}$.*

Proof If $\pi : G \to G/N$ is the natural homomorphism, then

$$(\gamma_i G)N/N = \pi(\gamma_i G) = \gamma_i(\pi(G)) = \gamma_i(G/N)$$

by Lemma 2.5.   □

The lower central subgroups of a group can always be generated by a certain collection of simple commutators.

Lemma 2.6 *Let G be any group. For any $n \in \mathbb{N}$, we have*

$$\gamma_n G = gp\left([g_1, \ldots, g_n] \mid g_i \in G\right). \tag{2.6}$$

Furthermore, if X is a generating set of G, then $\gamma_n G$ is generated by all simple commutators of weight n or more in the elements of X and their inverses.

Proof Corollary 1.6 immediately gives (2.6). Suppose that $G = gp(X)$. Each element of G can be written as a product of the elements of X and their inverses. In particular, we may replace each g_i in a simple commutator $[g_1, \ldots, g_n] \in G$ of weight n by such a product. Since $\gamma_n G$ is generated by such simple commutators, the result follows from repeatedly applying Lemma 1.4.   □

Example 2.6 Let G be a group generated by x, y, and z, and consider the simple commutator $\left[x^{-1}y^2, \ z \right]$ of weight 2. By Lemma 2.6, this commutator can be expressed as a product of simple commutators of weight 2 or more in the elements of the set $\{ x, \ x^{-1}, \ y, \ y^{-1}, \ z, \ z^{-1} \}$. To see how this is done, we use Lemma 1.4 (v) to (vi) and get

$$\left[x^{-1}y^2, \ z \right] = \left[x^{-1}, \ z \right]\left[x^{-1}, \ z, \ y^2 \right]\left[y^2, \ z \right]$$

$$= \left[x^{-1}, \ z \right]\left[x^{-1}, \ z, \ y \right]\left[x^{-1}, \ z, \ y \right]\left[x^{-1}, \ z, \ y, \ y \right]\left[y, \ z \right]\left[y, \ z, \ y \right]\left[y, \ z \right].$$

2.1.4   The Upper Central Series

The upper central series plays a key role in the study of nilpotent groups. This series is constructed as follows:

Let G be any group. Set $\zeta_1 G = Z(G)$, and let $\pi_1 : G \to G/\zeta_1 G$ be the natural homomorphism of G onto $G/\zeta_1 G$. Define

$$\zeta_2 G = \pi_1^{-1}(Z(G/\zeta_1 G)),$$

so that $\zeta_2 G/\zeta_1 G = Z(G/\zeta_1 G)$. Observe that $\zeta_2 G \trianglelefteq G$ by the Correspondence Theorem.

Next, take $\pi_2 : G \to G/\zeta_2 G$ to be the natural homomorphism of G onto $G/\zeta_2 G$, and define

$$\zeta_3 G = \pi_2^{-1}(Z(G/\zeta_2 G)).$$

Thus, $\zeta_3 G/\zeta_2 G = Z(G/\zeta_2 G)$. As before, $\zeta_3 G \trianglelefteq G$. Continuing in this way, we obtain the subgroups of the upper central series of G.

Definition 2.5 Let G be any group. The ascending series

$$1 = \zeta_0 G \leq \zeta_1 G \leq \cdots \tag{2.7}$$

recursively defined by $\zeta_{i+1} G/\zeta_i G = Z(G/\zeta_i G)$ for $i \geq 0$ is called the *upper central series* of G, and its terms are called the *upper central subgroups* of G.

If $\pi_i : G \to G/\zeta_i G$ is the natural homomorphism of G onto $G/\zeta_i G$, then

$$\zeta_{i+1}G = \pi_i^{-1}(Z(G/\zeta_i G))$$
$$= \{g \in G \mid g\zeta_i G \text{ is central in } G/\zeta_i G\}$$
$$= \{g \in G \mid (g\zeta_i G)(h\zeta_i G) = (h\zeta_i G)(g\zeta_i G) \text{ for all } h \in G\}$$
$$= \{g \in G \mid [g, h] \in \zeta_i G \text{ for all } h \in G\}.$$

In particular, $\zeta_1 G$ is the center of G. By taking $N = \zeta_i G$ and $H = \zeta_{i+1}G$ in Lemma 1.9, we find that $[\zeta_{i+1}G, G] \le \zeta_i G$.

Remark 2.3 Let G be any group.

(i) If $\zeta_i G = G$ for some $i \ge 0$, then

$$\zeta_{i+1}G = \{g \in G \mid [g, h] \in \zeta_i G \text{ for all } h \in G\}$$
$$= \{g \in G \mid [g, h] \in G \text{ for all } h \in G\}$$
$$= G.$$

It follows by induction on j that $\zeta_j G = G$ for $j \ge i$. In this situation, the upper central series of G is a central series in the sense of Definition 2.3.

(ii) If $Z(G) = 1$, then

$$\zeta_2 G = \{g \in G \mid [g, h] \in Z(G) \text{ for all } h \in G\}$$
$$= \{g \in G \mid [g, h] = 1 \text{ for all } h \in G\}$$
$$= Z(G).$$

Thus, $\zeta_2 G = 1$. Continuing in this way, we find that $\zeta_j G = 1$ for $j \ge 0$.

We provide the upper central subgroups of some groups.

Example 2.7 If G is an abelian group, then $\zeta_1 G = G$. Thus, $\zeta_i G = G$ for all $i \ge 1$ by Remark 2.3 (i).

Example 2.8 If $n \ge 3$, then the upper central subgroups of S_n are trivial from Example 1.2 and Remark 2.3 (ii). The same is true for the upper central subgroups of A_n when $n > 3$ (see Example 1.3). This illustrates that the upper central series of a group does not necessarily ascend to the group.

Example 2.9 We find the upper central subgroups of D_n. The last two cases rely on the fact that $D_{2n}/Z(D_{2n})$ is isomorphic to D_n. See [4] for details.

• For all $i \ge 1$, $\zeta_i D_1 = D_1$ and $\zeta_i D_2 = D_2$ since D_1 and D_2 are abelian. This follows from Remark 2.3 (i).
• If $n \ge 3$ is odd, then D_n has trivial center (see Example 1.4). By Remark 2.3 (ii), $\zeta_i D_n = 1$ for all $i \ge 0$.

- If $n = 2^k m$, where $m \geq 3$ is odd and $k \geq 1$, then $\zeta_i D_{2^k m} = gp\left(x^{n/2^i}\right)$ for $1 \leq i \leq k$. In particular,

$$\zeta_k D_{2^k m} = gp\left(x^{n/2^k}\right) = gp\left(x^m\right).$$

It follows that $\zeta_i D_{2^k m} = gp\left(x^m\right)$ for $i \geq k$.

- If $n = 2^k$ for some $k > 1$, then $\zeta_i D_{2^k} = gp\left(x^{n/2^i}\right)$ for $1 \leq i \leq k-1$. In particular,

$$\zeta_{k-1} D_{2^k} = gp\left(x^{n/2^{k-1}}\right) = gp\left(x^2\right).$$

For $i = k$, we get $\zeta_k D_{2^k} = D_{2^k}$, and consequently, $\zeta_i D_{2^k} = D_{2^k}$ for $i \geq k$. Thus, the upper central series of D_{2^k} is central in the sense of Definition 2.3.

Example 2.10 We find the upper central subgroups of the Heisenberg group. By Example 1.10, we know that $Z(\mathcal{H}) = [\mathcal{H}, \mathcal{H}]$. Consequently,

$$\zeta_2 \mathcal{H} = \{g \in \mathcal{H} \mid [g, h] \in Z(\mathcal{H}) \text{ for all } h \in \mathcal{H}\} = \mathcal{H}.$$

By Remark 2.3, $\zeta_i \mathcal{H} = \mathcal{H}$ for $i \geq 2$. We conclude that $\gamma_1 \mathcal{H} = \mathcal{H} = \zeta_2 \mathcal{H}$, $\gamma_2 \mathcal{H} = \zeta_1 \mathcal{H}$, and $\gamma_3 \mathcal{H} = I_3 = \zeta_0 \mathcal{H}$. Thus, the upper and lower central series of $\mathcal{H}$ coincide.

The next lemma deals with epimorphic images of the upper central subgroups of a group.

Lemma 2.7 *If G and H are any groups and $\varphi : G \to H$ is an epimorphism, then $\varphi(\zeta_i G) \leq \zeta_i H$ for $i \geq 0$.*

Proof The proof is done by induction on i. The result is obviously true when $i = 0$. Suppose that $\varphi(\zeta_{i-1} G) \leq \zeta_{i-1} H$ for $i > 0$, and let $g \in \varphi(\zeta_i G)$. We claim that $g \in \zeta_i H$. Since $g \in \varphi(\zeta_i G)$, there exists $x \in \zeta_i G$ such that $g = \varphi(x)$. Suppose that h is any element of H. Since φ is an epimorphism, there exists $y \in G$ such that $h = \varphi(y)$. Now,

$$[g, h] = [\varphi(x), \varphi(y)] = \varphi([x, y])$$

and

$$[x, y] \in [\zeta_i G, G] \leq \zeta_{i-1} G.$$

By induction,

$$\varphi([x, y]) \in \varphi(\zeta_{i-1} G) \leq \zeta_{i-1} H.$$

Thus, $[g, h] \in \zeta_{i-1} H$ and $g \in \zeta_i H$. $\square$

If we put $G = H$ in Lemma 2.7, then we obtain:

Corollary 2.2 *The upper central subgroups of a group are characteristic.*

Remark 2.4 In contrast to the lower central subgroups, the upper central subgroups of a group are not necessarily fully invariant. For instance, let G be a nontrivial abelian group, and let H be a nontrivial group with trivial center. Suppose, in addition, that H contains a subgroup K which is isomorphic to G. We claim that the center of $G \times H$ is not fully invariant. Let $\pi : G \times H \to G$ be the standard projection map, and suppose that α is an isomorphism from G to K. By Lemma 1.2, $Z(G \times H) = G$. However, the endomorphism $\alpha \circ \pi$ of $G \times H$ clearly does not map G to itself. As a particular example, take $G = \mathbb{Z}_2$, $H = S_3$, and $K = \{e, (1\ \ 2)\} \cong G$.

2.1.5   Comparing Central Series

The upper central series of a nilpotent group ascends to the group faster than any other central series, whereas its lower central series descends to the identity faster than any other central series. This is highlighted in the next theorem.

Theorem 2.1 *If G is a nilpotent group with a (descending) central series*

$$G = G_1 \geq G_2 \geq \cdots \geq G_n \geq G_{n+1} = 1,$$

then $\gamma_i G \leq G_i$ and $G_{n-j+1} \leq \zeta_j G$ for $1 \leq i \leq n+1$ and $0 \leq j \leq n$.

Proof First, we prove that $\gamma_i G \leq G_i$. If $i = 1$, then $\gamma_1 G = G = G_1$. Let $i > 1$ and assume that $\gamma_{i-1} G \leq G_{i-1}$. By Proposition 1.1 (iii) and Lemma 2.1,

$$\gamma_i G = [\gamma_{i-1} G,\ G] \leq [G_{i-1},\ G] \leq G_i.$$

Next, we show that $G_{n-j+1} \leq \zeta_j G$. If $j = 0$, then $G_{n+1} = 1 = \zeta_0 G$. Let $j > 0$ and assume the result holds for $j - 1$. Lemma 2.1 now gives

$$\left[G_{n-j+1},\ G \right] \leq G_{n-j+2} \leq \zeta_{j-1} G.$$

By setting $H = G_{n-j+1}$ and $N = \zeta_{j-1} G$ in Lemma 1.9, we obtain

$$\left(G_{n-j+1} \zeta_{j-1} G \right) / \zeta_{j-1} G \leq Z \left(G / \zeta_{j-1} G \right) = \zeta_j G / \zeta_{j-1} G.$$

Thus, $G_{n-j+1} \leq \zeta_j G$. $\square$

Remark 2.5 By Theorem 2.1, we have

$$\gamma_{i+1}G \leq \zeta_{n-i}G \text{ for } 0 \leq i \leq n. \tag{2.8}$$

If G has nilpotency class c and we set $i = c-1$ and $n = c$ in (2.8), then $\gamma_c G \leq Z(G)$.

Remark 2.6 The proof of Theorem 2.1 shows that if

$$G = G_1 \geq G_2 \geq \cdots \geq G_n \geq \cdots$$

is any descending series such that $[G_i, G] \leq G_{i+1}$ for $i = 1, 2, \ldots$, then $\gamma_i G \leq G_i$.

Corollary 2.3 *Let G be a group. The following are equivalent:*

  (i) G is nilpotent of class at most c;
  (ii) $\gamma_{c+1}G = 1$;
  (iii) $\zeta_c G = G$;
  (iv) $[g_1, \ldots, g_{c+1}] = 1$ for all $g_i \in G$.

Proof The result follows from Theorem 2.1 and Lemma 2.6.   □

The next theorem is another consequence of Theorem 2.1. It shows that the lengths of the upper and lower central series (when finite) coincide with the nilpotency class of the group, and no other central series has smaller length.

Theorem 2.2 *Let G be a group. The following are equivalent:*

  (i) G is nilpotent of class $c \geq 1$;
  (ii) $\gamma_{c+1}G = 1$ and $\gamma_c G \neq 1$;
  (iii) $\zeta_c G = G$ and $\zeta_{c-1}G \neq G$.

Example 2.11 The dihedral group D_{2^k} ($k > 1$) has nilpotency class k (see Examples 2.4 and 2.9).

Example 2.12 The Heisenberg group has nilpotency class 2 (see Example 2.10).

We have seen that some groups coincide with their derived subgroup (refer to Example 1.7). This never happens for nontrivial nilpotent groups.

Corollary 2.4 *If G is a nontrivial nilpotent group, then $\gamma_2 G$ is a proper subgroup of G.*

Proof The proof is done by contradiction. If $\gamma_2 G = G$, then $\gamma_i G = G$ for all $i \geq 2$ by Remark 2.2 (ii). However, Theorem 2.2 implies that $\gamma_i G = 1$ for some i since G is nilpotent. Consequently, G must be trivial.   □

Remark 2.7 In fact, if G is a nontrivial nilpotent group and $N \neq 1$ is a normal subgroup of G, then $[N, G]$ is a proper subgroup of G.

2.2   Examples of Nilpotent Groups

In this section, we give more examples of nilpotent groups.

2.2.1   Finite p-Groups

A classical result in finite group theory is that finite p-groups are nilpotent.

Theorem 2.3 *Every finite p-group is nilpotent, where p is any prime.*

Proof We use the fact that the center of a finite p-group is itself a finite p-group. Let G be a finite p-group of order p^n for some $n \in \mathbb{N}$. By Theorem 1.2, $Z(G)$ is nontrivial, and thus $G/Z(G)$ is a finite p-group of order p^r for some $r \in \mathbb{N}$ with $r < n$. Invoking Theorem 1.2 again, we have that $G/Z(G)$ has nontrivial center. Hence, $Z(G/Z(G)) = \zeta_2 G/Z(G)$ is a p-group of order p^s for some $s \in \mathbb{N}$ with $s < r$. This means that $|Z(G)| < |\zeta_2 G|$, so $Z(G) < \zeta_2 G$. By iterating this procedure, we see that $|\zeta_i G| < |\zeta_{i+1} G|$ for $i \geq 0$, and thus $\zeta_i G$ is a proper subgroup of $\zeta_{i+1} G$ for $i \geq 0$. And so, the upper central series for G is strictly increasing. Since G is finite, the series must terminate at $\zeta_k G = G$ for some $k \in \mathbb{N}$. Therefore, G is nilpotent. $\square$

Example 2.13 The dihedral groups D_{2^n} for $n \geq 1$ are finite 2-groups, and thus nilpotent.

Example 2.14 The *quaternion group* Q is the group with presentation

$$Q = \left\langle x, y \mid x^4 = 1, \ x^2 = y^2, \ y^{-1}xy = x^{-1} \right\rangle.$$

The elements of Q are 1, x, x^2, x^3, y, xy, x^2y, and x^3y. Since Q has order $8 = 2^3$, it is nilpotent.

Example 2.15 If G and H are finite groups of orders m and n respectively, then both the direct product $G \times H$ and semi-direct product $G \rtimes_\varphi H$ by φ have order mn. In particular, the direct and semi-direct product of any two finite p-groups is itself a finite p-group.

An important construction of groups is the wreath product. Let A and T be any two groups. For each $s \in T$, let A_s be an isomorphic copy of A, and let a_s denote the isomorphic image of $a \in A$ in A_s. Consider the direct product $B = \prod_{s \in T} A_s$, and define the *standard (or restricted) wreath product* of A by T as

$$W = A \wr T = B \rtimes_\varphi T,$$

where $\varphi : T \to Aut(B)$ is the homomorphism that maps each $t \in T$ to $\varphi(t)$, where $\varphi(t)$ is the automorphism of B induced by the mapping

$$a_s \mapsto a_{st} \quad \text{for all } a \in A \text{ and } s, t \in T.$$

Thus, T acts on B by permuting its factors. This action can be realized as conjugation, so that $t^{-1}a_s t = a_{st}$ in W for all $a \in A$ and $s, t \in T$. A presentation for W is

$$W = A \wr T = \langle B, T \mid t^{-1}a_s t = a_{st} \ (a \in A, \ s, t \in T) \rangle.$$

We call W the *unrestricted wreath product* of A by T in case B is an unrestricted direct product of the A_s. In both situations, B is called the *base group*, A is the *bottom group*, and T is the *top group*.

Example 2.16 Suppose that A and T are finite groups of orders m and n, respectively. Using the notation above, we have that $B = \prod_{s \in T} A_s$ is a finite group of order m^n, and thus $A \wr T$ has order nm^n. In particular, if $|A| = p^m$ and $|T| = p^n$ for some prime p, then

$$|A \wr T| = p^n(p^m)^{p^n} = p^{n+mp^n}.$$

Thus, the wreath product of any two finite p-groups is a finite p-group.

Remark 2.8 In contrast to Theorem 2.3, an infinite p-group does not have to be nilpotent. In [2], G. Baumslag showed how to construct infinite p-groups which are not nilpotent using wreath products. Take a nontrivial p-group A and an infinite p-group B, and form the wreath product $W = A \wr B$. Clearly, W is an infinite group. By Corollary 3.2 of [2], W must have trivial center, and thus fails to be nilpotent by Lemma 2.2. Furthermore, W is a p-group since it is the wreath product of two p-groups (see [11]). Thus, W is an infinite p-group that is not nilpotent.

Two groups which are infinite p-groups that are not nilpotent are the wreath products $\mathbb{Z}_p \wr \mathbb{Z}_{p^\infty}$ and $\mathbb{Z}_{p^\infty} \wr \mathbb{Z}_{p^\infty}$, where $\mathbb{Z}_p$ is the cyclic group of order p and

$$\mathbb{Z}_{p^\infty} = \langle x_1, x_2, \ldots \mid px_1 = 0, \ px_{n+1} = x_n \text{ for } n = 1, 2, \ldots \rangle \qquad (2.9)$$

is an additively written presentation for the *Prüfer p-group* (or *p-quasicyclic group*).

2.2.2 An Example Involving Rings

Nilpotency in ring theory relates to nilpotency in group theory in a natural way.

Definition 2.6 Let R be a ring with unity 1, and let T be a subring of R. For any $k \in \mathbb{N}$, let T^k be the subring of T consisting of all finite sums of the form

$$\sum a_{p_1 \cdots p_k} x_{p_1} \cdots x_{p_k} \quad \left(a_{p_1 \cdots p_k} \in \mathbb{Z}, \ x_{p_1}, \ldots, x_{p_k} \in T \right).$$

If there exists a natural number m such that $T^m = \{0\}$, then T is termed a *nilpotent subring* of R.

Let R be as in Definition 2.6, and suppose that S is a nilpotent subring of R with $S^n = \{0\}$. Define

$$G = 1 + S = \{1 + x \mid x \in S\}.$$

Clearly, G is closed under multiplication since

$$(1 + x)(1 + y) = 1 + y + x + xy \in G$$

for all $x, y \in S$, and it is closed under inverses because

$$\left(1 + x\right)\left(1 - x + x^2 - x^3 + \cdots + (-1)^{n-1} x^{n-1}\right) = 1$$

for all $x \in S$. Thus, G is a subgroup of the group of units of R.

We claim that G is nilpotent of class at most $n - 1$. Let

$$G_i = 1 + S^i = \left\{1 + x \mid x \in S^i\right\} \text{ for } i = 1, 2, \ldots, n.$$

By the same argument as before, we find that G_i is a subgroup of G.

Consider the (descending) series

$$G = G_1 \geq G_2 \geq \cdots \geq G_n = 1. \tag{2.10}$$

We claim that (2.10) is a central series for G. By Lemma 2.1, it suffices to show that $[G_i, G] \leq G_{i+1}$ for $i = 1, 2, \ldots, n - 1$. Let $g = 1 + x \in G_i$ and $h = 1 + y \in G$, where $x \in S^i$ and $y \in S$. A straightforward computation gives

$$gh - hg = (1 + x)(1 + y) - (1 + y)(1 + x)$$
$$= xy - yx \in S^{(i+1)}.$$

Thus,

$$[g, h] = g^{-1} h^{-1} (gh - hg) + 1 \in S^{(i+1)} + 1 = G_{i+1},$$

and consequently, G is nilpotent of class at most $n - 1$ as claimed.

Example 2.17 Let R be a commutative ring with unity, and let T be the ring of $n \times n$ matrices over R. Let S be the subring of T consisting of all $n \times n$ matrices over R whose entries on and below the main diagonal are equal to zero. Thus,

$$S = \left\{ \begin{pmatrix} 0 & b_{12} & \ldots & b_{1n} \\ 0 & 0 & \ldots & b_{2n} \\ \vdots & \vdots & \ddots & \vdots \\ 0 & 0 & \ldots & 0 \end{pmatrix} \middle| b_{ij} \in R \right\}.$$

A direct computation shows that S^p consists of all elements of S whose first $p-1$ superdiagonals have zero entries. Thus, a typical matrix in S^p has the form

$$
\begin{pmatrix}
0 \cdots\cdots 0 & c_{1\,p+1} & c_{1\,p+2} & \cdots & c_{1n} \\
0 \cdots\cdots 0 & 0 & c_{2\,p+2} & \cdots & c_{2n} \\
0 \cdots\cdots 0 & 0 & \ddots & \ddots & \vdots \\
\vdots \cdots\cdots & \ddots & \ddots & \ddots & c_{p+1\,n} \\
\vdots \cdots\cdots & \ddots & \ddots & 0 & 0 \\
\vdots \cdots\cdots & \ddots & \ddots & \vdots & \vdots \\
0 \cdots\cdots & \cdots & \cdots & 0 & 0
\end{pmatrix},
$$

where $c_{ij} \in R$. In particular, S^n is the $n \times n$ zero matrix. Let

$$
UT_n(R) = \{I_n + M \mid M \in S\},
$$

where I_n is the $n \times n$ identity matrix (we use this notation throughout the book). It follows from the above that $UT_n(R)$ is a nilpotent group of class less than n, called the *(upper) unitriangular group of degree n over R*. A typical element of $UT_n(R)$ is an $n \times n$ upper unitriangular matrix of the form

$$
\begin{pmatrix}
1 & a_{12} & a_{13} & \cdots & \cdots & a_{1n} \\
0 & 1 & a_{23} & \cdots & \cdots & a_{2n} \\
0 & 0 & 1 & \cdots & \cdots & a_{3n} \\
\vdots & \vdots & \ddots & \ddots & \ddots & \vdots \\
\vdots & \vdots & \cdots & \cdots & \cdots & a_{n-1\,n} \\
0 & \cdots & \cdots & \cdots & 0 & 1
\end{pmatrix},
$$

where $a_{ij} \in R$. In particular, $UT_3(\mathbb{Z})$ is the Heisenberg group $\mathcal{H}$.

For more on nilpotent rings and nilpotent groups which arise from them, see [6].

2.3 Elementary Properties of Nilpotent Groups

In this section, we take a look at some fundamental results on nilpotent groups. The first one deals with subgroups and homomorphic images of nilpotent groups.

Theorem 2.4 *If G is a nilpotent group of class c, then every subgroup and homomorphic image of G is nilpotent of class at most c.*

Proof Suppose that H is a subgroup of G. By Lemma 2.4, $\gamma_i H \leq \gamma_i G$ for each $i \in \mathbb{N}$. Since G has nilpotency class c, $\gamma_{c+1} G = 1$ by Theorem 2.2. Thus, $\gamma_{c+1} H = 1$ and H is nilpotent of class at most c by Corollary 2.3.

Let K be any group and $\varphi \in Hom(G, K)$. By Lemma 2.5, $\varphi(\gamma_i G) = \gamma_i(\varphi(G))$ for each $i \in \mathbb{N}$. Since $\gamma_{c+1} G = 1$ and φ is a homomorphism,

$$1 = \varphi(\gamma_{c+1} G) = \gamma_{c+1}(\varphi(G)).$$

It follows from Corollary 2.3 that $\varphi(G)$ is nilpotent of class at most c. $\qquad\qquad$ □

Corollary 2.5 *If G is a nilpotent group of class c and $N \trianglelefteq G$, then G/N is nilpotent of class at most c.*

This is immediate from Theorem 2.4 since G/N is a homomorphic image of G. Note that Corollary 2.5 is also a consequence of Corollaries 2.1 and 2.3.

2.3.1 Establishing Nilpotency by Induction

Many of the theorems on nilpotent groups are proven using induction on the nilpotency class. The next few results are commonly used.

Lemma 2.8 *If G is a nilpotent group of class $c \geq 1$, then $G/\gamma_c G$ is nilpotent of class $c - 1$.*

Proof Let $\pi : G \to G/\gamma_c G$ be the natural homomorphism. By Corollary 2.5, $G/\gamma_c G$ is a nilpotent group of class at most c. Furthermore, for any $n \in \mathbb{N}$,

$$\gamma_n(G/\gamma_c G) = \pi(\gamma_n G) = \gamma_n G/\gamma_c G$$

by Lemma 2.5. In particular, $\gamma_{c-1}(G/\gamma_c G) = \gamma_{c-1} G/\gamma_c G \neq 1$ and $\gamma_c(G/\gamma_c G) = 1$. Thus, $G/\gamma_c G$ has nilpotency class $c - 1$ by Theorem 2.2. $\qquad\qquad$ □

Lemma 2.9 *Let G be a nilpotent group of class $c \geq 2$. For any element $g \in G$, the subgroup $H = gp(g, \gamma_2 G)$ is nilpotent of class less than c.*

Proof We prove that $\gamma_i H \leq \gamma_{i+1} G$ for $i \geq 2$ by induction on i. If $i = 2$, then

$$\gamma_2 H = gp\left([g^m h, g^n k] \mid h, k \in \gamma_2 G \text{ and } m, n \in \mathbb{Z}\right).$$

By Lemmas 1.1 and 1.4 (iv), (v), and (vi),

$$[g^m h, g^n k] = [g^m, g^n k]^h [h, g^n k]$$

$$= \left([g^m, k][g^m, g^n]^k\right)^h [h, g^n k]$$

$$= [g^m, k]^h [g^m, g^n]^{kh} [h, g^n k].$$

Now, $[g^m, k]^h$ and $[h, g^n k]$ are contained in $\gamma_3 G$ and $[g^m, g^n] = 1$. Therefore, $[g^m h, g^n k] \in \gamma_3 G$, and consequently, $\gamma_2 H \leq \gamma_3 G$.

If we assume that $\gamma_{i-1} H \leq \gamma_i G$ for $i > 2$, then

$$\gamma_i H = [\gamma_{i-1} H, H] \leq [\gamma_i G, H] \leq [\gamma_i G, G] = \gamma_{i+1} G.$$

Thus, $\gamma_i H \leq \gamma_{i+1} G$. In particular, $\gamma_c H \leq \gamma_{c+1} G = 1$. By Corollary 2.3, H has nilpotency class less than c. □

Lemma 2.10 *If G is any group, then $\zeta_n G/Z(G) \cong \zeta_{n-1}(G/Z(G))$ for any $n \in \mathbb{N}$.*

Proof The proof is done by induction on n. If $n = 1$, then the result is obviously true. Suppose that $\zeta_i G/Z(G) \cong \zeta_{i-1}(G/Z(G))$ for $2 \leq i \leq n - 1$. We claim that $\zeta_n G/Z(G) \cong \zeta_{n-1}(G/Z(G))$. By definition, $\zeta_n G/\zeta_{n-1} G = Z(G/\zeta_{n-1} G)$. By the Third Isomorphism Theorem,

$$\frac{\zeta_n G/Z(G)}{\zeta_{n-1} G/Z(G)} \cong Z\left(\frac{G/Z(G)}{\zeta_{n-1} G/Z(G)}\right). \tag{2.11}$$

By induction, $\zeta_{n-1} G/Z(G) \cong \zeta_{n-2}(G/Z(G))$. Substituting this in (2.11) yields

$$\frac{\zeta_n G/Z(G)}{\zeta_{n-2}(G/Z(G))} \cong Z\left(\frac{G/Z(G)}{\zeta_{n-2}(G/Z(G))}\right) = \frac{\zeta_{n-1}(G/Z(G))}{\zeta_{n-2}(G/Z(G))}.$$

The result follows. □

More generally, we have the next result of P. Hall.

Lemma 2.11 *If G is any group, then $\zeta_i(G/\zeta_j G) \cong \zeta_{i+j} G/\zeta_j G$ for $i, j \geq 0$.*

Proof The proof is done by induction on j. Lemma 2.10 settles the case for $j = 1$. Suppose that the lemma is true for $j > 1$. By the Third Isomorphism Theorem,

$$\frac{\zeta_{i+j+1} G}{\zeta_{j+1} G} \cong \frac{\zeta_{(i+1)+j} G/\zeta_j G}{\zeta_{j+1} G/\zeta_j G}.$$

By induction, $\zeta_{(i+1)+j} G/\zeta_j G \cong \zeta_{i+1}(G/\zeta_j G)$. Since $\zeta_{j+1} G/\zeta_j G$ is just $Z(G/\zeta_j G)$, we have

$$\frac{\zeta_{i+j+1} G}{\zeta_{j+1} G} \cong \frac{\zeta_{i+1}(G/\zeta_j G)}{Z(G/\zeta_j G)} \cong \zeta_i\left(\frac{G/\zeta_j G}{Z(G/\zeta_j G)}\right)$$

by Lemma 2.10. However,

$$\zeta_i\left(\frac{G/\zeta_j G}{Z(G/\zeta_j G)}\right) \cong \zeta_i\left(\frac{G}{\zeta_{j+1} G}\right)$$

by the Third Isomorphism Theorem. This completes the proof. □

Theorem 2.5 *Let G be a group, and suppose that $N \trianglelefteq G$. If $N \leq \zeta_i G$ for some $i \in \mathbb{N}$ and G/N is nilpotent, then G is nilpotent.*

Proof Consider the upper central series

$$1 = \zeta_0(G/\zeta_i G) \leq \zeta_1(G/\zeta_i G) \leq \cdots \qquad (2.12)$$

for $G/\zeta_i G$. By Lemma 2.11, $\zeta_k(G/\zeta_i G) \cong \zeta_{k+i} G/\zeta_i G$ for $k \geq 0$. Thus, (2.12) becomes

$$1 = \zeta_i G/\zeta_i G \leq \zeta_{i+1} G/\zeta_i G \leq \cdots . \qquad (2.13)$$

Since $G/\zeta_i G \cong (G/N)/(\zeta_i G/N)$ by the Third Isomorphism Theorem and G/N is nilpotent, then $G/\zeta_i G$ is nilpotent by Corollary 2.5. Thus, the series (2.13) terminates at $G/\zeta_i G$. Therefore, there exists an integer $n \geq i$ such that $\zeta_n G/\zeta_i G = G/\zeta_i G$, and hence, $\zeta_n G = G$. By Theorem 2.3, G is nilpotent. $\qquad\qquad\square$

If $N \leq Z(G)$ in Theorem 2.5, then the next theorem gives information about the nilpotency class of G.

Theorem 2.6 *Let G be a group, and suppose that $N \leq Z(G)$. If G/N is nilpotent of class c, then G is nilpotent of class either c or $c + 1$.*

Proof We first prove that if $gN \in \zeta_n(G/N)$ for any $g \in G$ and $n \geq 0$, then $g \in \zeta_{n+1} G$. If $n = 0$, then $\zeta_0(G/N) = N$. In this case, $gN \in \zeta_0(G/N) = N$, and thus $g \in N$. And so, g is central because $N \leq Z(G)$ by the hypothesis. Assume that $hN \in \zeta_{k-1}(G/N)$ implies $h \in \zeta_k G$ for $2 \leq k \leq n$, and let $gN \in \zeta_n(G/N)$. Since

$$[\zeta_n(G/N), \; G/N] \leq \zeta_{n-1}(G/N),$$

we have $[gN, \; hN] \in \zeta_{n-1}(G/N)$ for all $h \in G$. Thus, $[g, \; h] \in \zeta_n G$ by the induction hypothesis. Consequently, $g \in \zeta_{n+1} G$ as claimed.

Next, we prove that $G = \zeta_{c+1} G$. If $g \in G$, then $gN \in G/N = \zeta_c(G/N)$ by Theorem 2.2. This implies that $g \in \zeta_{c+1} G$ by our discussion above. Hence $G = \zeta_{c+1} G$. Now, if $\zeta_c G \neq G$, then G has nilpotency class $c + 1$ by Theorem 2.2. Suppose that $\zeta_c G = G$. If $\zeta_{c-1} G = G$, then G is of class $d \leq c - 1$ by Theorem 2.3. By Corollary 2.5, G/N is of class at most d. However, G/N is of class c by hypothesis. Thus, $c \leq d \leq c - 1$, which is false. It follows from Theorem 2.2 that $\zeta_{c-1} G \neq G$, and thus G is of nilpotency class c. $\qquad\qquad\square$

If $N = Z(G)$ in Theorem 2.6, then the nilpotency class can be determined.

Lemma 2.12 *A group G is nilpotent of class $c \geq 1$ if and only if $G/Z(G)$ is nilpotent of class $c - 1$.*

Proof We invoke Theorem 2.2 (iii). If G is nilpotent of class c, then $\zeta_c G = G$ and $\zeta_{c-1} G \neq G$. Thus,

$$\zeta_{c-1}(G/Z(G)) \cong \zeta_c G/Z(G) = G/Z(G)$$

and

$$\zeta_{c-2}(G/Z(G)) \cong \zeta_{c-1}G/Z(G) \neq G/Z(G)$$

by Lemma 2.10. Therefore, $G/Z(G)$ is of class $c - 1$. The converse is similar.   □

2.3.2 A Theorem on Root Extraction

We illustrate how Lemma 2.9 is used to prove a theorem on the extraction of roots in nilpotent groups by induction on the nilpotency class.

Definition 2.7 Let G be a group, and let P be a set of primes. A natural number n is called a *P-number* if every prime divisor of n belongs to P.

By convention, 1 is a P-number for any set of primes P. If P happens to be the empty set, then the only P-number is 1.

Definition 2.8 Let G be a group, and let P be a set of primes.

1. An element of G is called a *P-torsion element* if its order is a P-number. The set of P-torsion elements of G is denoted by $\tau_P(G)$. Thus,

$$\tau_P(G) = \{g \in G \mid g^n = 1 \text{ for some } P\text{-number } n\}.$$

2. If every element of G is P-torsion, then G is called a *P-torsion group.*
3. If G has no P-torsion elements other than the identity, then G is *P-torsion-free.*

If $P = \{p\}$, then a P-torsion group is just a p-group by Definition 1.7. If P is the set of all primes, then $\tau_P(G)$ is the set of all elements of finite order of G and is written as $\tau(G)$. Note that G is P-torsion-free whenever P is empty.

An element of $\tau(G)$ is called a *torsion element* of G, and G is a *torsion* (or *periodic*) group if $\tau(G) = G$. We say that G is *torsion-free* if it has no torsion elements other than the identity element.

The group properties "P-torsion" and "P-torsion-free" are preserved under extensions.

Definition 2.9 Let G, H, and N be groups.

(i) If $N \trianglelefteq G$ and $G/N \cong H$, then G is called an *extension* of H by N. Thus, there exists a short exact sequence

$$1 \to N \to G \to H \to 1.$$

(ii) An extension G of H by N is called *central* if $N \leq Z(G)$.
(iii) Let $N \trianglelefteq G$, and suppose that G is an extension of H by N. A property $\mathscr{Q}$ of groups is said to be *preserved under extensions* if G has property $\mathscr{Q}$ whenever both N and H have property $\mathscr{Q}$.

Lemma 2.13 *If P is a set of primes, then "P-torsion" and "P-torsion-free" are preserved under extensions.*

Proof Let G be a group with $N \trianglelefteq G$.

- Suppose that N and G/N are P-torsion, and let $g \in G$. Since G/N is P-torsion, the element $gN \in G/N$ has order a P-number n. Thus, $(gN)^n = N$, or equivalently, $g^n \in N$. Since N is also P-torsion, there exists a P-number m such that $(g^n)^m = 1$; that is, $g^{nm} = 1$. Since nm is a P-number, G is P-torsion.
- Suppose that N and G/N are P-torsion-free. Let $g \in G$ such that $g^n = 1$ for some P-number n. Then $(gN)^n = N$ in G/N. Since G/N is P-torsion-free, $gN = N$; that is, $g \in N$. Therefore, $g = 1$ because N is P-torsion-free. $\quad\square$

We now prove a classical result on extraction of roots in nilpotent groups. If G is any group and $g \in G$, then $h \in G$ is an *nth root* of g if $h^n = g$ for some natural number $n > 1$.

Theorem 2.7 (S. N. Černikov, A. I. Mal'cev) *Let P be a nonempty set of primes. A nilpotent group G is P-torsion-free if and only if the following condition holds:*

$$\text{if } g, \ h \in G \text{ and } g^n = h^n \text{ for some P-number } n, \text{ then } g = h. \tag{2.14}$$

Equation (2.14) is equivalent to the condition that every element of G has at most one nth root for every P-number n.

Proof Suppose that G is P-torsion-free, and assume that $g^n = h^n$ for some $g, h \in G$ and P-number n. We prove that $g = h$ by induction on the class c of G. If $c = 1$, then G is abelian. In this case, $g^n = h^n$ for some P-number n implies that $\left(gh^{-1}\right)^n = 1$. Since G is P-torsion-free, $gh^{-1} = 1$ and $g = h$.

Suppose that $c > 1$, and assume that the result holds for all P-torsion-free nilpotent groups of class less than c. By Lemma 2.9, $H = gp(g, \ \gamma_2 G)$ is nilpotent of class less than c. It is clear that $h^{-1}gh \in H$ because $h^{-1}gh = g[g, \ h]$. Now, $g^n = h^n$ is the same as $g^n = h^{-1}h^n h$ which, after replacing h^n by g^n, becomes

$$g^n = h^{-1}g^n h = \left(h^{-1}gh\right)^n.$$

By induction, $g = h^{-1}gh$, so g and h commute. Hence, the equality $g^n = h^n$ can be expressed as $\left(gh^{-1}\right)^n = 1$. Since G is P-torsion-free, $gh^{-1} = 1$, and thus $g = h$.

Conversely, suppose that G is *any* group such that (2.14) is satisfied for any elements g and h in G. If we take $h = 1$, then $g^n = 1^n = 1$ implies $g = 1$. And so, G is P-torsion-free. $\quad\square$

Example 2.18 The Heisenberg group is torsion-free. To see this, suppose that

$$\begin{pmatrix} 1 & a & b \\ 0 & 1 & c \\ 0 & 0 & 1 \end{pmatrix}^n = \begin{pmatrix} 1 & 0 & 0 \\ 0 & 1 & 0 \\ 0 & 0 & 1 \end{pmatrix} \tag{2.15}$$

for some a, b, $c \in \mathbb{Z}$ and $n \in \mathbb{N}$. We use the Binomial Theorem to compute the left-hand side of (2.15):

$$
\begin{pmatrix} 1 & a & b \\ 0 & 1 & c \\ 0 & 0 & 1 \end{pmatrix}^n = \left(\begin{pmatrix} 1 & 0 & 0 \\ 0 & 1 & 0 \\ 0 & 0 & 1 \end{pmatrix} + \begin{pmatrix} 0 & a & b \\ 0 & 0 & c \\ 0 & 0 & 0 \end{pmatrix} \right)^n
$$

$$
= \begin{pmatrix} 1 & 0 & 0 \\ 0 & 1 & 0 \\ 0 & 0 & 1 \end{pmatrix} + n \begin{pmatrix} 0 & a & b \\ 0 & 0 & c \\ 0 & 0 & 0 \end{pmatrix} + \binom{n}{2} \begin{pmatrix} 0 & a & b \\ 0 & 0 & c \\ 0 & 0 & 0 \end{pmatrix}^2 + \cdots
$$

$$
= \begin{pmatrix} 1 & 0 & 0 \\ 0 & 1 & 0 \\ 0 & 0 & 1 \end{pmatrix} + n \begin{pmatrix} 0 & a & b \\ 0 & 0 & c \\ 0 & 0 & 0 \end{pmatrix} + \binom{n}{2} \begin{pmatrix} 0 & 0 & ac \\ 0 & 0 & 0 \\ 0 & 0 & 0 \end{pmatrix} + \cdots
$$

$$
= \begin{pmatrix} 1 & na & nb + \binom{n}{2}ac \\ 0 & 1 & nc \\ 0 & 0 & 1 \end{pmatrix}.
$$

Therefore,

$$
\begin{pmatrix} 1 & na & nb + \binom{n}{2}ac \\ 0 & 1 & nc \\ 0 & 0 & 1 \end{pmatrix} = \begin{pmatrix} 1 & 0 & 0 \\ 0 & 1 & 0 \\ 0 & 0 & 1 \end{pmatrix},
$$

and thus $a = b = c = 0$.

Since $\mathscr{H}$ is torsion-free, (2.14) must hold in $\mathscr{H}$. Indeed, suppose that

$$
\begin{pmatrix} 1 & a_1 & b_1 \\ 0 & 1 & c_1 \\ 0 & 0 & 1 \end{pmatrix}^n = \begin{pmatrix} 1 & a_2 & b_2 \\ 0 & 1 & c_2 \\ 0 & 0 & 1 \end{pmatrix}^n
$$

for some a_1, a_2, b_1, b_2, c_1, $c_2 \in \mathbb{Z}$ and $n \in \mathbb{N}$. The same computation used above gives

$$
\begin{pmatrix} 1 & na_1 & nb_1 + \binom{n}{2}a_1c_1 \\ 0 & 1 & nc_1 \\ 0 & 0 & 1 \end{pmatrix} = \begin{pmatrix} 1 & na_2 & nb_2 + \binom{n}{2}a_2c_2 \\ 0 & 1 & nc_2 \\ 0 & 0 & 1 \end{pmatrix}.
$$

Therefore, $a_1 = a_2$, $b_1 = b_2$, and $c_1 = c_2$. Hence, $\begin{pmatrix} 1 & a_1 & b_1 \\ 0 & 1 & c_1 \\ 0 & 0 & 1 \end{pmatrix} = \begin{pmatrix} 1 & a_2 & b_2 \\ 0 & 1 & c_2 \\ 0 & 0 & 1 \end{pmatrix}$ as claimed.

2.3.3 The Direct Product of Nilpotent Groups

The direct product of finitely many nilpotent groups is again nilpotent. This is the point behind the next theorem.

Theorem 2.8 *If* $\{H_1, \ldots, H_n\}$ *is a set of nilpotent groups of class* $c_1, \ldots, c_n$ *respectively, then the direct product* $H_1 \times \cdots \times H_n$ *is nilpotent of class* $max\{c_1, \ldots, c_n\}$.

Proof We prove the theorem for $n = 2$. Assume that H_1 and H_2 are nontrivial groups of nilpotency classes c_1 and c_2 respectively, and suppose that $c_1 \geq c_2 > 0$. The proof is done by induction on c_1. If $c_1 = 1$, then H_1 and H_2 are abelian, and thus $H_1 \times H_2$ is abelian.

Suppose that $c_1 > 1$. By Lemma 1.3,

$$\frac{H_1 \times H_2}{Z(H_1 \times H_2)} \cong \frac{H_1}{Z(H_1)} \times \frac{H_2}{Z(H_2)}. \tag{2.16}$$

Note that the right side of (2.16) is a direct product of nilpotent groups of classes less than c_1. By Lemma 2.12, the class of $H_1/Z(H_1)$ is $c_1 - 1$. By induction, $(H_1 \times H_2)/Z(H_1 \times H_2)$ is a nilpotent group of class $c_1 - 1$. The result follows from Lemma 2.12.   □

Remark 2.9 It is not always the case that the direct product of an arbitrary number of nilpotent groups is nilpotent. For example, suppose that $\{G_1, G_2, \ldots\}$ is an infinite set of nilpotent groups, and assume that G_i has nilpotency class at least i for each $i = 1, 2, \ldots$. We claim that the infinite direct product of the groups $G_1, G_2, \ldots$ is not nilpotent. Assume, on the contrary, that this direct product is nilpotent of class c. By Theorem 2.4, each of its subgroups is of class at most c. Consequently, every G_i is of class at most c. This contradicts the fact that G_j is of class at least j whenever $j > c$.

On the other hand, if the nilpotency class of each G_i is bounded above, then their direct product is nilpotent. The proof of this is analogous to that of Theorem 2.8.

2.3.4 Subnormal Subgroups

Subgroups of nilpotent groups enjoy several noteworthy properties, one of which is subnormality.

Definition 2.10 A subgroup H of a group G is called *subnormal* if there is a subnormal series of subgroups of G beginning at H and terminating at G.

Theorem 2.9 *Every subgroup of a nilpotent group is subnormal.*

Proof Let H be a subgroup of a nilpotent group G of class c, and consider the subgroups $H\zeta_i G$ of G for $i = 1, 2, \ldots, c$. Since the upper central series of G is normal, we have

$$H = H\zeta_0 G \leq H\zeta_1 G \leq \cdots \leq H\zeta_c G = G. \tag{2.17}$$

We claim that (2.17) is a subnormal series. If $h \in H$ and $z \in \zeta_{i+1}G$, then

$$z^{-1}hz = h[h, z] \in H[H, \zeta_{i+1}G] = H\zeta_i G.$$

Therefore, $z \in N_G(H\zeta_i G)$, and thus $\zeta_{i+1}G < N_G(H\zeta_i G)$. Since $H < N_G(H\zeta_i G)$ as well, $H\zeta_{i+1}G < N_G(H\zeta_i G)$ and the claim is proved. Thus, (2.17) is a subnormal series from H to G in c steps.   □

Remark 2.10 Another subnormal series from H to G can be constructed using successive normalizers. Put $H_0 = H$, and recursively define $H_{i+1} = N(H_i)$. It is simple to verify that the series

$$H = H_0 < H_1 < \cdots < H_c = G$$

is, indeed, subnormal.

Corollary 2.6 *If G is a nilpotent group and $H < G$ with $[G : H] = n$, then $g^n \in H$ for all $g \in G$.*

Proof Suppose that G has nilpotency class c. If H is a normal subgroup of G, then $|G/H| = [G : H] = n$. Hence, $(gH)^n = H$ for all $g \in G$, and thus $g^n \in H$.

Assume that H is any subgroup of G. By Theorem 2.9, there is a subnormal series

$$H = H_0 \lhd H_1 \lhd \cdots \lhd H_c = G.$$

Furthermore, each H_i is nilpotent by Theorem 2.4. If we put $[H_{i+1} : H_i] = m_i$, so that $n = m_{c-1}m_{c-2}\cdots m_0$, then we obtain

$$g^n = ((g^{m_{c-1}})^{m_{c-2}})^{\cdots m_0}.$$

Since each H_i is normal in G, we have

$$g^{m_{c-1}} \in H_{c-1}, \ (g^{m_{c-1}})^{m_{c-2}} \in H_{c-2}, \ \ldots.$$

Continuing in this way leads to $g^n \in H_0 = H$.   □

2.3.5   The Normalizer Condition

An important feature of nilpotent groups is that all of their maximal subgroups are normal. In fact, this property leads to a structure theorem for finite nilpotent groups which will be proven in the next section. Groups whose maximal subgroups are normal satisfy the so-called *normalizer condition*.

Definition 2.11 A group G satisfies the *normalizer condition* if H is a proper subgroup of $N_G(H)$ whenever H is a proper subgroup of G.

Lemma 2.14 *If a group G satisfies the normalizer condition, then every maximal subgroup of G is normal.*

Proof Let M be a maximal subgroup of G. By hypothesis, M is a proper subgroup of $N_G(M)$. Thus, $N_G(M) = G$ because M is maximal. And so, $M \lhd G$.   □

Lemma 2.15 *If every subgroup of a group G is subnormal, then G satisfies the normalizer condition.*

Proof Suppose that H is a proper subgroup of G. Since H is subnormal, there exists a subnormal series

$$H = H_0 \lhd H_1 \lhd \cdots \lhd H_n = G$$

for some $n \in \mathbb{N}$. Clearly, H_1 properly contains and normalizes H since $H \lhd H_1$.   □

Theorem 2.10 *Every nilpotent group satisfies the normalizer condition.*

Proof This is a consequence of Theorem 2.9 and Lemma 2.15.   □

Corollary 2.7 *Every maximal subgroup of a nilpotent group is normal.*

Proof The result follows at once from Theorem 2.10 and Lemma 2.14.   □

2.3.6   Products of Normal Nilpotent Subgroups

We prove a theorem pertaining to the product of normal nilpotent subgroups of an arbitrary group.

Theorem 2.11 (H. Fitting) *Let G be any group, and suppose that H and K are normal nilpotent subgroups of G of classes c and d respectively. Then HK is a normal nilpotent subgroup of G of class at most $c + d$.*

Proof By Theorem 2.2, $\gamma_{c+1}H = 1$ and $\gamma_{d+1}K = 1$. The result will follow at once from Theorem 2.3 once we prove that $\gamma_{c+d+1}(HK) = 1$. By repeatedly applying Lemma 1.11, we get

$$\gamma_{c+d+1}(HK) = \underbrace{[HK, \ HK, \ \cdots, \ HK]}_{c+d+1}$$

$$= \underbrace{[H, \ HK, \ \cdots, \ HK]}_{c+d+1}\underbrace{[K, \ HK, \ \cdots, \ HK]}_{c+d+1}$$

$$= \cdots$$

Thus, $\gamma_{c+d+1}(HK)$ is a product of commutators of the form

$$[X_1, \ X_2, \ \ldots, \ X_{c+d+1}],$$

where X_j is either H or K for $1 \le j \le c+d+1$. Let $Y = [X_1, \ X_2, \ \ldots, \ X_{c+d+1}]$ be one of the commutators arising in this product. Since Y contains $(c+d+1)$ X_j's, either H appears at least $(c+1)$ times in Y or K appears at least $(d+1)$ times in Y. Now, $\gamma_m H \trianglelefteq G$ and $\gamma_n K \trianglelefteq G$ for each $m, \ n > 0$ by Corollary 1.3 because both H and K are normal in G. By Theorem 1.4,

$$[\gamma_m H, \ K] \le \gamma_m H \quad \text{and} \quad [\gamma_n K, \ H] \le \gamma_n K. \tag{2.18}$$

Hence, if s of the X_j's in the commutator Y equal H, then $Y \le \gamma_{s+1}H$ by (2.18). Similarly, if t of the X_j's in the commutator Y equal K, then $Y \le \gamma_{t+1}K$. It follows that if H occurs at least $(c+1)$ times in Y, then $Y \le \gamma_{c+1}H$. However, if K occurs at least $(d+1)$ times in Y, then $Y \le \gamma_{d+1}K$. In either case, we obtain $Y = 1$. Therefore, $\gamma_{c+d+1}(HK) = 1$. $\qquad\qquad\square$

2.4 Finite Nilpotent Groups

In this section, we give a characterization of finite nilpotent groups. We begin by mentioning some of the well-known Sylow theorems and consequences of them. These play a fundamental role in the study of finite groups, and their proofs can be found in various places in the literature (see [3, 9], or [10] for instance).

Definition 2.12 Let G be a finite group of order $p^n k$, where p is a prime, $k \in \mathbb{N}$, and p doesn't divide k. A subgroup of G whose order is exactly p^n is called a *Sylow p-subgroup* of G.

A subgroup H of a finite group G is called a *Sylow subgroup* of G if it is a Sylow p-subgroup of G for some prime p. The fact that a finite group has Sylow subgroups is contained in the next fundamental theorem.

Theorem 2.12 (Sylow) *Let G be a finite group of order $p^n k$, where p is a prime, $k \in \mathbb{N}$, and p doesn't divide k.*

  (i) *G has at least one subgroup of order p^i for each $i = 1, 2, \ldots, n$.*
 (ii) *If $H \leq G$ and $|H| = p^n$, then H is contained in some Sylow p-subgroup.*
(iii) *Any two Sylow p-subgroups of G are conjugate.*

A consequence of Theorem 2.12 (iii) is:

Corollary 2.8 *Let p be a prime, and suppose that P is a Sylow p-subgroup of a finite group G. Then $P \trianglelefteq G$ if and only if P is the unique Sylow p-subgroup of G.*

Another result which will be needed later is:

Lemma 2.16 *Let P be a Sylow p-subgroup of a finite group G.*

 (i) *If $K \leq G$ and K contains $N_G(P)$, then $K = N_G(K)$.*
(ii) *If $N \trianglelefteq G$, then $P \cap N$ is a Sylow p-subgroup of N and PN/N is a Sylow p-subgroup of G/N.*

The proof of Lemma 2.16 (i) relies on the so-called Frattini Argument.

Lemma 2.17 (Frattini Argument) *Let G be a finite group and $H \trianglelefteq G$. If P is a Sylow p-subgroup of H for some prime p, then $G = HN_G(P)$.*

We now prove the main theorem of this section.

Theorem 2.13 *Let G be a finite group. The following are equivalent:*

  (i) *G is nilpotent.*
 (ii) *Every subgroup of G is subnormal.*
(iii) *G satisfies the normalizer condition.*
(iv) *Every maximal subgroup of G is normal.*
 (v) *Every Sylow subgroup of G is normal.*
(vi) *G is a direct product of its Sylow subgroups.*
(vii) *Elements of coprime order commute.*

Proof (i) $\Rightarrow$ (ii) by Theorem 2.9, (ii) $\Rightarrow$ (iii) by Lemma 2.15, and (iii) $\Rightarrow$ (iv) by Lemma 2.14.

We prove (iv) $\Rightarrow$ (v) by contradiction. Let P be a Sylow subgroup of G, and assume that P is not normal in G. Then $N_G(P) < G$, and consequently, $N_G(P) < M$ for some maximal subgroup M of G. Since $M \triangleleft G$, we have $N_G(M) = G$. This contradicts Lemma 2.16 (i).

Next, we prove (v) $\Rightarrow$ (vi). Suppose that G has order $p_1^{r_1} p_2^{r_2} \cdots p_n^{r_n}$, where the p_i's are distinct primes and $r_i \in \mathbb{N}$. Assume that each Sylow subgroup of G is normal.

By Corollary 2.8, there is a unique Sylow p_i-subgroup P_i of order $p_i^{r_i}$ for each p_i. We claim that G is the direct product of the P_i's. Observe that if $g_i \in P_i$ and $g_j \in P_j$ for $i \neq j$, then

$$[g_i, g_j] \in P_i \cap P_j = 1$$

by Lagrange's Theorem and normality of P_i and P_j. Thus, the elements of P_i commute with the elements of P_j whenever $i \neq j$. Now, define the map

$$\varphi : P_1 \times \cdots \times P_n \to G \text{ by } \varphi(g_1, \ldots, g_n) = g_1 \cdots g_n.$$

By the observation above, we have that φ is a homomorphism. We claim that φ is injective. Suppose that

$$\varphi(h_1, \ldots, h_n) = h_1 \cdots h_n = 1$$

for some $h_i \in P_i$. Since the h_i and h_j commute and have coprime order when $i \neq j$, we have

$$|h_1 h_2 \cdots h_n| = |h_1||h_2| \cdots |h_n| = 1.$$

This means that $|h_1| = |h_2| = \cdots = |h_n| = 1$, and thus $h_1 = h_2 = \cdots = h_n = 1$. And so, *ker* φ is trivial. This proves the claim. Since φ is an injective map between finite groups of equal order, it is an isomorphism. Therefore, G is a direct product of its Sylow subgroups.

Next, we prove (vi) $\Leftrightarrow$ (vii). Suppose that $G = P_1 \times \cdots \times P_n$ for Sylow p_i-subgroups P_i (here, of course, the p_i are distinct primes). Let $g = g_1 \cdots g_n$ and $h = h_1 \cdots h_n$ be elements of coprime order in G, where $g_i, h_i \in P_i$. Since

$$[g_i, g_j] = [h_i, h_j] = 1$$

when $i \neq j$, we have $|g| = |g_1| \cdots |g_n|$ and $|h| = |h_1| \cdots |h_n|$. Now, $|g|$ and $|h|$ are coprime only if one of the g_i or h_i equals 1 for each $i = 1, 2, \ldots, n$. We conclude that $gh = hg$.

Conversely, suppose that the elements of coprime order commute. Let $p_1, \ldots, p_n$ be the distinct prime divisors of $|G|$, and let $P_1, \ldots, P_n$ be corresponding Sylow subgroups associated with these primes. We assert that $G \cong P_1 \times \cdots \times P_n$. Let $g \in G$ and $h \in P_i$ for some $1 \leq i \leq n$. Clearly, $h^g \in P_i$ if $g \in P_i$. If $g \notin P_i$, then $|g|$ is coprime to $|h|$. By assumption, $[g, h] = 1$, and thus $h^g = h \in P_i$. And so, $P_i \trianglelefteq G$. Furthermore, $G = P_1 P_2 \cdots P_n$ because P_i and P_j are commuting subgroups for $i \neq j$. Finally, we find that

$$gp(P_1, \ldots, \widehat{P_i}, \ldots, P_n) \cap P_i = 1$$

for any $1 \leq i \leq n$ by Lagrange's Theorem. Here, $\widehat{P_i}$ means that P_i is omitted from the collection $P_1, \ldots, P_n$. This proves the assertion.

It remains to prove that (vii) $\Rightarrow$ (i). Suppose that the elements of coprime order in G commute. By (vii) $\Rightarrow$ (vi), G is a direct product of its Sylow subgroups. Since the Sylow subgroups have prime power order, each of them is nilpotent by Theorem 2.3. The result follows from Theorem 2.8. $\square$

2.5   The Tensor Product of the Abelianization

Tensor products serve as a useful tool in the study of nilpotent groups. In this section, we discuss the connection between the factors $\gamma_i G / \gamma_{i+1} G$ of the lower central series of a group G and the i-fold tensor product of $Ab(G)$, the abelianization of G. In particular, we demonstrate that certain properties of a nilpotent group are inherited from its abelianization.

2.5.1   The Three Subgroup Lemma

We begin with a result of P. Hall and L. Kalužnin.

Lemma 2.18 (Three Subgroup Lemma)   *Let G be a group with subgroups H, K, and L. If $N \trianglelefteq G$ and any two of the following subgroups $[H, K, L]$, $[K, L, H]$, $[L, H, K]$ are subgroups of N, then the third subgroup is also a subgroup of N.*

Proof Let h, k, and l be any elements of the subgroups H, K, and L respectively. By Corollary 1.5, the groups $[H, K, L]$, $[K, L, H]$, and $[L, H, K]$ are generated by conjugates of commutators of the forms $\left[h, k^{-1}, l\right]$, $\left[k, l^{-1}, h\right]$, and $\left[l, h^{-1}, k\right]$ respectively. By Lemma 1.5,

$$\left[h, k^{-1}, l\right]^k \left[k, l^{-1}, h\right]^l \left[l, h^{-1}, k\right]^h = 1.$$

Without loss of generality, suppose that $[H, K, L]$ and $[K, L, H]$ are contained in N. Since $N \trianglelefteq G$, we have $\left[h, k^{-1}, l\right]^k \in N$ and $\left[k, l^{-1}, h\right]^l \in N$. Hence,

$$\left[l, h^{-1}, k\right] = \left[k, l^{-1}, h\right]^{-l} \left(\left[h, k^{-1}, l\right]^{-k} \right)^{h^{-1}}$$

belongs to N, and consequently, $[L, H, K]$ is contained in N. $\square$

Corollary 2.9 *If H, K, and L are normal subgroups of a group G, then*

$$[H, K, L] \le [K, L, H][L, H, K].$$

Proof The result follows from Corollary 1.3 by putting $N = [K, L, H][L, H, K]$ in Lemma 2.18. □

The Three Subgroup Lemma plays a fundamental role in establishing certain connections between the commutators of the upper and lower central subgroups.

Theorem 2.14 (P. Hall) *Let G be any group and $i, j \in \mathbb{N}$.*

(i) $[\gamma_i G, \gamma_j G] \le \gamma_{i+j} G$;
(ii) $\gamma_i(\gamma_j G) \le \gamma_{ij} G$;
(iii) *If $j \ge i$, then $[\gamma_i G, \zeta_j G] \le \zeta_{j-i} G$.*

Proof The proofs of (i), (ii), and (iii) are done by induction on i.

(i) If $i = 1$, then $[\gamma_1 G, \gamma_j G] = \gamma_{1+j} G$ by Definition 2.4. Assume that $i > 1$ and the result holds for $i - 1$. By definition,

$$[\gamma_i G, \gamma_j G] = [[\gamma_1 G, \gamma_{i-1} G], \gamma_j G] = [\gamma_1 G, \gamma_{i-1} G, \gamma_j G].$$

We examine the subgroups obtained by permuting the entries of $[\gamma_1 G, \gamma_{i-1} G, \gamma_j G]$. Observe that

$$[\gamma_{i-1} G, \gamma_j G, \gamma_1 G] = [[\gamma_{i-1} G, \gamma_j G], \gamma_1 G] \le [\gamma_{i-1+j} G, \gamma_1 G] = \gamma_{i+j} G$$

and

$$[\gamma_j G, \gamma_1 G, \gamma_{i-1} G] = [[\gamma_j G, \gamma_1 G], \gamma_{i-1} G] = [\gamma_{j+1} G, \gamma_{i-1} G] \le \gamma_{i+j} G.$$

Setting $N = \gamma_{i+j} G$ in Lemma 2.18 gives
$[\gamma_i G, \gamma_j G] = [\gamma_1 G, \gamma_{i-1} G, \gamma_j G] \le \gamma_{i+j} G.$

(ii) The result is obvious when $i = 1$. Suppose that $i > 1$, and assume that the result holds for $i - 1$. By (i), we have

$$\gamma_i(\gamma_j G) = [\gamma_{i-1}(\gamma_j G), \gamma_j G] \le [\gamma_{(i-1)j} G, \gamma_j G] \le \gamma_{(i-1)j+j} G = \gamma_{ij} G.$$

(iii) If $i = 1$, then $[\gamma_1 G, \zeta_j G] = [G, \zeta_j G] \le \zeta_{j-1} G$ and the result holds by Lemma 2.1. Let $j \ge i > 1$, and suppose that the result is true for $i - 1$. By induction and Lemma 2.1, we have

$$[G, \zeta_j G, \gamma_{i-1} G] = [[G, \zeta_j G], \gamma_{i-1} G] \le [\zeta_{j-1} G, \gamma_{i-1} G] \le \zeta_{j-i} G$$

and

$$[\zeta_j G,\ \gamma_{i-1}G,\ G] = [[\zeta_j G,\ \gamma_{i-1}G],\ G] = [\zeta_{j-i+1}G,\ G] \le \zeta_{j-i}G.$$

Lemma 2.18 ultimately gives

$$[\gamma_i G,\ \zeta_j G] = [\gamma_{i-1}G,\ G,\ \zeta_j G] \le \zeta_{j-i}G.$$

This completes the proof. □

2.5.2 The Epimorphism $\bigotimes_{\mathbb{Z}}^n Ab(G) \to \gamma_n G / \gamma_{n+1} G$

We illustrate how the abelianization of a group influences the factors of its lower central series.

Definition 2.13 Suppose that A, B, and M are R-modules. A function $\varphi : A \times B \to M$ is called *bilinear* if, for all a, a_1, $a_2 \in A$, b, b_1, $b_2 \in B$, and $r \in R$, we have:

$$\varphi(a_1 + a_2,\ b) = \varphi(a_1,\ b) + \varphi(a_2,\ b);$$
$$\varphi(a,\ b_1 + b_2) = \varphi(a,\ b_1) + \varphi(a,\ b_2);$$
$$\varphi(ra,\ b) = \varphi(a,\ rb) = r\varphi(a,\ b).$$

If the R-modules are written using multiplicative notation, then the conditions above become:

$$\varphi(a_1 a_2,\ b) = \varphi(a_1,\ b)\varphi(a_2,\ b);$$
$$\varphi(a,\ b_1 b_2) = \varphi(a,\ b_1)\varphi(a,\ b_2);$$
$$\varphi(a^r,\ b) = \varphi(a,\ b^r) = (\varphi(a,\ b))^r.$$

In this case, φ is said to be *multiplicative in each variable*. In what follows, all $\mathbb{Z}$-modules (equivalently, abelian groups) are written multiplicatively.

Theorem 2.15 (D. J. S. Robinson) *Let G be any group. For each integer $n > 1$, the mapping*

$$\psi : \gamma_{n-1}G/\gamma_n G \bigotimes_{\mathbb{Z}} Ab(G) \to \gamma_n G/\gamma_{n+1}G$$

defined by

$$\psi(x\gamma_n G \otimes y\gamma_2 G) = [x,\ y]\gamma_{n+1}G \quad (x \in \gamma_{n-1}G,\ y \in G)$$

is a well-defined $\mathbb{Z}$-module epimorphism.

Proof Consider the function

$$\varphi_n : \gamma_{n-1}G/\gamma_n G \times Ab(G) \rightarrow \gamma_n G/\gamma_{n+1}G$$

defined by

$$(x\gamma_n G, \, y\gamma_2 G) \mapsto [x, \, y]\gamma_{n+1}G \quad (x \in \gamma_{n-1}G, \, y \in G).$$

We claim that φ_n is well defined and multiplicative in each variable.

- φ_n is well defined.

  (i) Let $g \in G$, $g_{n-1} \in \gamma_{n-1}G$, and $g_n \in \gamma_n G$. By Theorem 2.14 (i), the commutators $[g_n, \, g]$ and $[g_{n-1}, \, g, \, g_n]$ are contained in $\gamma_{n+1}G$. By Lemma 1.4 (v), we have

$$\begin{aligned}
\varphi_n(g_{n-1}g_n\gamma_n G, \, g\gamma_2 G) &= [g_{n-1}g_n, \, g]\gamma_{n+1}G \\
&= [g_{n-1}, \, g][g_{n-1}, \, g, \, g_n][g_n, \, g]\gamma_{n+1}G \\
&= [g_{n-1}, \, g]\gamma_{n+1}G \\
&= \varphi_n(g_{n-1}\gamma_n G, \, g\gamma_2 G).
\end{aligned}$$

  (ii) Let $g \in G$, $g_{n-1} \in \gamma_{n-1}G$, and $g_2 \in \gamma_2 G$. The commutators $[g_{n-1}, \, g_2]$ and $[g_{n-1}, \, g, \, g_2]$ are elements of $\gamma_{n+1}G$ by Theorem 2.14 (i). An application of Lemma 1.4 (vi) gives

$$\begin{aligned}
\varphi_n(g_{n-1}\gamma_n G, \, gg_2\gamma_2 G) &= [g_{n-1}, \, gg_2]\gamma_{n+1}G \\
&= [g_{n-1}, \, g_2][g_{n-1}, \, g][g_{n-1}, \, g, \, g_2]\gamma_{n+1}G \\
&= [g_{n-1}, \, g]\gamma_{n+1}G \\
&= \varphi_n(g_{n-1}\gamma_n G, \, g\gamma_2 G).
\end{aligned}$$

  Hence, φ_n is well defined. Consequently, φ_n naturally extends to a $\mathbb{Z}$-module homomorphism from the free $\mathbb{Z}$-module on $\gamma_{n-1}G/\gamma_n G \times Ab(G)$ to $\gamma_n G/\gamma_{n+1}G$.

- φ_n is multiplicative in each variable.

  (i) Let $a_1, \, a_2 \in \gamma_{n-1}G$. By Theorem 2.14 (i), $[a_1, \, g, \, a_2] \in \gamma_{n+1}G$. Thus,

$$\begin{aligned}
\varphi_n(a_1 a_2 \gamma_n G, \, g\gamma_2 G) &= [a_1 a_2, \, g]\gamma_{n+1}G \\
&= [a_1, \, g][a_1, \, g, \, a_2][a_2, \, g]\gamma_{n+1}G \\
&= [a_1, \, g][a_2, \, g]\gamma_{n+1}G \\
&= \varphi_n(a_1 \gamma_n G, \, g\gamma_2 G)\varphi_n(a_2 \gamma_n G, \, g\gamma_2 G).
\end{aligned}$$

54   2 Introduction to Nilpotent Groups

(ii) Let b_1, $b_2 \in G$. Since $[g_{n-1}, b_1, b_2] \in \gamma_{n+1}G$ by Theorem 2.14 (i), we have

$$
\begin{aligned}
\varphi_n(g_{n-1}\gamma_n G, \, b_1 b_2 \gamma_2 G) &= [g_{n-1}, \, b_1 b_2] \gamma_{n+1}G \\
&= [g_{n-1}, \, b_2][g_{n-1}, \, b_1][g_{n-1}, \, b_1, \, b_2] \gamma_{n+1}G \\
&= [g_{n-1}, \, b_2][g_{n-1}, \, b_1] \gamma_{n+1}G \\
&= [g_{n-1}, \, b_1][g_{n-1}, \, b_2] \gamma_{n+1}G \\
&= \varphi_n(g_{n-1}\gamma_n G, \, b_1 \gamma_2 G)\varphi_n(g_{n-1}\gamma_n G, \, b_2 \gamma_2 G).
\end{aligned}
$$

This shows that φ_n is multiplicative in each variable.

By the Universal Mapping Property of the Tensor Product, there is an induced $\mathbb{Z}$-module homomorphism from the tensor product $\gamma_{n-1}G/\gamma_n G \otimes_{\mathbb{Z}} Ab(G)$ to $\gamma_n G/\gamma_{n+1}G$ given by

$$
x\gamma_n G \otimes y\gamma_2 G \mapsto [x, \, y]\gamma_{n+1}G \quad (x \in \gamma_{n-1}G, \, y \in G).
$$

This map is an epimorphism since $\gamma_n G = [\gamma_{n-1}G, \, G]$. □

Remark 2.11 Theorem 2.15 also holds for groups which come equipped with operator domains. See [7].

In the next few results, we exploit Theorem 2.15. Some notation is needed. If M is an R-module, then the n-fold tensor product of M is written as

$$
\bigotimes_R^n M = \underbrace{M \otimes_R \cdots \otimes_R M}_{n}.
$$

By convention, we set $\bigotimes_R^1 M = M$.

Corollary 2.10 *Let G be any group. For each $n \in \mathbb{N}$, the mapping*

$$
\varphi_n : \bigotimes_{\mathbb{Z}}^n Ab(G) \to \gamma_n G/\gamma_{n+1}G
$$

defined by

$$
\varphi_n(x_1\gamma_2 G \otimes \cdots \otimes x_n\gamma_2 G) = [x_1, \, \ldots, \, x_n]\gamma_{n+1}G
$$

is a $\mathbb{Z}$-module epimorphism.

Proof This easily follows by induction on n. □

Corollary 2.11 *Suppose that G is a finitely generated group with generating set $X = \{x_1, \, \ldots, \, x_k\}$. For each $n \in \mathbb{N}$, the factor group $\gamma_n G/\gamma_{n+1}G$ is finitely* generated, modulo $\gamma_{n+1}G$, by the simple commutators of weight n of the form $[x_{i_1}, \ldots, x_{i_n}]$, where the x_{i_j}'s vary over all elements of X and are not necessarily distinct.

Proof Since G is finitely generated by X, $Ab(G)$ is finitely generated by the elements $x_1\gamma_2G, \ldots, x_k\gamma_2G$. Hence, $\bigotimes_{\mathbb{Z}}^{n} Ab(G)$ is finitely generated by the k^n n-fold tensor products of the form

$$x_{i_1}\gamma_2G \otimes \cdots \otimes x_{i_n}\gamma_2G,$$

where the x_{i_j} vary over X. It follows from Corollary 2.10 that $\gamma_nG/\gamma_{n+1}G$ is finitely generated by the simple commutators, modulo $\gamma_{n+1}G$, of the form $[x_{i_1}, \ldots, x_{i_n}]$, where the x_{i_j}'s vary over all elements of X. $\qquad\square$

Remark 2.12 Corollary 2.11 could also be proven using Lemma 2.6. Notice however, that Lemma 2.6 allows inverses of elements of the generating set in the simple commutators, whereas the corollary does not. This issue can be resolved by a repeated application of Lemmas 1.4 and 1.13.

Example 2.19 Let G be a group generated by $X = \{x_1, x_2, x_3\}$. If $g = x_2^3x_1^{-1}$ and $h = x_1x_2^{-4}x_3^2$ are elements of G, then $[g, h]\gamma_3G \in \gamma_2G/\gamma_3G$. Using Lemmas 1.4 and 1.13, together with the fact that all simple commutators of weight 2 are central, modulo γ_3G, we have

$$[g, h]\gamma_3G = \left[x_2^3x_1^{-1}, x_1x_2^{-4}x_3^2\right]\gamma_3G$$

$$= \left[x_2^3, x_1x_2^{-4}x_3^2\right]\left[x_1^{-1}, x_1x_2^{-4}x_3^2\right]\gamma_3G$$

$$= \left[x_2^3, x_1\right]\left[x_2^3, x_2^{-4}\right]\left[x_2^3, x_3^2\right]\left[x_1^{-1}, x_1\right]\left[x_1^{-1}, x_2^{-4}\right]\left[x_1^{-1}, x_3^2\right]\gamma_3G$$

$$= [x_2, x_1]^3[x_2, x_2]^{-12}[x_2, x_3]^6[x_1, x_1]^{-1}[x_1, x_2]^4[x_1, x_3]^{-2}\gamma_3G$$

$$= [x_2, x_1]^3[x_2, x_3]^6[x_1, x_2]^4[x_3, x_1]^2\gamma_3G,$$

which illustrates that $[g, h]$ modulo γ_3G is expressible as a product of commutators of weight 2 in the elements of X.

Corollary 2.10 can be used to prove that a nilpotent group is finitely generated whenever its abelianization is finitely generated. We need some preliminary material.

Definition 2.14 A group G is said to satisfy condition *Max* (the *maximal condition on subgroups*) if every subgroup of G is finitely generated.

A group in which every ascending series of subgroups stabilizes is said to satisfy the *Noetherian condition*.

Theorem 2.16 *A group G satisfies Max if and only if it satisfies the Noetherian condition.*

Proof Suppose that G satisfies Max, and let

$$H_1 < H_2 < H_3 < \cdots$$

be an ascending series of subgroups of G. We assert that this series stabilizes. Put $H = \bigcup_{i=1}^{\infty} H_i$. Clearly, H is a subgroup of G and is finitely generated by hypothesis. Let $X = \{h_1, \ldots, h_k\}$ be a set of generators of H. It is evident that each element of X is contained in some H_i since X generates H. Thus, there exists $n \in \mathbb{N}$ such that $X \subset H_n$. It follows that $H \le H_n$. Since $H_n \le H$, we have $H = H_n$ and the series stabilizes.

Conversely, suppose that every ascending series of subgroups stabilizes. Let H be a subgroup of G, and choose an element $h_1 \in H$. If $H = gp(h_1)$, then H is finitely generated. Otherwise, there exists an element $h_2 \in H$ such that $h_2 \notin gp(h_1)$. Now, if $H = gp(h_1, h_2)$, then H is finitely generated. If $H \ne gp(h_1, h_2)$, then we continue this argument to obtain an ascending series of subgroups

$$gp(h_1) \le gp(h_1, h_2) \le \cdots$$

which stabilizes by assumption. Hence, $H = gp(h_1, h_2, \ldots, h_n)$ for some $n \in \mathbb{N}$. And so, H is finitely generated. $\square$

Groups which satisfy Max must be finitely generated. There are finitely generated groups, however, which do not satisfy Max. For example, let $F = \langle x, y \rangle$ be the free group of rank two, and let

$$G_i = gp\left(x, \, yxy^{-1}, \, \ldots, \, y^i x y^{-i}\right).$$

Every element of G_i can be written as

$$y^{m_1} x^{n_1} y^{m_2 - m_1} x^{n_2} y^{m_3 - m_2} \cdots y^{-m_k} \quad (0 \le m_r \le i).$$

Thus, $y^{i+1} x y^{-(i+1)}$ is not an element of G_i. This implies that the ascending sequence of subgroups

$$G_1 < G_2 < G_3 < \cdots$$

does not stabilize. By Theorem 2.16, F does not satisfy Max.

Lemma 2.19 *Max is preserved under extensions.*

Proof Let G be a group with $N \trianglelefteq G$, and suppose that G/N and N satisfy Max. Let H be any subgroup of G. Clearly, $H \cap N$ is finitely generated since $H \cap N < N$ and N satisfies Max. By the Second Isomorphism Theorem,

$$H/(H \cap N) \cong HN/N < G/N.$$

This implies that $H/(H \cap N)$ is finitely generated because G/N satisfies Max. It follows that H is finitely generated.   □

Theorem 2.17 *Every finitely generated abelian group satisfies Max.*

Proof Let G be a finitely generated abelian group with generating set $\{x_1, \ldots, x_k\}$. The proof is done by induction on k. If $k = 1$, then G is cyclic. In this case, it is easy to show that $[G : H] < \infty$ for every nontrivial subgroup H of G. Hence, H must be finitely generated.

Suppose that the theorem is true for $1 \leq i \leq k - 1$, and consider the subgroup $H = gp(x_1, \ldots, x_{k-1})$ of G. Since H is finitely generated and abelian, H satisfies Max by induction. Furthermore, $G/H \cong gp(x_k)$ is cyclic, and thus satisfies Max. The result follows from Lemma 2.19.   □

Theorem 2.18 (R. Baer) *Every finitely generated nilpotent group satisfies Max.*

Proof Let G be a finitely generated nilpotent group of class c, and let $H \leq G$. Set $H_i = H \cap \gamma_i G$ for $1 \leq i \leq c$. It follows from Lemma 2.1 that the series

$$H = H_1 \geq H_2 \geq \cdots \geq H_c \geq H_{c+1} = 1$$

is a central series for H. Furthermore, the Second Isomorphism Theorem gives

$$\frac{H_i}{H_{i+1}} = \frac{H \cap \gamma_i G}{H \cap \gamma_{i+1} G} = \frac{H \cap \gamma_i G}{(H \cap \gamma_i G) \cap \gamma_{i+1} G} \cong \frac{\gamma_{i+1} G (H \cap \gamma_i G)}{\gamma_{i+1} G}$$

for $1 \leq i \leq c$. Therefore, each H_i/H_{i+1} is isomorphic to a subgroup of $\gamma_i G/\gamma_{i+1} G$. Since $\gamma_i G/\gamma_{i+1} G$ is finitely generated and abelian by Corollary 2.11, so is H_i/H_{i+1} by Theorem 2.17. In particular, $H_c = H_c/H_{c+1}$ is finitely generated. Thus, H_{c-1} is finitely generated since both H_{c-1}/H_c and H_c are finitely generated. Repeating this argument gives that H_i is finitely generated for $1 \leq i \leq c - 2$. In particular, $H_1 = H$ is finitely generated.   □

We now prove that nilpotent groups with finitely generated abelianization must be finitely generated.

Corollary 2.12 *If G is a nilpotent group and $Ab(G)$ is finitely generated, then G satisfies Max. Hence, G is finitely generated.*

Proof The proof is done by induction on the class c of G. Theorem 2.17 takes care of the case $c = 1$. Assume that the corollary is true for nilpotent groups of class less than c, and let $n \in \{1, \ldots, c\}$. The tensor product $\bigotimes_{\mathbb{Z}}^n Ab(G)$ is finitely generated because it involves a finite number of finitely generated abelian groups. By Corollary 2.10, each $\gamma_n G/\gamma_{n+1} G$ is finitely generated abelian, and thus satisfies Max by Theorem 2.17. In particular, $\gamma_c G$ satisfies Max. By the induction hypothesis, $G/\gamma_c G$ also satisfies Max. The result now follows from Lemma 2.19.   □

2.5.3 Property $\mathscr{P}$

The proof of Corollary 2.12 shows that certain properties of the abelianization of a nilpotent group can be passed on to the group itself. This is the substance of the next result.

Definition 2.15 A group-theoretical property is called *property $\mathscr{P}$* if it satisfies the following criteria:

1. Property $\mathscr{P}$ is preserved under extensions.
2. If G is an abelian group having property $\mathscr{P}$ and $k \in \mathbb{N}$, then any homomorphic image of the k-fold tensor product $\bigotimes_{\mathbb{Z}}^{k} G$ has property $\mathscr{P}$.

It is clear that finiteness is a property $\mathscr{P}$. Other possibilities for property $\mathscr{P}$ include finite generation, P-torsion for a set of primes P (see Lemma 2.13), and Max (see Lemma 2.19 and the proof of Corollary 2.12).

Theorem 2.19 (D. J. S. Robinson) *If G is nilpotent and $Ab(G)$ has property $\mathscr{P}$, then G has property $\mathscr{P}$.*

Proof Suppose that G is of class c, and let $k > 0$. By Corollary 2.10, $\gamma_k G / \gamma_{k+1} G$ is an image of the k-fold tensor product $\bigotimes_{\mathbb{Z}}^{k} Ab(G)$. Thus, each $\gamma_k G / \gamma_{k+1} G$ has property $\mathscr{P}$ because $Ab(G)$ does. Now, $\gamma_{c+1} G = 1$ by Theorem 2.3. This means that $\gamma_c G$ has property $\mathscr{P}$. Since $\gamma_{c-1} G / \gamma_c G$ has property $\mathscr{P}$ and $\gamma_{c-1} G$ is an extension of $\gamma_{c-1} G / \gamma_c G$ by $\gamma_c G$, we have that $\gamma_{c-1} G$ also has property $\mathscr{P}$. We continue this argument to conclude that G has property $\mathscr{P}$. $\square$

Definition 2.16 The *exponent* of a torsion group G is the smallest natural number m, if it exists, satisfying $g^m = 1$ for every $g \in G$. If no such m exists, then G has *infinite exponent*.

Every finite group has finite exponent dividing the order of the group. For any prime p, both the infinite direct product

$$\mathbb{Z}_p \times \mathbb{Z}_{p^2} \times \mathbb{Z}_{p^3} \times \cdots$$

and the p-quasicyclic group are infinite torsion groups with infinite exponent. Thus, torsion groups need not be finite nor have finite exponent. A group with infinite exponent is necessarily infinite. However, the infinite direct product of cyclic groups of order p is an example of an infinite group with finite exponent.

Theorem 2.20 (S. Dixmier) *Let G be a nilpotent group of class c. If $Ab(G)$ has finite exponent m, then G has finite exponent dividing m^c.*

Proof The exponent of $\bigotimes_{\mathbb{Z}}^{i} Ab(G)$ divides m for $1 \leq i \leq c$ because $Ab(G)$ has exponent m. Thus, $\gamma_i G / \gamma_{i+1} G$ also has exponent dividing m by Corollary 2.10. In particular, $\gamma_c G = \gamma_c G / \gamma_{c+1} G$ has exponent dividing m. This, combined with the fact that $\gamma_{c-1} G / \gamma_c G$ also has exponent dividing m, gives that the exponent of $\gamma_{c-1} G$ divides m^2. We iterate this process to finally obtain that the exponent of $G = \gamma_{c-(c-1)} G$ divides m^c. $\square$

2.5.4 The Hirsch-Plotkin Radical

We end this section with an important result whose proof depends on Theorem 2.18. Motivated by Theorem 2.11, it is natural to ask whether or not a group has a *maximal* normal nilpotent subgroup.

Definition 2.17 A maximal normal nilpotent subgroup of a group is called a *nilpotent radical* of the group.

One attempt to construct a nilpotent radical is by trying to use Zorn's Lemma. Suppose that

$$N_1 < N_2 < N_3 < \cdots$$

is an ascending chain of normal nilpotent subgroups of a group G, where N_i is of class c_i for $i = 1, 2, \ldots$. A nilpotent radical would exist if $\cup_{i=1}^{k} N_i$ were normal and nilpotent for all $k \geq 1$. However, it is not nilpotent since the class of

$$\bigcup_{i=1}^{k} N_i = N_1 \cdots N_k,$$

which is $c_1 + \cdots + c_k$ according to Theorem 2.11, becomes unbounded as k approaches infinity. Hence, Zorn's Lemma does not apply.

Even though the nilpotent radical doesn't always exist, one can always find a *locally nilpotent radical*. This is the basis of our next discussion.

Definition 2.18 A group G is called *locally nilpotent* if every finitely generated subgroup of G is nilpotent.

Clearly, every nilpotent group is locally nilpotent. If $G = \prod_{i=1}^{\infty} G_i$, where each G_i is nilpotent of class c_i and $c_k < c_{k+1}$ for $k \geq 1$, then G is locally nilpotent. In particular, $\prod_{i=1}^{\infty} \mathbb{Z}_{p^i}$ and $\prod_{i=1}^{\infty} UT_i(\mathbb{Z})$ are locally nilpotent.

Lemma 2.20 *(i) Every nilpotent group is locally nilpotent.*
(ii) Every subgroup of a locally nilpotent group is locally nilpotent.
(iii) Every homomorphic image of a locally nilpotent group is locally nilpotent.

Proof

(i) This is immediate from Theorem 2.4.
(ii) Let G be a locally nilpotent group, and suppose that $H < G$. If K is a finitely generated subgroup of H, then it is also a finitely generated subgroup G. Since G is locally nilpotent, K is nilpotent, and thus K is a nilpotent subgroup of H. This means that H is locally nilpotent.
(iii) Let G be a locally nilpotent group, and suppose that $\varphi \in Hom(G, H)$ for some group H. Let K be a finitely generated subgroup of $\varphi(G)$ with finite generating set $\{x_1, \ldots, x_m\}$. There exist elements $g_1, \ldots, g_m$ in G such that $\varphi(g_i) = x_i$

for $1 \leq i \leq m$. Consider the subgroup $L = gp(g_1, \ldots, g_m)$ of G. It is finitely generated, and thus nilpotent since G is locally nilpotent. By Theorem 2.4, $\varphi(L) = K$ is also nilpotent. And so, $\varphi(G)$ is locally nilpotent.   □

Theorem 2.21 (K. Hirsch, B. Plotkin) *If H and K are normal locally nilpotent subgroups of a group G, then HK is a normal locally nilpotent subgroup of G.*

Proof We adopt the proof given by D.J.S. Robinson in [8]. Clearly, $HK \trianglelefteq G$ since $H \trianglelefteq G$ and $K \trianglelefteq G$. We claim that HK is locally nilpotent. Let

$$\{h_1, \ldots, h_m\} \subset H \ \text{and} \ \{k_1, \ldots, k_m\} \subset K.$$

Then $\{h_1 k_1, \ldots, h_m k_m\} \subset HK$. Define the subgroups

$$A = gp(h_1, \ldots, h_m) \leq H \ \text{and} \ B = gp(k_1, \ldots, k_m) \leq K,$$

and set $C = gp(A, B)$ and $S = gp(h_1 k_1, \ldots, h_m k_m)$. In order to prove the claim, we need to establish that S is nilpotent. Since $S \leq C$, it suffices to show that C is nilpotent.

Define the set $T = \{[h_i, k_j] \mid i, j = 1, \ldots, m\}$, and observe that $T \subseteq H \cap K$ since $H \trianglelefteq G$ and $K \trianglelefteq G$. Clearly, both A and T are finitely generated and contained in H. Thus, $gp(A, T)$ is a finitely generated subgroup of H. Since H is locally nilpotent, $gp(A, T)$ is also nilpotent. By Theorems 2.4 and 2.18, the normal closure T^A of T in $gp(A, T)$ is finitely generated and nilpotent. Furthermore, $T^A \leq H \cap K$, and consequently, $gp(B, T^A) \leq K$. Therefore, $gp(B, T^A)$ is finitely generated and nilpotent. By Corollary 1.5, we have $[A, B] = (T^A)^B$. Hence,

$$gp(B, T^A) = gp\left(B, (T^A)^B\right) = gp(B, [A, B]) = B^A.$$

It follows that B^A is nilpotent, and similarly, A^B is nilpotent. By Theorem 2.11, $A^B B^A = C$ is nilpotent.   □

Corollary 2.13 *Every group G has a unique maximal normal locally nilpotent subgroup containing all normal locally nilpotent subgroups of G.*

This subgroup is called the *Hirsch-Plotkin radical* of G.

Proof If $N_1 < N_2 < \cdots$ is a chain of locally nilpotent subgroups of G, then $\cup_{i=1}^{\infty} N_i$ is locally nilpotent. By Zorn's Lemma, each normal locally nilpotent subgroup of G is contained in a maximal normal locally nilpotent subgroup of G.

We establish uniqueness. Suppose that M_1 and M_2 are both maximal normal locally nilpotent subgroups of G. By Theorem 2.21, the product $M_1 M_2$ is locally nilpotent. The maximality of M_1 and M_2 implies that $M_1 = M_1 M_2 = M_2$.   □

The Hirsch-Plotkin radical is a valuable tool for studying various generalized nilpotent groups. We refer the reader to [8] for a discussion of such groups.

2.5.5   An Extension Theorem for Nilpotent Groups

The symmetric group S_3 is an extension of S_3/A_3 by A_3, groups of order 2 and 3 respectively. Both of these groups are cyclic (hence, nilpotent). However, S_3 is not nilpotent. This illustrates that nilpotency is not preserved under extensions. The next theorem addresses the following question: when is an extension of a nilpotent group by another group again nilpotent?

Theorem 2.22 (P. Hall, A. G. R. Stewart) *Let G be any group, and suppose that* $N \lhd G$. *If N is nilpotent of class c and* $G/\gamma_2 N$ *is nilpotent of class d, then G is nilpotent of class at most* $cd + (c - 1)(d - 1)$.

In [5], P. Hall initially found the bound on the class of G to be at most

$$\binom{c+1}{2} d - \binom{c}{2}.$$

A. G. R. Stewart improved on this in [12] and obtained the bound to be at most

$$cd + (c - 1)(d - 1).$$

In the same paper, he provided an example to illustrate that this bound cannot be improved. We give A. G. R. Stewart's proof below. In what follows, we define

$$[N, \underbrace{G, \ldots, G}_{0}] = N.$$

Lemma 2.21 *Let G be any group. If* $N \unlhd G$, *then*

$$[\gamma_2 N, \underbrace{G, G, \ldots, G}_{s}] \leq \prod_{k=1}^{m} S_k$$

for some $m \in \mathbb{N}$, *where*

$$S_k = [[N, \underbrace{G, \ldots, G}_{i}], [N, \underbrace{G, \ldots, G}_{s-i}]]$$

for some $i \in \{1, 2, \ldots, s\}$.

Proof The proof is done by induction on s. Suppose that $s = 1$. By Proposition 1.1 (i) and Corollary 2.9, we have

$$
\begin{aligned}
[\gamma_2 N, G] &\le [N, G, N][G, N, N] \\
&= [N, G, N][N, G, N] \\
&= [N, G, N] \\
&= [[N, G], [N, \underbrace{G, \ldots, G}_{0}]]
\end{aligned}
$$

and the lemma holds. Next, assume that the lemma is true for $s - 1$:

$$
[\gamma_2 N, \underbrace{G, \ldots, G}_{s-1}] \le \prod_{k=1}^{n} T_k
$$

for some $n \in \mathbb{N}$, where

$$
T_k = [[N, \underbrace{G, \ldots, G}_{i}], [N, \underbrace{G, \ldots, G}_{s-i-1}]]
$$

for some $i \in \{1, 2, \ldots, s-1\}$. Notice that

$$
[\gamma_2 N, \underbrace{G, \ldots, G}_{s}] = [\gamma_2 N, \underbrace{G, \ldots, G}_{s-1}, G] \le \left[\prod_{k=1}^{n} T_k, G\right] = \prod_{k=1}^{n} [T_k, G],
$$

where the last equality follows from Lemma 1.10. By applying Proposition 1.1 (i) and Corollary 2.9, we get

$$
\begin{aligned}
[T_k, G] &= [[N, \underbrace{G, \ldots, G}_{i}], [N, \underbrace{G, \ldots, G}_{s-i-1}], G] \\
&\le [[N, \underbrace{G, \ldots, G}_{(s-1)-i}], G, [N, \underbrace{G, \ldots, G}_{i}]][G, [N, \underbrace{G, \ldots, G}_{i}], [N, \underbrace{G, \ldots, G}_{s-i-1}]] \\
&= [[N, \underbrace{G, \ldots, G}_{s-i}], [N, \underbrace{G, \ldots, G}_{i}]][[N, \underbrace{G, \ldots, G}_{i+1}], [N, \underbrace{G, \ldots, G}_{s-(i+1)}]] \\
&= [[N, \underbrace{G, \ldots, G}_{i}], [N, \underbrace{G, \ldots, G}_{s-i}]][[N, \underbrace{G, \ldots, G}_{i+1}], [N, \underbrace{G, \ldots, G}_{s-(i+1)}]]
\end{aligned}
$$

and the result follows.   □

We now prove Theorem 2.22. First, note that $\gamma_{c+1} N = 1$ and $\gamma_{d+1} G \le \gamma_2 N$ by Theorem 2.2 because the classes of N and $G/\gamma_2 N$ are c and d respectively. The proof is done by induction on c. If $c = 1$, then N is abelian. In this case, $\gamma_2 N = 1$, and thus $G/\gamma_2 N \cong G$ is nilpotent of class d.

Next, suppose that $c > 1$, and assume that the theorem is true for $c - 1$. For any $r \in \{1, 2, \ldots, c\}$, $M_r = N/\gamma_{r+1}N$ is a normal subgroup of $H_r = G/\gamma_{r+1}N$, where M_r is of class r and $H_r/\gamma_2 M_r$ is of class d by the Third Isomorphism Theorem. Thus, we may assume by induction that

$$\gamma_{2rd-r-d+2}G \leq \gamma_{r+1}N \tag{2.19}$$

for all $r \in \{1, 2, \ldots, c - 1\}$. We invoke Lemma 2.21 to find that

$$\gamma_{2cd-c-d+2}G = [\gamma_{d+1}G, \underbrace{G, \ldots, G}_{2cd-2d-c+1}] \leq [\gamma_2 N, \underbrace{G, \ldots, G}_{2cd-2d-c+1}] \leq \prod_{k=1}^{m} S_k$$

for some $m \in \mathbb{N}$, where

$$S_k = [[N, \underbrace{G, \ldots, G}_{i}], [N, \underbrace{G, \ldots, G}_{2cd-2d-c+1-i}]]$$

for some $i \in \{1, 2, \ldots, (2cd - 2d - c + 1)\}$. Now, each

$$i \in \{1, 2, \ldots, (2cd - 2d - c + 1)\}$$

is contained in one of the following sets:

$$2(j - 1)d - d - (j - 1) + 1 \leq i \leq 2jd - d - j + 1, \quad \text{where } j \in \{1, 2, \ldots, c\}.$$

For arbitrary j,

$$[[N, \underbrace{G, \ldots, G}_{i}], [N, \underbrace{G, \ldots, G}_{2cd-2d-c+1-i}]] \leq [\gamma_j N, \gamma_w G], \tag{2.20}$$

where

$$w = 2d(c - j) - d - (c - j) + 2 + 2dj - d - j - i.$$

The result follows from the fact that $[N, \underbrace{G, \ldots, G}_{t}] \leq \gamma_{t+1} G$. Since $2dj - d - j \geq 1$ and $\gamma_{r+s}G \leq \gamma_r G$ for all $s \geq 0$, we find that

$$[\gamma_j N, \gamma_w G] \leq \left[\gamma_j N, \gamma_{2d(c-j)-d-(c-j)+2}G\right]. \tag{2.21}$$

Substituting r by $(c - j)$ in (2.19) shows that

$$\left[\gamma_j N, \gamma_{2d(c-j)-d-(c-j)+2}G\right] \leq \left[\gamma_j N, \gamma_{c-j+1}N\right].$$

By Theorem 2.14 (i), $\left[\gamma_j N, \ \gamma_{c-j+1} N\right] \leq \gamma_{c+1} N$. We conclude that for all possible k, $S_k \leq \gamma_{c+1} N = 1$, and thus $\prod_{k=1}^{m} S_k = 1$. This completes the proof of Theorem 2.22.

2.6 Finitely Generated Torsion Nilpotent Groups

In [1], R. Baer proved that every finitely generated torsion nilpotent group is finite. This allows one to answer certain questions involving torsion in a nilpotent group by passing to a finite group. In this section, we focus on some of these questions. We begin with a result due to A. I. Mal'cev which contains R. Baer's theorem as a special case.

Theorem 2.23 (A. I. Mal'cev) *Let G be a finitely generated nilpotent group, and let $H \leq G$. If G has a finite set of generators X such that some positive power of each element of X is contained in H, then a positive power of every element of G is contained in H. Furthermore, H is of finite index in G.*

Proof The proof is done by induction on the class c of G. If $c = 1$, then G is a finitely generated abelian group and the result is clear.

Suppose that $c > 1$, and assume that the lemma is true for all finitely generated nilpotent groups of class less than c. By Lemma 2.8, $G/\gamma_c G$ is finitely generated nilpotent of class $c - 1$. By induction, $H\gamma_c G$ has finite index in G and a positive power of every element of G is contained in $H\gamma_c G$. We claim that a positive power of every element of G is contained in H and $[G : H] < \infty$.

Let $G = gp(g_1, g_2, \ldots, g_s)$ such that $g_i^{m_i} \in H$, where $m_i > 0$ and $1 \leq i \leq s$. By Theorem 2.18, $\gamma_{c-1} G$ is finitely generated. Suppose that $\gamma_{c-1} G = gp(x_1, x_2, \ldots, x_t)$ such that $x_j^{n_j} \in H\gamma_c G$, where $n_j > 0$ and $1 \leq j \leq t$. By Lemmas 1.4 and 1.13, together with Remark 2.5, we have

$$\gamma_c G = gp\left([x_j, \ g_i] \mid 1 \leq i \leq s, \ 1 \leq j \leq t\right)$$

and

$$\left[x_j, \ g_i\right]^{n_j m_i} = \left[x_j^{n_j}, \ g_i^{m_i}\right] \in [H\gamma_c G, \ H] = [H, \ H] \leq H$$

for $1 \leq i \leq s$ and $1 \leq j \leq t$. Since $\gamma_c G \leq Z(G)$, a positive power of every element of $\gamma_c G$ lies in H. If $g \in G$, then there exists $m \in \mathbb{N}$ such that $g^m = hz$, where $h \in H$ and $z \in \gamma_c G$. Furthermore, there exists $n \in \mathbb{N}$ such that $z^n \in H$. Thus,

$$g^{mn} = (hz)^n = h^n z^n \in H$$

since z is central. This means that a positive power of every element of G is contained in H.

Next, we show that $H\gamma_c G/H$ is a finite abelian group. This, together with the
fact that $[G : H\gamma_c G] < \infty$, will give $[G : H] < \infty$ as claimed. By the Second
Isomorphism Theorem, $H\gamma_c G/H$ is abelian since it is isomorphic to $\gamma_c G/(H \cap \gamma_c G)$,
a quotient of the abelian group $\gamma_c G$. It is finite because it has a finite set of
generators, each having finite order. More precisely, $\gamma_c G/(H \cap \gamma_c G)$ is finitely
generated because $\gamma_c G$ is finitely generated, and each generator $[x_j, g_i](H \cap \gamma_c G)$
of $\gamma_c G/(H \cap \gamma_c G)$ has finite order since $[x_j, g_i]^{n_j m_i} \in H \cap \gamma_c G$. $\qquad\square$

An analogue of Theorem 2.23 for a given nonempty set of primes is:

Theorem 2.24 *Let P be a nonempty set of primes. Suppose that G is a finitely
generated nilpotent group and $H \leq G$. If G has a finite set of generators X such that
some P-number power of each element of X is contained in H, then each element of
G has a P-number power contained in H. Furthermore, $[G : H]$ is a P-number.*

The proof is the same as for Theorem 2.23.

Theorem 2.25 (R. Baer) *Let P be a nonempty set of primes. If there is a finite set
of generators X of a finitely generated nilpotent group G for which each element of
X has order a P-number, then G is a finite P-torsion group. In particular, finitely
generated torsion nilpotent groups are finite.*

Proof Set $H = 1$ in Theorem 2.24. $\qquad\square$

We point out that the finiteness of G in Theorem 2.25 is a consequence of the
fact that the trivial subgroup $H = 1$ must be of finite index in G according to
Theorem 2.24.

Corollary 2.14 *The elements of coprime order in any locally nilpotent group
commute.*

Proof Let G be a locally nilpotent group, and suppose that g and h are elements of
coprime order in G. The subgroup $H = gp(g, h)$ of G is finitely generated, and
thus nilpotent. Since each generator g and h has finite order, H must be finite by
Theorem 2.25. Therefore, g and h commute by Theorem 2.13. $\qquad\square$

2.6.1   The Torsion Subgroup of a Nilpotent Group

If P is a nonempty set of primes and G is a group, then the set $\tau_P(G)$ of P-torsion
elements of G is not necessarily a subgroup of G. For example, consider the (non-
nilpotent) *infinite dihedral group*

$$D_\infty = \left\langle x, y \,\middle|\, x^2 = 1, \, y^2 = 1 \right\rangle.$$

Clearly, xy is not a torsion element, even though x and y are torsion elements. For nilpotent groups, however, we have:

Theorem 2.26 (R. Baer, K. A. Hirsch) *If G is a nilpotent group and P is any nonempty set of primes, then $\tau_P(G)$ is a normal subgroup of G. Furthermore, if $\mathbb{P}$ denotes the set of all prime numbers, then*

$$\tau(G) = \prod_{p \in \mathbb{P}} \tau_p(G).$$

This coincides with Theorem 2.13 in the case when G is finite.

Proof Let g and h be P-torsion elements. By Theorem 2.25, $gp(g, h)$ is a finite P-torsion group. Hence, $g^{-1}h$ is a P-torsion element, and thus $\tau_P(G)$ is a subgroup of G. It is easy to see that $\tau_P(G)$ is, in fact, normal in G. In particular, $\tau_p(G)$ is a normal p-subgroup of G for any prime p. Moreover, if q is a prime different from p, then $\big[\tau_p(G),\ \tau_q(G)\big] = 1$ by Corollary 2.14. Thus,

$$\prod_{p \in \mathbb{P}} \tau_p(G) = gp\big(\tau_p(G) \mid p \text{ varies over all of } \mathbb{P}\big). \qquad (2.22)$$

We claim that the right-hand side of (2.22) is just $\tau(G)$. It is clearly contained in $\tau(G)$ by the previous discussion. We establish the reverse inclusion. Let $g \in \tau(G)$ be a torsion element of order $d = p_1^{m_1} \cdots p_n^{m_n}$ for some $m_1, \ldots, m_n \in \mathbb{N}$ and distinct primes $p_1, \ldots, p_n$. Define $a_i = d/p_i^{m_i}$ for $i = 1, \ldots, n$. Since $(g^{a_i})^{p_i^{m_i}} = 1$, we have $g^{a_i} \in \tau_{p_i}(G)$. Furthermore, the greatest common divisor of $a_1, \ldots, a_n$ is 1 because they are pairwise relatively prime. Thus, there are integers $s_1, \ldots, s_n$ such that $\sum_{i=1}^{n} a_i s_i = 1$. Hence,

$$g = g^{a_1 s_1 + \cdots + a_n s_n}$$
$$= (g^{a_1})^{s_1} \cdots (g^{a_n})^{s_n},$$

which is contained in $\tau_{p_1}(G)\tau_{p_2}(G) \cdots \tau_{p_n}(G)$. This proves the claim. $\square$

Corollary 2.15 *Let P be a nonempty set of primes. If G is a nilpotent group, then $G/\tau_P(G)$ is P-torsion-free.*

Proof By Theorem 2.26, $\tau_P(G) \trianglelefteq G$. Suppose that $(g\tau_P(G))^n = \tau_P(G)$ for some $g\tau_P(G) \in G/\tau_P(G)$ and P-number n. We need to show that $g\tau_P(G) = \tau_P(G)$. Since $(g\tau_P(G))^n = \tau_P(G)$, we have $g^n \in \tau_P(G)$. Thus, there is a P-number m such that $g^{nm} = (g^n)^m = 1$. Since mn is a P-number, $g \in \tau_P(G)$; that is, $g\tau_P(G) = \tau_P(G)$. $\square$

Corollary 2.16 *Let P be a nonempty set of primes. If G is a finitely generated nilpotent group, then $\tau_P(G)$ is a finite P-torsion group.*

Proof By Theorems 2.18 and 2.26, $\tau_P(G)$ is a finitely generated P-torsion nilpotent group. The result follows from Theorem 2.25.   □

Theorem 2.26 holds for locally nilpotent groups as well.

Theorem 2.27 *If G is a locally nilpotent group and P is any nonempty set of primes, then $\tau_P(G) \trianglelefteq G$. If $\mathbb{P}$ denotes the set of all prime numbers, then*

$$\tau(G) = \prod_{p \in \mathbb{P}} \tau_p(G).$$

Proof Let g, $h \in \tau_P(G)$, and put $H = gp(g, h)$. Since H is a finitely generated subgroup of G, it is nilpotent. By Theorem 2.26, $\tau_P(H) \trianglelefteq H$. Therefore, $gh \in \tau_P(H)$, and thus $gh \in \tau_P(G)$. The rest of the proof is the same as for Theorem 2.26.   □

An analogue of Corollary 2.15 clearly holds for locally nilpotent groups.

Corollary 2.17 *If P is a nonempty set of primes and G is a locally nilpotent group, then $G/\tau_P(G)$ is P-torsion-free.*

2.7 The Upper Central Subgroups and Their Factors

In this section, we focus our attention on some properties of the upper central subgroups and their factors.

2.7.1 Intersection of the Center and a Normal Subgroup

We begin by proving that every nontrivial normal subgroup of a nilpotent group contains a nonidentity central element.

Theorem 2.28 (K. A. Hirsch) *If G is a nilpotent group and N is a nontrivial normal subgroup of G, then $N \cap Z(G) \neq 1$.*

Proof If $N \leq Z(G)$, then the result is immediate. Suppose that $N \nleq Z(G)$. Since G is nilpotent, there exists $i \in \mathbb{N}$ such that $N \cap \zeta_i G \neq 1$. If $i = 1$, then the result is immediate. Assume that $i > 1$, and let $n \in N \cap \zeta_i G$ for some $n \neq 1$. If $n \in Z(G)$, then we have the result. If $n \notin Z(G)$, then there exists $g \in G$ such that $[n, g] \neq 1$. Observe that $[n, g] \in [\zeta_i G, G] \leq \zeta_{i-1} G$ and $[n, g] \in N$ since $N \trianglelefteq G$. Thus, if $N \cap \zeta_i G \neq 1$, then $N \cap \zeta_{i-1} G \neq 1$ for $i > 1$. It follows that $N \cap Z(G) \neq 1$.   □

Theorem 2.28 has several consequences.

Lemma 2.22 *Every maximal normal abelian subgroup of a nilpotent group G coincides with its centralizer in G.*

Proof The proof is done by contradiction. Let M be a maximal normal abelian subgroup of G, and assume that $M \neq C_G(M)$. Clearly, $M \leq C_G(M)$ and $C_G(M)/M$ is a nontrivial normal subgroup of G/M. By Theorem 2.28, there exists an element

$$gM \in Z(G/M) \cap (C_G(M)/M)$$

such that $g \notin M$. Now, $gp(g, M)$ is abelian because $g \in C_G(M)$. Moreover, $gp(g, M)$ is normal in G. To see this, let $g^k m \in gp(g, M)$ for some $k \in \mathbb{Z}$ and $m \in M$, and let $h \in G$. Since $g^k M \in Z(G/M)$, we have

$$h^{-1} g^k m h = g^k m_1 \in gp(g, M)$$

for some $m_1 \in M$. By the maximality of M, we have $g \in M$, a contradiction. $\square$

Corollary 2.18 *Let G be a nilpotent group, and let K be any group. A homomorphism $\varphi \in Hom(G, K)$ is a monomorphism if and only if $\varphi|_{Z(G)}$, the restriction of φ to $Z(G)$, is a monomorphism.*

Proof Suppose that $\varphi|_{Z(G)}$ is a monomorphism. Assume, on the contrary, that φ is not a monomorphism. Then *ker* φ is a nontrivial normal subgroup of G. By Theorem 2.28, *ker* $\varphi \cap Z(G) \neq 1$, and thus $\varphi|_{Z(G)}$ also has a nontrivial kernel. Thus, $\varphi|_{Z(G)}$ is not a monomorphism, a contradiction. The converse is clear. $\square$

Definition 2.19 A nontrivial normal subgroup N of a group G is termed a *minimal normal subgroup* if there is no normal subgroup M of G such that $1 < M < N$.

Thus, if N is a minimal normal subgroup of G and $M \trianglelefteq N$, then either $M = 1$ or $M = N$.

Corollary 2.19 *If G is a nilpotent group, then every minimal normal subgroup of G is contained in $Z(G)$.*

Proof Let N be a minimal normal subgroup of G. Clearly, $N \cap Z(G) \trianglelefteq N$. By minimality, either $N \cap Z(G) = 1$ or $N \cap Z(G) = N$. However, $N \cap Z(G) \neq 1$ by Theorem 2.28. Thus, $N \cap Z(G) = N$ and the result follows. $\square$

Corollary 2.19 allows us to characterize a finite nilpotent group in terms of a certain type of series. Our discussion that follows is based on [9].

Definition 2.20 Let G be a group. A normal series

$$1 = G_0 \leq G_1 \leq \cdots \leq G_n = G$$

of G is called a *chief series* if each factor group G_{i+1}/G_i for $i = 0, 1, \ldots, n-1$ is a minimal normal subgroup of G/G_i. The factor groups G_{i+1}/G_i are called the *chief factors* of G.

Every finite group has a chief series. By the Correspondence Theorem, the condition that G_{i+1}/G_i is a minimal normal subgroup of G/G_i is equivalent to the condition that if $N \triangleleft G$ and $G_i \leq N \leq G_{i+1}$, then either $N = G_i$ or $N = G_{i+1}$.

Lemma 2.23 *Let G be a group with normal subgroups M and N, and suppose that N < M. Further suppose that G has a chief series. The factor M/N is a minimal normal subgroup of G/N if and only if it is a chief factor of G.*

Proof Suppose that M/N is a minimal normal subgroup of G/N. Since G has a chief series, every proper normal series of G can be refined to a chief series of G. In particular, G has a chief series containing M and N as two of its terms. We conclude that M/N is a chief factor of G. The converse is trivial. $\qquad\square$

Theorem 2.29 *A finite group G is nilpotent if and only if every chief factor of G is central.*

Proof If G is nilpotent, then so is any factor group of G by Corollary 2.5. In view of Lemma 2.23, it suffices to show that every minimal normal subgroup of G is in $Z(G)$. This was done in Corollary 2.19.

Conversely, suppose that every chief factor of G is central. This implies that every chief series of G is also a central series of G. Therefore, G is nilpotent. $\qquad\square$

Lemma 2.24 *If G is any group with a chief series, then any central factor of the series is finite and has prime order.*

Proof In light of Lemma 2.23, it suffices to consider a minimal normal subgroup N of G such that $N \leq Z(G)$ and to prove that $|N| = p$ for some prime p. Clearly, every subgroup of N is normal in G because $N \leq Z(G)$. By the minimality of N, the only normal subgroups of N are 1 and N. It follows that $|N| = p$ for some prime p. $\qquad\square$

Remark 2.13 By Theorem 2.29 and Lemma 2.24, every factor of a chief series in a finite nilpotent group is central and has prime order. The converse need not be true (consider S_3).

2.7.2 Separating Points in a Group

Certain properties of the upper central subgroups of a group, as well as their factors, are inherited from the center of the group. These properties allow one to understand the structure of the group, especially when it is nilpotent. The next definition can be found in [13] for abelian groups.

Definition 2.21 Let G and H be any pair of nontrivial groups. We say that H *separates* G if for each element $g \neq 1$ in G, there exists $\varphi \in Hom(G, H)$ such that $\varphi(g) \neq 1$. Such elements of $Hom(G, H)$ are said to *separate points* in G.

Lemma 2.25 *Let P be a nonempty set of primes. Suppose that G and H are groups and H separates G.*

(i) If H is P-torsion-free, then G is P-torsion-free.
(ii) If H has finite exponent m, then G has finite exponent dividing m.

In particular, if H is torsion-free and H separates G, then G is torsion-free.

Proof Both results are proven by contradiction.

(i) Suppose that $1 \neq g \in G$ is a P-torsion element. There exists $\varphi \in Hom(G, H)$
    such that $\varphi(g) \neq 1$. If $g^n = 1$ for some P-number n, then $\varphi(g^n) = (\varphi(g))^n = 1$;
    that is, $\varphi(g)$ is a P-torsion element of H. This contradicts the P-torsion-freeness
    of H. Hence, G is P-torsion-free.

(ii) Assume that there exists $g \in G$ such that $g^m \neq 1$. There exists $\varphi \in Hom(G, H)$
    such that $\varphi(g^m) \neq 1$; that is, $(\varphi(g))^m \neq 1$. However, $\varphi(g) \in H$ and H has
    exponent m. Therefore, $g^m = 1$ for every $g \in G$. Thus, G has exponent
    dividing m.   □

Theorem 2.30 *If G is any group, then $Z(G)$ separates $\zeta_i G/\zeta_{i-1} G$.*

Here of course, we are assuming that $Z(G)$ and $\zeta_i G/\zeta_{i-1} G$ are nontrivial.
In particular, if both $Z(G)$ and $Z(G/Z(G))$ are nontrivial, then there exists a
homomorphism of G onto a nontrivial subgroup of $Z(G)$. This is the case for $i = 2$
and it is due to O. Grün.

Proof The proof is done by induction on i. The case for $i = 1$ is obviously true.
Suppose $i = 2$. We prove that $Z(G)$ separates $\zeta_2 G/Z(G)$. For any element $g \in G$,
consider the map

$$\psi_g : \zeta_2 G \to Z(G) \text{ defined by } \psi_g(x) = [x, g].$$

This map makes sense since $[\zeta_2 G, G] \leq Z(G)$. By Lemma 1.12, ψ_g is a
homomorphism whose kernel clearly contains $Z(G)$. Thus, ψ_g induces a well-
defined homomorphism

$$\overline{\psi}_g : \zeta_2 G/Z(G) \to Z(G) \text{ given by } \overline{\psi}_g(xZ(G)) = [x, g].$$

Let $hZ(G)$ be a nonidentity element of $\zeta_2 G/Z(G)$, so that $h \in \zeta_2 G$ and $h \notin Z(G)$.
There exists some element $g \in G$ such that $[h, g] \neq 1$. This means that $\psi_g(h) \neq 1$,
and consequently, $\overline{\psi}_g(hZ(G)) \neq 1$. Therefore, $Z(G)$ separates $\zeta_2 G/Z(G)$.

Assume that $Z(G)$ separates $\zeta_i G/\zeta_{i-1} G$ for $i > 2$. In order to prove that $Z(G)$
separates $\zeta_{i+1} G/\zeta_i G$, it is enough to show that $\zeta_i G/\zeta_{i-1} G$ separates $\zeta_{i+1} G/\zeta_i G$. By
Lemma 2.11 and the Third Isomorphism Theorem,

$$\frac{\zeta_{i+1} G}{\zeta_i G} \cong \frac{\zeta_{i+1} G/\zeta_{i-1} G}{\zeta_i G/\zeta_{i-1} G} \cong \frac{\zeta_2(G/\zeta_{i-1} G)}{Z(G/\zeta_{i-1} G)}. \tag{2.23}$$

It follows from the previous case that $\zeta_i G/\zeta_{i-1} G$ separates $\zeta_{i+1} G/\zeta_i G$.   □

By Lemma 2.25 (i) and Theorem 2.30, we have:

Corollary 2.20 (D. H. McLain) *Let G be any group, and let P be a nonempty
set of primes. If $Z(G)$ is P-torsion-free, then $\zeta_{i+1} G/\zeta_i G$ is P-torsion-free for each
integer $i \geq 0$.*

We mention that A. I. Mal'cev and S. N. Černikov proved Corollary 2.20 for the
case when P is the set of all primes.

We offer another proof of Corollary 2.20 which uses Lemma 1.13. Let $g \neq 1$ be an element of $\zeta_2 G$ such that $(gZ(G))^n = Z(G)$ in $\zeta_2 G/Z(G)$, where n is a P-number. This means that $g^n \in Z(G)$. If $h \in G$, then

$$[g, h]^n = [g^n, h] = 1$$

by Lemma 1.13 because $[g, h] \in Z(G)$. Since $Z(G)$ is P-torsion-free, $[g, h] = 1$. Therefore, $g \in Z(G)$ and $\zeta_2 G/Z(G)$ is P-torsion-free. The rest now follows by induction on i.

Corollary 2.21 *Let P be a nonempty set of primes. A nilpotent group is P-torsion-free if and only if its center is P-torsion-free.*

Proof Suppose that G is nilpotent of class c and $Z(G)$ is P-torsion-free. By Corollary 2.20, $\zeta_{i+1}G/\zeta_i G$ is P-torsion-free for $0 \leq i \leq c - 1$. Let $1 \neq g \in G$, and let n be any P-number. Since $g \neq 1$, there exists an integer $i \in \{0, \ldots, c - 1\}$ such that $g \in \zeta_{i+1}G \smallsetminus \zeta_i G$. Now, $(g\zeta_i G)^n \neq \zeta_i G$ because $\zeta_{i+1}G/\zeta_i G$ is P-torsion-free. Hence, $g^n \notin \zeta_i G$. This means that $g^n \neq 1$, and thus G is P-torsion-free. The converse is obvious.   □

By Corollaries 2.20 and 2.21, we see that each upper central factor of a torsion-free nilpotent group must be torsion-free abelian.

Corollary 2.22 *Let P be a nonempty set of primes. If G is a P-torsion-free nilpotent group, then so is $G/Z(G)$.*

Proof The center of G is P-torsion-free since G is. By Corollary 2.20, $\zeta_2 G/Z(G)$ is P-torsion-free as well. The result follows from Corollary 2.21 since $\zeta_2 G/Z(G)$ is the center of $G/Z(G)$.   □

Remark 2.14 The lower central factors of a torsion-free nilpotent group are not necessarily torsion-free. For example, fix a positive integer $n > 1$, and let

$$G = \left\{ \begin{pmatrix} 1 & xn & y \\ 0 & 1 & z \\ 0 & 0 & 1 \end{pmatrix} \;\middle|\; x, y, z \in \mathbb{Z} \right\}.$$

It is easy to see that G is a subgroup of the Heisenberg group, which is torsion-free and nilpotent (see Example 2.18). Thus, G is also torsion-free and nilpotent. Now,

$$\gamma_2 G = \left\{ \begin{pmatrix} 1 & 0 & wn \\ 0 & 1 & 0 \\ 0 & 0 & 1 \end{pmatrix} \;\middle|\; w \in \mathbb{Z} \right\}.$$

It follows that $G/\gamma_2 G$ is isomorphic to the direct sum $\mathbb{Z} \oplus \mathbb{Z} \oplus \mathbb{Z}_n$, and this group has torsion.

Corollary 2.23 (S. Dixmier) *Let G be a nilpotent group of class c. If $Z(G)$ has finite exponent m, then $\zeta_{i+1}G/\zeta_iG$ has exponent dividing m for $0 \leq i \leq c$. Consequently, G has exponent dividing m^c.*

Proof By Theorem 2.30 and Lemma 2.25 (ii), each $\zeta_{i+1}G/\zeta_iG$ has exponent dividing m. Let $1 \neq g \in G$. For some $i \in \{0, \ldots, c-1\}$, we have $g \in \zeta_{i+1}G \setminus \zeta_iG$. Since every upper central quotient has exponent dividing m, we have

$$g^m \in \zeta_iG, \; g^{m^2} \in \zeta_{i-1}G, \; \ldots, \; g^{m^{i+1}} \in \zeta_0G = 1.$$

Thus, $g^{m^c} = 1$. □

Lemma 2.26 *Let P be a nonempty set of primes. If G is a finitely generated nilpotent group and $Z(G)$ is a P-torsion group, then G is a finite P-torsion group.*

In particular, every finitely generated nilpotent group with finite center is finite.

Proof By Theorem 2.18, $Z(G)$ is finitely generated. Since $Z(G)$ is also P-torsion and abelian, it must be finite with exponent a P-number. Thus, G has finite exponent which is a P-number by Corollary 2.23. The result follows from Theorem 2.25. □

The center of any torsion group is obviously a torsion group. There are nilpotent groups which are torsion-free, yet their center is a torsion group. This is illustrated in the next example.

Example 2.20 Suppose that A is an additive abelian torsion group with infinite exponent, and let $\vartheta \in Aut(A \oplus A)$ be defined by $\vartheta(x, y) = (x + y, y)$. For each $m \in \mathbb{N}$, set

$$\vartheta^{\circ m} = \underbrace{\vartheta \circ \cdots \circ \vartheta}_{m}.$$

Since $\vartheta^{\circ m}(x, y) = (x + my, y)$ for every $m \in \mathbb{N}$, ϑ has infinite order. Define a mapping

$$\varphi : \mathbb{Z} \to Aut(A \oplus A) \text{ by } \varphi(k) = \vartheta^{\circ k},$$

and let $G = (A \oplus A) \rtimes_\varphi \mathbb{Z}$. Observe that

$$(i, (x, y))(j, (\tilde{x}, \tilde{y})) = (i + j, (\varphi(j))(x, y) + (\tilde{x}, \tilde{y}))$$
$$= (i + j, (x + jy, y) + (\tilde{x}, \tilde{y}))$$
$$= (i + j, (x + \tilde{x} + jy, y + \tilde{y})).$$

It is easy to check that G is torsion-free and

$$Z(G) = \{(0, (x, 0)) \mid x \in A\} \cong A.$$

It follows that G is nilpotent of class 2.

We end this section with a lemma which will be useful later.

Lemma 2.27 *Every infinite finitely generated nilpotent group contains a central element of infinite order.*

Proof Let G be an infinite finitely generated nilpotent group. If G has no central elements of infinite order, then $Z(G)$ is a torsion group. By Lemma 2.26, G must be finite, a contradiction. Therefore, G has a central element of infinite order. □

References

1. R. Baer, Nilpotent groups and their generalizations. Trans. Am. Math. Soc. **47**, 393–434 (1940). MR0002121
2. G. Baumslag, Wreath products and p-groups. Proc. Camb. Philos. Soc. **55**, 224–231 (1959). MR0105437
3. B. Baumslag, B. Chandler, *Schaum's Outline of Theory and Problems of Group Theory* (McGraw-Hill, New York, 1968)
4. K. Conrad, http://www.math.uconn.edu/~kconrad/blurbs/grouptheory/subgpseries1.pdf
5. P. Hall, Some sufficient conditions for a group to be nilpotent. Ill. J. Math. **2**, 787–801 (1958). MR0105441
6. M.I. Kargapolov, J.I. Merzljakov, *Fundamentals of the Theory of Groups*. Graduate Texts in Mathematics, vol. 62 (Springer, New York, 1979). MR0551207. Translated from the second Russian edition by Robert G. Burns
7. D.J.S. Robinson, A property of the lower central series of a group. Math. Z. **107**, 225–231 (1968). MR0235043
8. D.J.S. Robinson, *A Course in the Theory of Groups*. Graduate Texts in Mathematics, vol. 80, 2nd edn. (Springer, New York, 1996). MR1357169
9. J.S. Rose, *A Course on Group Theory* (Dover Publications, New York, 1994). MR1298629. Reprint of the 1978 original [Dover, New York; MR0498810 (58 #16847)]
10. J.J. Rotman, *Advanced Modern Algebra* (Prentice Hall, Upper Saddle River, NJ, 2002). MR2043445
11. W.R. Scott, *Group Theory* (Prentice-Hall, Inc., Englewood Cliffs, NJ, 1964). MR0167513
12. A.G.R. Stewart, On the class of certain nilpotent groups. Proc. R. Soc. Ser. A **292**, 374–379 (1966). MR0197573
13. R.B. Warfield, Jr., *Nilpotent Groups*. Lecture Notes in Mathematics, vol. 513 (Springer, Berlin, 1976). MR0409661

Chapter 3
The Collection Process and Basic Commutators

The goal of this chapter is to determine normal forms for elements in finitely generated free groups and free nilpotent groups. This is done by using a collection process which we discuss in Section 3.1. A study of weighted commutators and basic commutators in a group relative to a given generating set also appears in this section. The highlight of Section 3.1 is a fundamental result stating that if G is any group generated by a set X and $\gamma_i G$ denotes the ith lower central subgroup, then each quotient $\gamma_n G / \gamma_{n+1} G$ is generated, modulo $\gamma_{n+1} G$, by a sequence of basic commutators on X of weight n. Section 3.2 is devoted to the so-called collection formula. This formula expresses a positive power of a product of elements $x_1, \ldots, x_r$ of a group as a product of positive powers of basic commutators in $x_1, \ldots, x_r$. The collection process developed in Section 3.1 plays a key role here. In Section 3.3, we investigate basic commutators in finitely generated free groups and free nilpotent groups. We prove a major result which states that a finitely generated free nilpotent group, freely generated by a set X, has a "basis" consisting of basic commutators in X. The techniques used in this section involve groupoids, Lie rings, and the Magnus embedding. We end the chapter with Section 3.4, which is devoted to the rather technical proof of the collection formula obtained in Section 3.2.

3.1 The Collection Process

Let G be an abelian group generated by $X = \{x_1, \ldots, x_k\}$. Since any two elements of an abelian group commute, it is clear that every element of G can be written in the (not necessarily unique) form $x_1^{e_1} \cdots x_k^{e_k}$ for some integers $e_1, \ldots, e_k$.

© Springer International Publishing AG 2017
A.E. Clement et al., *The Theory of Nilpotent Groups*,
DOI 10.1007/978-3-319-66213-8_3

Next, suppose that G is *any* group generated by $X = \{x_1, \ldots, x_k\}$, and choose $n \in \mathbb{N}$. Our goal in this section is to prove that every element of G can be expressed, modulo $\gamma_{n+1}G$, as a (not necessarily unique) product of the form

$$x_1^{e_1} \cdots x_k^{e_k} c_{k+1}^{e_{k+1}} \cdots c_t^{e_t},$$

where $e_1, \ldots, e_t$ are integers and $c_{k+1}, \ldots, c_t$ are certain commutators in the elements of X. In order to establish this, we introduce a collection process. Our discussion is based on the work of M. Hall [7].

3.1.1   Weighted Commutators

Definition 3.1 Let G be a group generated by $X = \{x_1, \ldots, x_k\}$. A *commutator* c_j of *weight* $w(c_j)$ is defined as follows:

1. The commutators of weight one are the elements of X:

$$c_1 = x_1, \; c_2 = x_2, \; \ldots, \; c_k = x_k.$$

2. If $i \neq j$ and c_i and c_j are commutators of weights $w(c_i)$ and $w(c_j)$ respectively, then $[c_i, c_j]$ is a commutator of weight $w(c_i) + w(c_j)$.

Definition 3.1 is relative to a given generating set X. Moreover, every commutator with an assigned weight is built upon the elements of the generating set, but not their inverses. If $X = \{x_1, x_2, x_3, x_4\}$, for example, then

$$[[x_1, x_3], [x_4, x_2]]$$

is a commutator of weight 4, whereas the commutator $\left[x_1, x_4^{-1}\right]$ does not have an assigned weight.

We impose an ordering on the weighted commutators $c_1, c_2, \ldots$ by their subscripts in the following manner:

(i) $c_1 = x_1, \; c_2 = x_2, \; \ldots, \; c_k = x_k$;
(ii) After c_k, we list $c_{k+1}, c_{k+2}, \ldots$ in order of their weight, arbitrarily ordering those commutators of equal weight.

Note that there are finitely many commutators of a given weight because X is a finite set. Thus, the ordering described above makes sense.

3.1.2   The Collection Process for Weighted Commutators

Every product of weighted commutators can be expressed in the form

$$c_{i_1} \cdots c_{i_m} c_{i_{m+1}} \cdots c_{i_n},$$ (3.1)

where $i_1 \leq \cdots \leq i_m$ and $i_m \leq i_s$ whenever $s = m + 1, \ldots, n$.

Definition 3.2 The product $c_{i_1} \cdots c_{i_m}$ is called the *collected part* and $c_{i_{m+1}} \cdots c_{i_n}$ is the *uncollected part*, provided that i_{m+1} is not a smallest subscript among $i_{m+1}, \ldots, i_n$.

Here it is assumed that the commutators $c_{i_{m+1}}, \ldots, c_{i_n}$ in the collected part do not appear in the prescribed order. Furthermore, the collected part in (3.1) may be empty.

Next, we discuss the collection process for a typical product of the form (3.1). The commutator identity

$$ab = ba[a, b]$$ (3.2)

is the key ingredient in the process. Let i_r be the smallest of the subscripts in the uncollected part (hence, $r \geq m + 1$). Note that there may be several occurrences of c_{i_r} among $c_{i_{m+1}}, \ldots, c_{i_n}$ in the uncollected part. Let c_{i_l} be the leftmost occurrence of c_{i_r} in the uncollected part. In order to *collect* c_{i_l} to the left in (3.1), we use (3.2) and replace $c_{i_{l-1}} c_{i_l}$ by $c_{i_l} c_{i_{l-1}} [c_{i_{l-1}}, c_{i_l}]$. As a result, (3.1) changes from

$$c_{i_1} \cdots c_{i_m} \cdots c_{i_{l-1}} c_{i_l} \cdots c_{i_n}$$

to

$$c_{i_1} \cdots c_{i_m} \cdots c_{i_l} c_{i_{l-1}} [c_{i_{l-1}}, c_{i_l}] \cdots c_{i_n}.$$

Realize that c_{i_l} has moved one step closer to the collected part, and a new commutator $[c_{i_{l-1}}, c_{i_l}]$ has been introduced. This new commutator appears later than c_{i_l} in the ordering since its weight exceeds the weight of c_{i_l}. This means that c_{i_l} is still an occurrence of the earliest commutator in the uncollected part.

If we repeat this process enough times, then c_{i_l} will be moved to the $(m + 1)$st position and become the last commutator of the collected part. Notice that this *collection process* may never end because a new commutator is introduced at each step. Furthermore, this process does not alter the element of G determined by (3.1).

3.1.3 Basic Commutators

An element g of a group G generated by X is called a *positive word* if it can be written as a product of the elements of X, but not their inverses. When the collection process described above is applied to g, only certain commutators will appear when g is in the collected form. For example, consider the positive word $g = x_1 x_2 x_1$. Collecting x_1 gives

$$g = x_1 x_2 x_1 = x_1 x_1 x_2 [x_2, \ x_1] = x_1^2 x_2 [x_2, \ x_1].$$

Note that $[x_2, \ x_1]$ occurs in the collected form of g, but $[x_1, \ x_2]$ does not. The weighted commutators which arise in the collected form of a positive word are called basic commutators.

Definition 3.3 Let G be a group generated by $X = \{x_1, \ \ldots, \ x_k\}$. A *basic commutator* b_j of weight $w(b_j)$ is defined as such:

1. The elements of X are the basic commutators of weight one. We impose an arbitrary ordering on these and relabel them as $b_1, \ b_2, \ \ldots, \ b_k$, where $b_i < b_j$ if $i < j$.
2. Suppose that we have defined and ordered the basic commutators of weight less than $l > 1$. The basic commutators of weight l are $[b_i, \ b_j]$, where

    (i) b_i and b_j are basic commutators and $w(b_i) + w(b_j) = l$,
    (ii) $b_i > b_j$, and
    (iii) if $b_i = [b_s, \ b_t]$, then $b_j \geq b_t$.

3. Basic commutators of weight l come after all basic commutators of weight less than l and are ordered arbitrarily with respect to one another.

The sequence $b_1, \ b_2, \ \ldots$ is called a *basic sequence of basic commutators* on X.

Example 3.1 We illustrate how to construct the basic commutators, up to weight 4, of a group with generating set $\{x_1, \ x_2, \ x_3\}$. This example also appears in Section 2.7 of [8]. Suppose that the ordering

$$x_1 < x_2 < x_3$$

is imposed on the generators, which are just the basic commutators of weight 1. The basic commutators of weight 2 are

$$[x_2, \ x_1], \ [x_3, \ x_1], \ \text{and} \ [x_3, \ x_2].$$

These commutators are ordered as such:

$$[x_2, \ x_1] < [x_3, \ x_1] < [x_3, \ x_2].$$

Now, the basic commutators of weight 3 are

$$[x_2, x_1, x_1], [x_2, x_1, x_2], [x_2, x_1, x_3], [x_3, x_1, x_1],$$

$$[x_3, x_1, x_2], [x_3, x_1, x_3], [x_3, x_2, x_2], \text{ and } [x_3, x_2, x_3].$$

Notice that $[x_3, x_2, x_1]$ doesn't occur as a basic commutator since, referring to Definition 3.3 (iii), $b_t = x_2$ and $b_j = x_1$, but $x_1 \not\geq x_2$. If the commutators of weight 3 are ordered as

$$[x_2, x_1, x_1] < [x_2, x_1, x_2] < [x_2, x_1, x_3] < [x_3, x_1, x_1] <$$

$$[x_3, x_1, x_2] < [x_3, x_1, x_3] < [x_3, x_2, x_2] < [x_3, x_2, x_3],$$

then the basic commutators of weight 4 will be

$$[x_2, x_1, x_1, x_1], [x_2, x_1, x_1, x_2], [x_2, x_1, x_1, x_3], [x_2, x_1, x_2, x_2], [x_2, x_1, x_2, x_3],$$

$$[x_2, x_1, x_3, x_3], [x_3, x_1, x_1, x_1], [x_3, x_1, x_1, x_2], [x_3, x_1, x_1, x_3], [x_3, x_1, x_2, x_2],$$

$$[x_3, x_1, x_2, x_3], [x_3, x_1, x_3, x_3], [x_3, x_2, x_2, x_2], [x_3, x_2, x_2, x_3], [x_3, x_2, x_3, x_3],$$

$$[[x_3, x_1], [x_2, x_1]], [[x_3, x_2], [x_2, x_1]], \text{ and } [[x_3, x_2], [x_3, x_1]].$$

Lemma 3.1 *Suppose that G is a finitely generated group, and let $g \in G$ be a positive word in the generating set. If commutators are ordered according to their weight, but those of the same weight are ordered arbitrarily, then the only commutators that are introduced when the collection process is applied to g are basic commutators.*

Proof Suppose that g has a subword of the form $b_i b_j$. If this subword is replaced by $b_j b_i [b_i, b_j]$, then b_j has been collected before b_i. This means that in the ordering, $b_i > b_j$. If it is the case that $b_i = [b_s, b_t]$, then $b_i b_j$ becomes

$$b_j b_i [b_i, b_j] = b_j [b_s, b_t][b_s, b_t, b_j].$$

Thus, b_t was collected before collecting this b_j. And so, $b_j \geq b_t$. □

3.1.4 The Collection Process For Arbitrary Group Elements

Our next goal is to apply the collection process to arbitrary words (not necessarily positive) in a group. Let $X = \{x_1, \ldots, x_k\}$ be a set of generators of any group G. Suppose that w is any word in $X \cup X^{-1}$, where $X^{-1} = \{x_1^{-1}, \ldots, x_k^{-1}\}$. We show that each step in the collection process when applied to w produces either a basic commutator or the inverse of a basic commutator.

Let a and b be basic commutators, and suppose that any of the expressions

$$b^{-1}a, \quad ba^{-1}, \quad \text{or } b^{-1}a^{-1}$$

occurs in the word w. We use the commutator identities in Lemma 1.4 to show that either a or a^{-1} can be collected in any one of these expressions in such a way that only basic commutators and their inverses appear in the resulting product (possibly modulo some term of the lower central series).

• Consider the expression $b^{-1}a$. This can be re-expressed as

$$b^{-1}a = a\left(a^{-1}b^{-1}ab\right)b^{-1} = a[a, b]b^{-1} = a[b, a]^{-1}b^{-1}. \tag{3.3}$$

Since a, b and $[b, a]$ are basic commutators, a has been collected in the desired way.

• Next, consider the expression ba^{-1}. By (3.2), $ba^{-1} = a^{-1}b\left[b, a^{-1}\right]$. Now, a and b are basic commutators, but $\left[b, a^{-1}\right]$ is not because it contains a^{-1}. This commutator needs to be rewritten so that a^{-1} is removed from the commutator. Observe that

$$1 = [b, 1] = \left[b, aa^{-1}\right] = \left[b, a^{-1}\right]\left[b, a\right]\left[b, a, a^{-1}\right], \tag{3.4}$$

and consequently,

$$\left[b, a^{-1}\right] = \left[b, a, a^{-1}\right]^{-1}\left[b, a\right]^{-1}. \tag{3.5}$$

Note that $[b, a]$ is a basic commutator, but $\left[b, a, a^{-1}\right]$ is not. We repeat the method used in (3.4) to rewrite $\left[b, a, a^{-1}\right]$ as

$$1 = [b, a, 1] = \left[b, a, aa^{-1}\right] = \left[b, a, a^{-1}\right]\left[b, a, a\right]\left[b, a, a, a^{-1}\right]. \tag{3.6}$$

Thus,

$$\left[b, a, a^{-1}\right] = \left[b, a, a, a^{-1}\right]^{-1}\left[b, a, a\right]^{-1}. \tag{3.7}$$

Once again, we get a basic commutator $[b, a, a]$ and a commutator $\left[b, a, a, a^{-1}\right]$ that is not basic.

This process is continued in a more systematic way. Set $b_0 = b$, and recursively define $b_{i+1} = [b_i, a]$. Since a and b are basic commutators, so is b_i for each $i \geq 0$. Generalizing our previous computations gives

$$\left[b_i, a^{-1}\right] = \left[b_{i+1}, a^{-1}\right]^{-1}\left[b_i, a\right]^{-1} = \left[b_{i+1}, a^{-1}\right]^{-1}b_{i+1}^{-1}, \tag{3.8}$$

and thus,

$$\left[b_i, a^{-1}\right]^{-1} = b_{i+1}\left[b_{i+1}, a^{-1}\right]. \tag{3.9}$$

Using (3.8) and (3.9) repeatedly yields

$$
\begin{aligned}
\left[b, a^{-1}\right] &= \left[b_1, a^{-1}\right]^{-1}b_1^{-1} \\
&= b_2\left[b_2, a^{-1}\right]b_1^{-1} \\
&= b_2\left[b_3, a^{-1}\right]^{-1}b_3^{-1}b_1^{-1} \\
&= b_2b_4\left[b_4, a^{-1}\right]b_3^{-1}b_1^{-1} \\
&= b_2b_4\left[b_5, a^{-1}\right]^{-1}b_5^{-1}b_3^{-1}b_1^{-1} \\
&= b_2b_4b_6\left[b_6, a^{-1}\right]b_5^{-1}b_3^{-1}b_1^{-1} \\
&\quad\vdots \\
&= b_2b_4b_6\cdots\left[b_s, a^{-1}\right]\cdots b_5^{-1}b_3^{-1}b_1^{-1}.
\end{aligned}
$$

Clearly, this procedure may never end because a commutator involving a^{-1} arises after each step. However, for any chosen $n \in \mathbb{N}$, there exists $s \in \mathbb{N}$ such the commutator $\left[b_s, a^{-1}\right]$ is of weight $(n + 1)$ or more. By Theorem 2.14, $\left[b_s, a^{-1}\right] \in \gamma_{n+1}G$. This means that, modulo $\gamma_{n+1}G$, the procedure does terminate and we get

$$\left[b, a^{-1}\right] = b_2b_4b_6\cdots b_5^{-1}b_3^{-1}b_1^{-1} \bmod \gamma_{n+1}G, \tag{3.10}$$

where only finitely many of the basic commutators b_i appear in (3.10). We conclude that

$$ba^{-1} = a^{-1}b\left[b, \ a^{-1}\right] = a^{-1}bb_2b_4b_6 \cdots b_5^{-1}b_3^{-1}b_1^{-1} \text{ mod } \gamma_{n+1}G \qquad (3.11)$$

and a has been collected.

- Lastly, we collect a^{-1} in the expression $b^{-1}a^{-1}$. Note that

$$b^{-1}a^{-1} = a^{-1}\left(ab^{-1}a^{-1}\right) = a^{-1}\left(aba^{-1}\right)^{-1}.$$

By (3.11), we obtain

$$aba^{-1} = bb_2b_4 \cdots b_5^{-1}b_3^{-1}b_1^{-1} \text{ mod } \gamma_{n+1}G.$$

Therefore,

$$b^{-1}a^{-1} = a^{-1}b_1b_3b_5 \cdots b_4^{-1}b_2^{-1}b^{-1} \text{ mod } \gamma_{n+1}G. \qquad (3.12)$$

We have shown that for any $n \in \mathbb{N}$, collecting in words involving inverses of basic commutators gives rise to a *finite* product of basic commutators and their inverses, modulo $\gamma_{n+1}G$. This leads to the next important result which can be found in [5].

Theorem 3.1 (P. Hall) *Let G be any finitely generated group with generating set $X = \{x_1, \ldots, x_k\}$, and choose $n \in \mathbb{N}$. The abelian group $\gamma_n G/\gamma_{n+1}G$ is generated, modulo $\gamma_{n+1}G$, by the basic commutators of weight n. Furthermore, every element of G can be expressed in the (not necessarily unique) form*

$$b_1^{e_1}b_2^{e_2} \cdots b_t^{e_t} \text{ mod } \gamma_{n+1}G,$$

where $e_1, \ldots, e_t$ are integers and $b_1, \ldots, b_t$ are the basic commutators of weights $1, 2, \ldots, n$.

Proof Apply (3.2), (3.3), (3.11), and (3.12). □

The next corollary characterizes nilpotent groups in terms of basic commutators.

Corollary 3.1 *If G is a finitely generated group, finitely generated by X, then G is nilpotent if and only if all but finitely many basic commutators on X equal the identity.*

Proof This is immediate from Theorems 2.2 (i) and 3.1. □

We conclude from Corollary 3.1 that if G is a finitely generated nilpotent group of class c, finitely generated by X, then every element of G can be written in the (not necessarily unique) form $b_1^{n_1} \cdots b_t^{n_t}$, where $n_1, \ldots, n_t$ are integers, and $b_1, \ldots, b_t$ are basic commutators on X of weights 1, 2, ..., c.

3.2   The Collection Formula

Let G be a group with $x_1, \ldots, x_r \in G$, and let $n \in \mathbb{N}$. The *collection formula* is a formula for expressing

$$(x_1 \cdots x_r)^n$$

as a product of positive powers of basic commutators in $x_1, \ldots, x_r$. In this section, we present the formula, along with some of its applications. Its derivation, which we give in the last section of this chapter, involves an application of the collection process which was discussed in Section 3.1. It appears in [3] for the case $r = 2$, and in [7] for the case $r \geq 2$.

3.2.1   Preliminary Examples

Before we state the collection formula, it is instructive to work out some examples for nilpotent groups. By convention, $\binom{n}{k} = 0$ whenever $n < k$.

Example 3.2 Let G be an abelian group with $x_1, \ldots, x_r \in G$, and let $n \in \mathbb{N}$. Using (3.2), we obtain

$$(x_1 \cdots x_r)^n = \underbrace{(x_1 \cdots x_r) \cdots (x_1 \cdots x_r)}_{n \text{ factors}} = x_1^n \cdots x_r^n$$

since $[x_i, x_j] = 1$ for $1 \leq i, j \leq r$.

Example 3.3 Suppose that G is a nilpotent group of class 2 with $x, y \in G$, and let $n \in \mathbb{N}$. We compute $(xy)^n$ by collecting terms to the left using (3.2) repeatedly. Since all commutators of G are central, every commutator of weight exceeding 2 equals the identity. Thus, we have

$$(xy)^2 = x \underline{yx} y = x^2 y^2 [y, x] \tag{3.13}$$

and

$$(xy)^3 = x\,\underline{yx}\,yxy = x^2y\,\underline{yx}\,y[y,\,x] = x^2\,\underline{yx}\,y^2[y,\,x]^2 = x^3y^3[y,\,x]^3.$$

In the above, we have applied (3.2) to the underlined products and moved all (central) basic commutators of weight 2 to the right. We could have used (3.13) to compute $(xy)^3$ just as well. By induction on n, it can be verified that

$$(xy)^n = x^ny^n[y,\,x]^{\binom{n}{2}}.$$

In general, if $x_1, \ldots, x_r$ are elements of G, then

$$(x_1\cdots x_r)^n = x_1^n\cdots x_r^n \prod [x_i,\,x_j]^{\binom{n}{2}}, \tag{3.14}$$

where the product is taken over all i and j satisfying the conditions $i > j$, $2 \le i \le r$, and $1 \le j \le r - 1$. Note that the exponents in (3.14) involve binomial coefficients containing n.

Example 3.4 Let G be a nilpotent group of class 3 with $x, y \in G$, and let $n \in \mathbb{N}$. Using the collection process, we compute $(xy)^2$ and $(xy)^3$ independently of one another. Since $\gamma_4 G = 1$, every basic commutator of weight 3 is central. Thus, every commutator of weight more than 3 equals the identity. Taking this into consideration, we get

$$(xy)^2 = x\,\underline{yx}\,y = x^2y\,\underline{[y,\,x]}y = x^2y^2[y,\,x][y,\,x,\,y] \tag{3.15}$$

and

$$
\begin{aligned}
(xy)^3 &= x\,\underline{yx}\,yxy \\
&= x^2y[y,\,x]\,\underline{yx}\,y \\
&= x^2y\,\underline{[y,\,x]x}\,y[y,\,x]y \\
&= x^2\,\underline{yx}\,[y,\,x]y[y,\,x]y[y,\,x,\,x] \\
&= x^3y[y,\,x]\,\underline{[y,\,x]y}\,[y,\,x]y[y,\,x,\,x] \\
&= x^3y\,\underline{[y,\,x]y}\,[y,\,x][y,\,x]y[y,\,x,\,x][y,\,x,\,y] \\
&= x^3y^2[y,\,x][y,\,x]\,\underline{[y,\,x]y}\,[y,\,x,\,x][y,\,x,\,y]^2 \\
&= x^3y^2[y,\,x]\,\underline{[y,\,x]y}\,[y,\,x][y,\,x,\,x][y,\,x,\,y]^3 \\
&= x^3y^2\,\underline{[y,\,x]y}\,[y,\,x][y,\,x][y,\,x,\,x][y,\,x,\,y]^4 \\
&= x^3y^3[y,\,x]^3[y,\,x,\,x][y,\,x,\,y]^5.
\end{aligned}
$$

Once again, we have applied (3.2) to the underlined products and all basic commutators of weight 3 have been moved to the right. It is also possible to get $(xy)^3$ directly from (3.15). One can show by induction on n that

$$(xy)^n = x^n y^n [y, x]^{\binom{n}{2}} [y, x, x]^{\binom{n}{3}} [y, x, y]^{\binom{n}{2}+2\binom{n}{3}}. \tag{3.16}$$

Once again, the exponents in (3.16) involve binomial coefficients containing n.

3.2.2 The Collection Formula and Applications

In Section 3.1, a certain ordering was imposed on the basic commutators. We need to make this ordering more specific by including an additional condition:

- The basic commutators of weight l come after all basic commutators of weight less than l in the ordering and satisfy the following condition:

$$[b_{11}, b_{12}] < [b_{21}, b_{22}] \text{ if either } b_{12} < b_{22}, \text{ or } b_{12} = b_{22} \text{ and } b_{11} < b_{21}.$$

Taking this new condition into account, we have:

Theorem 3.2 *Let G be a finitely generated group, generated by $X = \{x_1, \ldots, x_r\}$, and let $n \in \mathbb{N}$. Then*

$$(x_1 \cdots x_r)^n = x_1^n \cdots x_r^n b_{r+1}^{e_{r+1}} \cdots b_j^{e_j} d_1 \cdots d_t,$$

where $b_{r+1}, \ldots, b_j$ are basic commutators on X occurring in the prescribed order, and $d_1, \ldots, d_t$ are basic commutators occurring after b_j in the ordering. For each $i = r+1, \ldots, j$,

$$e_i = a_1 n + a_2 \binom{n}{2} + \cdots + a_m \binom{n}{m},$$

where $m = w(b_i)$, and $a_1, \ldots, a_m$ are nonnegative integers that depend on b_i but not on n.

The proof of Theorem 3.2 is rather technical and can be found in Section 3.4.

Corollary 3.2 *Suppose that G is a nilpotent group whose class c is less than a given prime p, and let $n \in \mathbb{N}$. If $g_1, \ldots, g_r$ are elements of G, then*

$$(g_1 \cdots g_r)^{p^n} = g_1^{p^n} \cdots g_r^{p^n} s_{r+1}^{p^n} \cdots s_k^{p^n},$$

where $s_{r+1}, \ldots, s_k$ are contained in the commutator subgroup of $gp(g_1, \ldots, g_r)$.

Proof By Theorem 3.2,

$$(g_1 \cdots g_r)^{p^n} = g_1^{p^n} \cdots g_r^{p^n} b_{r+1}^{e_{r+1}} \cdots b_j^{e_j} d_1 \cdots d_t,$$

where $b_{r+1}, \ldots, b_j$ are basic commutators on $\{g_1, \ldots, g_r\}$, occurring in the prescribed order, and $d_1, \ldots, d_t$ are basic commutators occurring after b_j in the ordering. Since G is nilpotent, there exists $k \in \mathbb{N}$ such that $b_k \neq 1$, $b_{k+1} = 1$, and

$$(g_1 \cdots g_r)^{p^n} = g_1^{p^n} \cdots g_r^{p^n} b_{r+1}^{e_{r+1}} \cdots b_k^{e_k}. \tag{3.17}$$

Furthermore,

$$e_i = a_1 p^n + a_2 \binom{p^n}{2} + \cdots + a_m \binom{p^n}{m}$$

for $i = r + 1, \ldots, k$, where $m = w(b_i)$ and $a_1, \ldots, a_m$ are nonnegative integers which are independent of p^n. Now, $m \leq c$ because b_i occurs in (3.17). Since $p > c$, we find that $p^n > m$. Let $t \in \mathbb{N}$ such that $1 \leq t \leq m$. Note that $\binom{p^n}{t} \neq 1$, and t and p^n are relatively prime since $t \leq m < p$. Consequently, p^n divides $\binom{p^n}{t}$, and thus divides each e_i. Setting $s_i = b_i^{e_i/p^n}$ for $r + 1 \leq i \leq k$ completes the proof.   □

We introduce some notation which will be used throughout. For any group G, put

$$G^n = gp(g^n \mid g \in G).$$

Thus, G^n is the subgroup of G generated by the nth powers of the elements of G.

Let G be an abelian group, and let g^n and h^n be elements of G^n for some g, $h \in G$. Since $g^n h^n = (gh)^n \in G^n$, we see that a product of nth powers of elements of G is again an nth power. Thus, $G^n \leq G$ in this case. Suppose on the other hand, that G is non-abelian. Then a product of elements of G^n does not have to be a power of n. For instance, let $n = 2$ and $G = \mathcal{H}$, a nilpotent group of class 2 by Example 2.12. Set

$$a = \begin{pmatrix} 1 & 1 & 0 \\ 0 & 1 & 0 \\ 0 & 0 & 1 \end{pmatrix} \text{ and } b = \begin{pmatrix} 1 & 0 & 0 \\ 0 & 1 & 1 \\ 0 & 0 & 1 \end{pmatrix}. \tag{3.18}$$

A direct calculation shows that there is no element $d \in \mathcal{H}$ such that $a^2 b^2 = d^2$.

Lemma 3.2 (P. Hall) *Let G be a nilpotent group of class c, and let p be a prime greater than c. If $n \in \mathbb{N}$, then every element of G^{p^n} is a p^nth power. Thus, a product of p^nth powers of elements of G is also a p^nth power.*

Proof We adopt the proof given in Section 2.2 of [9]. Let $g_1, \ldots, g_r$ be elements of G. The proof is done by induction on c. If $c = 1$, then G is abelian. In this case,

$$g_1^{p^n} \cdots g_r^{p^n} = (g_1 \cdots g_r)^{p^n}.$$

Suppose that the lemma holds for all nilpotent groups of class less than c, where $c > 1$. By Corollary 3.2,

$$(g_1 \cdots g_r)^{p^n} = g_1^{p^n} \cdots g_r^{p^n} s_1^{p^n} \cdots s_t^{p^n}$$

for $s_1, \ldots, s_t \in \gamma_2 G$. Thus,

$$g_1^{p^n} \cdots g_r^{p^n} = (g_1 \cdots g_r)^{p^n} s_t^{-p^n} \cdots s_1^{-p^n}. \tag{3.19}$$

We claim that the right-hand side of (3.19) is a p^nth power. Put

$$K = gp(g_1 \cdots g_r, \gamma_2 G) \quad \text{and} \quad L = gp(g_1 \cdots g_r, s_1, \ldots, s_t).$$

By Lemma 2.9, K is of class at most $c - 1$. Hence, L also is of class at most $c - 1$ since it is a subgroup of K. By induction,

$$(g_1 \cdots g_r)^{p^n} s_t^{-p^n} \cdots s_1^{-p^n} \in L$$

is a p^nth power. $\qquad\qquad\qquad\qquad\qquad\qquad\qquad\qquad\qquad\qquad\qquad\square$

Example 3.5 Let $G = \mathscr{H}$. Set $p = 3$ and $n = 1$, and consider the matrices a and b in (3.18). By Lemma 3.2, the product $a^3 b^3$ must be a third power. In fact,

$$\begin{pmatrix} 1 & 1 & 0 \\ 0 & 1 & 0 \\ 0 & 0 & 1 \end{pmatrix}^3 \begin{pmatrix} 1 & 0 & 0 \\ 0 & 1 & 1 \\ 0 & 0 & 1 \end{pmatrix}^3 = \begin{pmatrix} 1 & 1 & 2 \\ 0 & 1 & 1 \\ 0 & 0 & 1 \end{pmatrix}^3.$$

3.3 A Basis Theorem

According to Theorem 3.1, each abelian quotient $\gamma_n G / \gamma_{n+1} G$ of a finitely generated group G with finite generating set X is generated by the basic commutators on X of weight n, modulo $\gamma_{n+1} G$. In this section, we prove that if G happens to be a free group, freely generated by X, then these quotients are, in fact, free abelian on these basic commutators, modulo $\gamma_{n+1} G$.

The basic idea behind the proof is to examine the structure of certain Lie rings (see Definition 3.8). It turns out that the Lie bracket of elements in such a Lie ring resembles the commutator of group elements. Consequently, some of the information obtained from these rings transfers over to groups in a natural way. Many people have contributed toward solving this problem, including M. Hall, P. Hall, W. Magnus, and E. Witt. Our discussion is based on [1] and [5] (see also [4, 6, 7], and Chapter 5 of [11]).

3.3.1   Groupoids and Basic Sequences

We will develop certain machinery that will be applied to both groups and rings. First, we do this for groupoids. The reader can consult Chapter 2 in [2] for an introduction to groupoids.

Definition 3.4 A *groupoid* is a pair (G, μ) consisting of a nonempty set G and a binary operation $\mu : G \times G \to G$.

If (G, μ) is a groupoid and $g, h \in G$, then we write gh rather than $\mu(g, h)$ and refer to gh as the *product* of g and h. Note that G does not necessarily satisfy associativity, commutativity, or any other group-theoretic property except closure under μ. If μ is understood from the context, then we simply use the phrase "G is a groupoid."

Definition 3.5 A *free groupoid* G on a nonempty set X is the groupoid consisting of all elements of X, together with all bracketed products of elements of X. We say that G is *freely generated* by X.

Example 3.6 If G is a free groupoid, freely generated by $X = \{x_1, x_2, x_3\}$, then

$$x_1, \ (x_1x_2)x_1, \ (x_2(x_2x_3))((x_1x_3)x_3), \ \text{and} \ ((x_3x_1)x_1)(x_2x_3)$$

are elements of G.

Every element of a free groupoid G, freely generated by X, can be uniquely written as a bracketed product of elements of X. The *length* of $g \in G$, denoted $|g|$, is the number of elements of X occurring in g. For example, if $X = \{x_1, x_2, x_3\}$, then $|x_1| = 1$, $|x_1x_3| = 2$, and $|((x_3x_2)x_1)(x_1x_3)| = 5$.

Definition 3.6 Let G be a free groupoid, freely generated by a finite set X, and let

$$b_1, b_2, \ldots \tag{3.20}$$

be a sequence of elements in G. Then (3.20) is called a *basic sequence* in X if the following conditions are satisfied:

1. The elements of X appear in the sequence.
2. If $|b_i| < |b_j|$, then $i < j$.
3. If $u = vw \in G$ for some elements $v, w \in G$ and $|u| \geq 2$, then u belongs to (3.20) if and only if
   (i) $v = b_i$, $w = b_j$, and $i > j$, and
   (ii) either $|v| = 1$ or $v = b_kb_l$, where $l \leq j$.

The notion of a basic sequence of elements of a free groupoid extends to non-free groupoids in a natural way.

Definition 3.7 Let H be a groupoid generated by $M = \{\mu_1, \ldots, \mu_r\}$, and let G be a free groupoid, freely generated by $X = \{x_1, \ldots, x_r\}$. A sequence $\beta_1, \beta_2, \ldots$ of elements of H is called a *basic sequence* in M if and only if there exists a basic sequence $b_1, b_2, \ldots$ in X such that the groupoid homomorphism from G to H defined by $x_i \mapsto \mu_i$ maps b_i to β_i.

In order to construct a basic sequence, we use the so-called "rep" operation. Let G be any groupoid, and let A be a nonempty subset of G. Choose an element $a \in A$, and define the set A *rep* a to consist of all elements of $A \backslash \{a\}$, together with bracketed products of the form

$$((\cdots ((ba)a) \cdots)a)a,$$

where $b \in A \backslash \{a\}$. Such bracketed products are called *left-normed*. For instance, if $A = \{x, y, z\}$, then

$$A \ rep \ x = \{y, z, yx, (yx)x, ((yx)x)x, \ldots, zx, (zx)x, ((zx)x)x, \ldots\},$$

$$A \ rep \ y = \{x, z, xy, (xy)y, ((xy)y)y, \ldots, zy, (zy)y, ((zy)y)y, \ldots\},$$

$$A \ rep \ z = \{x, y, xz, (xz)z, ((xz)z)z, \ldots, yz, (yz)z, ((yz)z)z, \ldots\}.$$

Now, consider the groupoid G generated by $X = \{x_1, \ldots, x_r\}$. Set $X_1 = X$, and suppose that we have defined the set X_n for $n \geq 1$. Select an element $b_n \in X_n$ of minimal length, and put

$$X_{n+1} = X_n \ rep \ b_n.$$

The sequence $b_1, b_2, \ldots$ obtained in this way satisfies the definition of a basic sequence in X. In general, the first r terms of the sequence can be chosen to be $x_1, \ldots, x_r$.

3.3.2 Basic Commutators Revisited

The basic commutators of group elements form a basic sequence of elements in a certain groupoid. Suppose that G is a group generated by $Y = \{y_1, \ldots, y_r\}$. We introduce the binary operation $\mu : G \times G \to G$ defined by "commutation"

$$\mu(g, h) = [g, h] = g^{-1}h^{-1}gh.$$

This operation turns G into a groupoid relative to commutation. Now, Y generates a subgroupoid $G^\dagger$ of the groupoid G whose elements consist of the elements of Y, together with those obtained by performing iterated commutations of them. For example, if $Y = \{y_1, y_2, y_3\}$, then

$$y_2, \ [y_1, y_3], \ [y_2, [y_2, y_3]], \ \text{and} \ [[y_3, y_2], [y_1, y_2]]$$

are elements of $G^\dagger$. Using the "rep" operation, we can construct a basic sequence $c_1, c_2, \ldots$ in Y. The terms of this sequence are just basic commutators on Y, and the weight of c_i is the length of b_i, the canonical pre-image of c_i in the free groupoid on $X = \{x_1, \ldots, x_r\}$.

3.3.3 Lie Rings and Basic Lie Products

Next, we look at groupoids and basic sequences which arise in Lie rings.

Definition 3.8 A *Lie ring L* is an additive abelian group with a binary operation

$$[\ , \ ] : L \times L \to L$$

satisfying the following properties for $l, m, n \in L$:

(i) bi-linearity: $[l + m, n] = [l, n] + [m, n]$ and $[l, m + n] = [l, m] + [l, n]$;
(ii) skew symmetry: $[l, l] = 0$;
(iii) Jacobi identity: $[[l, m], n] + [[m, n], l] + [[n, l], m] = 0$.

The operation $[\ , \ ]$ is called the *Lie bracket*. Some properties satisfied by the Lie bracket are recorded in the next lemma.

Lemma 3.3 *Suppose that L is a Lie ring. For all $l, m, n \in L$, we have:*

(i) *anti-commutativity:* $[l, m] = -[m, l]$;
(ii) $[0, n] = [n, 0] = 0$;
(iii) $[-m, n] = [m, -n] = -[m, n]$.

Proof Notice that

$$0 = [l + m, l + m] = [l, l] + [l, m] + [m, l] + [m, m] = [l, m] + [m, l].$$

This gives (i). To obtain (ii), observe that

$$[0, n] = [0 + 0, n] = [0, n] + [0, n].$$

Thus, $[0, n] = 0$. Similarly, $[n, 0] = 0$ and (ii) is established. Finally, we get (iii) by noting that

$$0 = [0, n] = [m - m, n] = [m, n] + [-m, n].$$

And so, $[-m, n] = -[m, n]$. Similarly, $[m, -n] = -[m, n]$.   □

Definition 3.9 A nonempty subset X of a Lie ring L *generates* L if L is the smallest sub-Lie ring of L containing X. In this case, we write $L = lr(X)$.

We are interested in a certain groupoid whose underlying set comes from a Lie ring. Suppose that $L = lr(X)$ is a Lie ring with Lie bracket $[\ , \ ]$ for some subset $X \subseteq L$. If we ignore the fact that L has the addition operation, then L becomes a groupoid relative to the binary operation

$$\mu : L \times L \rightarrow L \ \text{ defined by } \ \mu(l_1, l_2) = [l_1, l_2] \quad (l_1, l_2 \in L).$$

Note that X generates a subgroupoid $L^{\dagger}$ of the groupoid L. The elements of $L^{\dagger}$ are called *Lie products*. The set of Lie products consists of the elements of X, together with those obtained by taking iterated Lie brackets of them. For example,

$$x_3, \ [x_1, x_2], \ [[x_3, x_2], x_1], \ \text{and} \ [[x_1, x_3], [x_2, x_3]]$$

are Lie products of the groupoid $L^{\dagger}$ on $X = \{x_1, x_2, x_3\}$.

Every element of the Lie ring L is a $\mathbb{Z}$-linear combination of the Lie products from the groupoid $L^{\dagger}$. Using the "rep" operation, we can form a basic sequence in X with respect to the Lie bracket. The elements arising in such a sequence are called *basic Lie products*. The next theorem shows that these basic Lie products span L as an additive abelian group.

Theorem 3.3 *Let L be a Lie ring generated by $X = \{x_1, \ldots, x_r\}$, and suppose that $b_1, b_2, \ldots$ is a basic sequence of basic Lie products in X. As an additive abelian group, L is generated by $b_1, b_2, \ldots$.*

The proof relies on Lemmas 3.4 and 3.5 below.

Lemma 3.4 *Let L be a Lie ring generated by a (finite or infinite) set Y. Choose an element $y \in Y$. If $u \in lr(Y \ rep \ y)$, then $[u, y] \in lr(Y \ rep \ y)$.*

Proof It suffices to prove the case when u is a monomial in the elements $Y \ rep \ y$; that is, u is obtained only from the application of the Lie bracket on the elements of $Y \ rep \ y$. In this case, u has formal length $||u||$, which we define to be the number of occurrences of elements of $Y \ rep \ y$ in u.

The proof is done by induction on the formal length of u. If $||u|| = 1$, then $u \in Y$ and $u \neq y$. It follows that $[u, y] \in Y \ rep \ y \subseteq lr(Y \ rep \ y)$. If $||u|| > 1$, then there are elements u_1, $u_2 \in lr(Y \, rep \, y)$ such that $u = [u_1, u_2]$, $||u_1|| < ||u||$, and $||u_2|| < ||u||$. An application of the Jacobi identity and Lemma 3.3 gives

$$
\begin{aligned}
[u, y] &= [[u_1, u_2], y] \\
&= -[[u_2, y], u_1] - [[y, u_1], u_2] \\
&= -[[u_2, y], u_1] - [-[u_1, y], u_2] \\
&= -[[u_2, y], u_1] + [[u_1, y], u_2].
\end{aligned}
$$

By induction, both $[u_2, y]$ and $[u_1, y]$ are contained in $lr(Y \, rep \, y)$. Thus, $[[u_2, y], u_1]$ and $[[u_1, y], u_2]$ also belong to $lr(Y \, rep \, y)$. The result follows. $\square$

In the next lemma, $gp(y)$ is the additive group generated by y.

Lemma 3.5 *Let L be a Lie ring generated by a (finite or infinite) set Y. For any $y \in Y$, we have $L = lr(Y \, rep \, y) + gp(y)$.*

Proof Set $M = lr(Y \, rep \, y)$. An element of L is a $\mathbb{Z}$-linear combination of monomials in Y. Let m be a typical monomial appearing in such a linear combination. There are three cases to consider:

  (i) m involves y and an element of Y different from y,
 (ii) m does not involve y, or
(iii) m involves only y; that is, m is an integral multiple of y.

In case (ii), $m \in M$ since $Y \setminus \{y\} \subseteq Y \, rep \, y$. In case (iii), $m \in gp(y)$. In light of this, it suffices to show that if m is as in case (i), then $m \in M$.

The proof is done by induction on the formal length of m as a monomial in the elements of Y. If the formal length of m is 1, then $m \in Y \setminus \{y\} \subseteq M$. Suppose that m has formal length greater than 1. Put $m = [m_1, m_2]$, where m_1 and m_2 are monomials in Y whose formal lengths are less than the formal length of m. By the hypothesis on m, either m_1 or m_2 contains an element of Y different from y. Since the Lie bracket satisfies anti-commutativity, we may as well assume that m_1 contains an element of Y different from y. By the induction hypothesis, $m_1 \in M$. If $m_2 = y$, then $m \in M$ by Lemma 3.4. On the other hand, if $m_2 \neq y$, then m_2 is a monomial in Y that involves elements different from y whose formal length is less than $|m|$. By induction, $m_2 \in M$, and thus $m \in M$. $\square$

We now give the proof Theorem 3.3, which is adopted from [1]. Put

$$
X_1 = X, \quad X_2 = X_1 \, rep \, b_1, \quad \ldots, \quad X_{n+1} = X_n \, rep \, b_n, \quad \ldots,
$$

where $b_i = x_i$ for $1 \leq i \leq r$. It suffices to show that any monomial m in the elements $x_1, \ldots, x_r$ can be expressed as a $\mathbb{Z}$-linear combination of the elements $b_1, b_2, \ldots$. By a repeated application of Lemma 3.5, we see that for any $n \geq 1$,

$$
L = gp(b_1) + gp(b_2) + \cdots + gp(b_n) + lr(X_{n+1}). \tag{3.21}
$$

Assume that m has formal length w. If s is chosen large enough so that every element of X_{s+1} has formal length more than w, then

$$m \in gp(b_1) + \cdots + gp(b_s).$$

The result follows from the fact that the monomials in $x_1, \ldots, x_r$ in the expansions found in (3.21) can be re-expressed as sums of monomials, each of which has the same length as the original monomials (as illustrated in the proofs of Lemmas 3.4 and 3.5). This completes the proof of Theorem 3.3.

3.3.4   The Commutation Lie Ring

Let $Y = \{y_i \mid i \in I\}$ for some nonempty index set I, and let R_0 be a ring with unity. For each $j \in \mathbb{N}$, let R_j be a free R_0-module, freely generated by all formal products (or *monomials*) of j elements from Y. Note that these j elements need not be distinct. The elements of R_j are said to be *homogeneous* of *degree j*. In particular, the elements of the ring R_0 are homogeneous of degree 0.

Example 3.7 If $Y = \{y_1, y_2, y_3\}$, then $y_3 y_2$ is a monomial in R_2, $y_1 y_3 y_3$ is a monomial in R_3, and

$$r_1(y_2 y_1 y_3 y_2) + r_2(y_3 y_2 y_3 y_1)$$

is an R_0-linear combination of monomials in R_4, where $r_1, r_2 \in R_0$.

Consider the direct sum

$$R = \bigoplus_{j=0}^{\infty} R_j. \qquad (3.22)$$

Every nonzero element of R can be uniquely written as a formal power series

$$\sum_{i=0}^{\infty} r_i = r_0 + r_1 + \cdots,$$

where $r_j \in R_j$ for $j = 0, 1, \ldots$, and all but finitely many of the r_j equal 0. We make R into a *free associative ring with unity*, freely generated by Y, by defining multiplication in R as such:

- If u and v are monomials and $t_1, t_2 \in R_0$, then

$$(t_1 u)(t_2 v) = t_1 t_2 (uv).$$

- If $\sum_{i=0}^{\infty} s_i$ and $\sum_{i=0}^{\infty} r_i$ are elements of R, where s_i, $r_i \in R_i$, then

$$\left(\sum_{i=0}^{\infty} s_i\right) \cdot \left(\sum_{i=0}^{\infty} r_i\right) = \sum_{i=0}^{\infty} \left(\sum_{j+k=i} s_j r_k\right). \tag{3.23}$$

Notice that R becomes an R_0-algebra whenever R_0 is commutative. In this situation, the ring and module operations are compatible:

$$r(ab) = (ra)b = a(rb) \quad \text{for all} \ \ r \in R_0 \ \ \text{and} \ \ a, \ b \in R.$$

We turn our attention to the case $R_0 = \mathbb{Z}$. Each R_j in the direct sum R defined in (3.22) is a free abelian group. We make R into a Lie ring by introducing the bracket operation

$$(\ , \ ) : R \times R \to R \ \text{ defined by } \ (a, \ b) = ab - ba.$$

This operation is referred to as *(ring) commutation*. We use "$(\ , \ )$" rather than "$[\ , \ ]$" in order to avoid confusion with commutation of group elements. More generally, an *m-fold commutator* of elements of R is recursively defined as

$$(a_1, \ a_2, \ \ldots, \ a_{m-1}, \ a_m) = ((a_1, \ a_2, \ \ldots, \ a_{m-1}), \ a_m), \ \text{ where } a_i \in R.$$

Definition 3.10 The ring R, equipped with addition and ring commutation, is termed the *commutation Lie ring* on R.

Our goal is to relate basic sequences of basic Lie products to basic sequences of basic commutators. This can be achieved by exploiting the algebraic structure of a commutation Lie ring. We proceed in this direction.

For the remainder of this section, all rings are with unity.

Theorem 3.4 *Let R be a free ring, freely generated by $X = \{x_1, \ \ldots, \ x_r\}$, and suppose that S is the Lie subring of the commutation Lie ring on R generated by X. If $b_1, \ b_2, \ \ldots$ is any basic sequence of basic Lie products in X, then the elements $b_1, \ b_2, \ \ldots$ are additively linearly independent.*

In order to prove this theorem, we need the next lemma. Refer to Lemma 5.6 in [11] for a more detailed proof.

Lemma 3.6 *Let R be a free ring, freely generated by the set*

$$X = \{x\} \cup \{y_\lambda \mid \lambda \in \Lambda\}$$

for some indexing set Λ. Consider the set X rep x in the commutation Lie ring on R. If S is the subring of R generated by X rep x, then S is a free ring, freely generated by X rep x.

Proof Each element of $X \, rep \, x$ can be expressed as

$$y_{\lambda, \, i} = (y_\lambda, \underbrace{x, \, \ldots, \, x}_{i \text{ of these}}),$$

where $\lambda \in \Lambda$ and $i > 0$. For $i = 0$, we set $y_{\lambda, \, 0} = y_\lambda$.

Let S be the subring of R generated by $X \, rep \, x$. The elements of S are $\mathbb{Z}$-linear combinations of monomials of the form

$$y_{\lambda_1, \, i_1} \cdots y_{\lambda_r, \, i_r}. \tag{3.24}$$

Since R is a free ring, freely generated by X, it suffices to show that these monomials are (additively) linearly independent.

First, observe that each $y_{\lambda, \, i}$ can be expressed as a polynomial in X as

$$y_{\lambda, \, i} = y_\lambda x^i - \binom{i}{1} x y_\lambda x^{i-1} + \binom{i}{2} x^2 y_\lambda x^{i-2} - \cdots + (-1)^i x^i y_\lambda. \tag{3.25}$$

Note that each $y_{\lambda, \, i}$ is homogenous of degree $(i + 1)$. Thus, (3.24) can be written as a homogeneous X-polynomial of degree

$$(i_1 + 1) + \cdots + (i_r + 1).$$

Now, any nontrivial $\mathbb{Z}$-linear combination of monomials whose degrees are distinct is different from 0 because R is free on X. In light of this, it is enough to show that different monomials on $X \, rep \, x$ which are of the *same* degree are $\mathbb{Z}$-linearly independent.

We begin by imposing a total ordering on the elements of X by choosing x as the first element of the ordering:

$$x < y_1 < y_2 < \cdots$$

This induces a lexicographic ordering on the set of monomials in X of the same total degree which have the same degree on x. According to this ordering, the last monomial in the expression for $y_{\lambda, \, i}$ in (3.24) is precisely $y_\lambda x^i$. Thus, when we rewrite the $X \, rep \, x$-monomial (3.24) as a polynomial in X, we find that the "last" term will be

$$y_{\lambda_1} x^{i_1} \cdots y_{\lambda_r} x^{i_r}.$$

This last term determines (3.24) uniquely. For if $y_{\lambda_1, \, i_1} \cdots y_{\lambda_r, \, i_r}$ and $y_{\mu_1, \, i_1} \cdots y_{\mu_r, \, i_r}$ are two distinct $X \, rep \, x$-monomials of same degree, then their last terms as polynomials in X are $y_{\lambda_1} x^{i_1} \cdots y_{\lambda_r} x^{i_r}$ and $y_{\mu_1} x^{i_1} \cdots y_{\mu_r} x^{i_r}$ respectively. These terms are distinct since R is free on X. We conclude that any nontrivial linear combination of monomials of type (3.24) of the same degree is not zero.   □

We now prove Theorem 3.4. Let $X_1 = X$ and $X_{n+1} = X_n$ rep b_n for $n > 1$. As usual, we assume that $b_i = x_i$ for $1 \leq i \leq r$. Notice that $lr(b_1) = gp(b_1)$. By Lemma 3.5,

$$S = lr(b_1) + lr(X_2).$$

We show that S is a direct sum by illustrating that no nonzero multiple of b_1 is contained in $lr(X_2)$. Let I be the ideal of the ring R generated by $\{b_2, \ldots, b_r\}$. Note that $X_2 = X_1$ rep $b_1 \subseteq I$, and thus $lr(X_2) \subseteq I$. Furthermore, the quotient ring R/I can be viewed as a polynomial ring in $b_1 + I$ over the integers. Thus, if $s \neq 0$, then $s(b_1 + I) \neq I$; that is, $sb_1 \notin I$. This means that $sb_1 \notin lr(X_2)$ as claimed. And so,

$$S = lr(b_1) \oplus lr(X_2).$$

Let R_1 be the subring of R generated by X_2. By Lemma 3.6, R_1 is a free ring, freely generated by X_2. Put $S_1 = lr(X_2)$ and repeat the previous argument to obtain

$$S = lr(b_1) \oplus S_1 = lr(b_1) \oplus lr(b_2) \oplus lr(X_3).$$

By induction,

$$S = lr(b_1) \oplus \cdots \oplus lr(b_m) \oplus lr(X_{m+1}),$$

for every $m \in \mathbb{N}$. Consequently, $b_1, b_2, \ldots$ are additively linearly independent and the proof of Theorem 3.4 is complete.

Definition 3.11 Let L be a Lie ring, and suppose that X is some nonempty set and $\sigma : X \to L$ is an injective set map. The Lie ring $L = (L, \sigma)$ is termed a *free Lie ring* on X if each function $\mu : X \to K$, where K is any Lie ring, extends uniquely to a Lie ring homomorphism $\beta : L \to K$ such that $\mu = \beta \circ \sigma$. If $X \subset L$ and σ is the identity map, then L is *freely generated* by X.

Corollary 3.3 *Let R be a free ring, freely generated by $X = \{x_1, \ldots, x_r\}$. If S is the Lie subring of the commutation Lie ring on R generated by X, then S is a free Lie ring, freely generated by X.*

Proof Define F to be the free Lie ring, freely generated by $Y = \{y_1, \ldots, y_r\}$. Let $\varphi : F \to S$ be the Lie ring homomorphism induced by the mapping $y_i \mapsto x_i$ for $i = 1, \ldots, r$. We claim that φ is a Lie ring isomorphism.

Suppose that $b_1, b_2, \ldots$ is a basic sequence of basic Lie products in Y. The sequence $\varphi(b_1), \varphi(b_2), \ldots$ is a basic sequence of basic Lie products in X. By

Theorems 3.3 and 3.4, we know that b_1, b_2, ... are additively linearly independent and span F additively. The same holds for $\varphi(b_1)$, $\varphi(b_2)$, ... in S. Thus, if $s \in S$, then there exist integers m_1, ..., m_k and $k \geq 1$ such that

$$s = m_1\varphi(b_1) + \cdots + m_k\varphi(b_k)$$
$$= \varphi(m_1 b_1 + \cdots + m_k b_k).$$

Hence, φ is a Lie ring epimorphism.

Next, suppose that $a = n_1 b_1 + \cdots + n_k b_k \neq 0$ for some integers n_1, ..., n_k. Clearly, at least one of n_1, ..., n_k is nonzero. Since $\varphi(b_1)$, ..., $\varphi(b_k)$ are linearly independent, $\varphi(a) \neq 0$. Therefore, φ is a Lie ring monomorphism. Thus, F and S are isomorphic as Lie rings.   □

3.3.5   The Magnus Embedding

We establish a connection between free groups and commutation Lie rings. Let R_0 be a ring, and let $Y = \{y_1, \ldots, y_r\}$. Let R_j be the same as in (3.22), and consider the free associative ring

$$R = \bigoplus_{j=0}^{\infty} R_j,$$

freely generated by $Y = \{y_1, \ldots, y_r\}$. It is clear that $R_m R_n \subseteq R_{m+n}$ for all $m \geq 0$ and $n \geq 0$. This means that R is a *graded ring*. This being the case, one can form the completion of R and obtain the *Magnus power series ring*

$$\overline{R} = R_0[[y_1, \ldots, y_r]]$$

in the variables y_1, ..., y_r over R_0. This is just the unrestricted direct sum of the R_0-modules R_j for $j \geq 0$. A typical element of $\overline{R}$ is an infinite sum of the form

$$\sum_{i=0}^{\infty} r_i = r_0 + r_1 + \cdots,$$

where $r_j \in R_j$ for $j = 0$, 1, ... Multiplication in $\overline{R}$ is defined the same way as in (3.23). If R_0 is commutative, then $\overline{R}$ is an R_0-algebra known as the *Magnus power series algebra* in the variables y_1, ..., y_r over R_0.

Suppose that $R_0 = \mathbb{Z}$, and let U be the subset of $\overline{R}$ consisting of all elements of the form $1 + a$, where $a \in \sum_{j=1}^{\infty} R_j$. We claim that U is a group under the ring multiplication in $\overline{R}$. If $1 + a$ and $1 + b$ are elements of U, then so is

$$(1 + a)(1 + b) = 1 + a + b + ab.$$

Thus, U is multiplicatively closed. Furthermore, U contains a unity element. In $\overline{R}$, we have

$$(1 + a)(1 - a + a^2 - a^3 + \cdots) = 1. \tag{3.26}$$

Hence, every element of U is a unit of $\overline{R}$, and its inverse is contained in U. Finally, multiplication of elements of U satisfies the associative law since $\overline{R}$ is an associative ring. And so, U is a group with respect to the ring multiplication in $\overline{R}$.

Next, let F be the free group on $X = \{x_1, \ldots, x_r\}$. The mapping from F to $\overline{R}$ induced by

$$x_i \mapsto 1 + y_i \quad (i = 1, \ldots, r) \tag{3.27}$$

gives rise to an embedding $\Psi : F \to U$ referred to as the *Magnus embedding* (see [1] or [10]).

Lemma 3.7 *Let F, R, $\overline{R}$, and Ψ be as above, and let S be the Lie subring of the commutation Lie ring on R generated by Y. Suppose that c_1, c_2, $\ldots$ is a basic sequence of basic commutators in X, and let b_1, b_2, $\ldots$ be the basic sequence of basic Lie products in Y whose terms b_j are obtained by replacing each generator $x_i \in X$ in the terms c_j by the generator $y_i \in Y$ and reinterpreting the group commutator operation as ring commutation. For each n, we have*

$$\Psi(c_n) = 1 + b_n + \cdots .$$

(Here and in the proof, "$\cdots$" represents an additive linear combination of basic Lie products of larger degree.)

Proof The proof is done by induction on the length of the basic commutators on X. If $|c_n| = 1$, then c_n is one of the generators and the result follows by the definition of the Magnus embedding.

Suppose that $|c_n| > 1$. In this case, we can express c_n uniquely as

$$c_n = [c_i, c_j] = c_i^{-1} c_j^{-1} c_i c_j,$$

where $|c_i| < |c_n|$ and $|c_j| < |c_n|$. In $\overline{R}$, the corresponding basic Lie product is

$$b_n = (b_i, b_j) = b_i b_j - b_j b_i.$$

By induction, together with (3.26), we have

$$
\begin{aligned}
\Psi(c_n) &= \Psi\big([c_i,\ c_j]\big) \\
&= \big[\Psi(c_i),\ \Psi(c_j)\big] \\
&= \big[1 + b_i + \cdots,\ 1 + b_j + \cdots\big] \\
&= \big(1 + b_i + \cdots\big)^{-1}\big(1 + b_j + \cdots\big)^{-1}\big(1 + b_i + \cdots\big)\big(1 + b_j + \cdots\big) \\
&= 1 + \big(b_i b_j - b_j b_i\big) + \cdots \\
&= 1 + b_n + \cdots
\end{aligned}
$$

and the lemma is proved.   □

We are now ready to prove the main theorem of this section.

Theorem 3.5 *Let F be a free group, freely generated by $X = \{x_1, \ldots, x_k\}$. Suppose that c_1, c_2, ... is a basic sequence of basic commutators in X, and choose $n \in \mathbb{N}$. The basic commutators of weight n, modulo $\gamma_{n+1}F$, form a basis for the free abelian group $\gamma_n F/\gamma_{n+1}F$. Furthermore, every element of F can be uniquely expressed in the form*

$$
c_1^{e_1} c_2^{e_2} \cdots c_t^{e_t} \bmod \gamma_{n+1}F, \tag{3.28}
$$

where $e_1, \ldots, e_t$ are integers and $c_1, \ldots, c_t$ are basic commutators of weights $1, 2, \ldots, n$.

Proof Let R and $\overline{R}$ be as before. By Theorem 3.1, the basic commutators $c_1, \ldots, c_m$ of weight n, modulo $\gamma_{n+1}F$, generate $\gamma_n F/\gamma_{n+1}F$. It suffices to show that these commutators are linearly independent, modulo $\gamma_{n+1}F$. Let $b_1, \ldots, b_m$ be the corresponding basic sequence of basic Lie products in the Lie subring of the commutation Lie ring on R generated by Y. By Lemma 3.7,

$$
c_i \mapsto 1 + b_i + \cdots
$$

under the Magnus embedding. It follows that for any integers $t_1, \ldots, t_m$, we have

$$
c_1^{t_1} \cdots c_m^{t_m} \mapsto 1 + (t_1 b_1 + \cdots + t_m b_m) + \cdots,
$$

where "$\cdots$" represents an additive linear combination of basic Lie products whose degrees exceed n. Now, $b_1, b_2, \ldots$ are additively linearly independent by Theorem 3.4. This means that the linear combination

$$
t_1 b_1 + \cdots + t_m b_m
$$

cannot equal zero unless $t_1 = \cdots = t_m = 0$. We conclude that $c_1, \ldots, c_m$ are linearly independent, modulo $\gamma_{n+1}F$. The uniqueness of (3.28) follows at once.   □

Example 3.8 Let F be a free group, freely generated by a, b, and c. Consider the element $aba^2ca^{-1}b$ of F.

- Take $n = 1$ in Theorem 3.1. Any basic commutator of weight 2 or more is contained in $\gamma_2 F$. Thus, applying (3.2) to the underlined products below gives

$$aba^2ca^{-1}b = a\,\underline{ba}\,aca^{-1}b \bmod \gamma_2 F$$
$$= a^2\,\underline{ba}\,ca^{-1}b \bmod \gamma_2 F$$
$$= a^3 b\,\underline{ca^{-1}}\,b \bmod \gamma_2 F$$
$$= a^3\,\underline{ba^{-1}}\,cb \bmod \gamma_2 F$$
$$= a^3 a^{-1}b\,\underline{cb} \bmod \gamma_2 F$$
$$= a^2 b^2 c \bmod \gamma_2 F.$$

- Next, take $n = 2$. In this case, the basic commutators of weight greater than 2 are contained in $\gamma_3 F$. This, together with (3.5), gives

$$aba^2ca^{-1}b = a\,\underline{ba}\,aca^{-1}b \bmod \gamma_3 F$$
$$= a^2 b\,\underline{[b,\,a]a}\,ca^{-1}b \bmod \gamma_3 F$$
$$= a^2\,\underline{ba}\,[b,\,a]ca^{-1}b \bmod \gamma_3 F$$
$$= a^3 b[b,\,a][b,\,a]\,\underline{ca^{-1}}b \bmod \gamma_3 F$$
$$= a^3 b[b,\,a]\,\underline{[b,\,a]a^{-1}}\,c[c,\,a^{-1}]b \bmod \gamma_3 F$$
$$= a^3 b\,\underline{[b,\,a]a^{-1}}\,[b,\,a]c[c,\,a]^{-1}b \bmod \gamma_3 F$$
$$= a^3\,\underline{ba^{-1}}\,[b,\,a][b,\,a]c[c,\,a]^{-1}b \bmod \gamma_3 F$$
$$= a^2 b[b,\,a]c\,\underline{[c,\,a]^{-1}b} \bmod \gamma_3 F$$
$$= a^2 b[b,\,a]\,\underline{cb}\,[c,\,a]^{-1} \bmod \gamma_3 F$$
$$= a^2 b\,\underline{[b,\,a]b}\,c[c,\,b][c,\,a]^{-1} \bmod \gamma_3 F$$
$$= a^2 b^2\,\underline{[b,\,a]c}\,[c,\,b][c,\,a]^{-1} \bmod \gamma_3 F$$
$$= a^2 b^2 c[b,\,a][c,\,b][c,\,a]^{-1} \bmod \gamma_3 F.$$

Once again, (3.2) has been applied to the underlined products.

Remark 3.1 A formula due to E. Witt provides the rank of the lower central quotients of a finitely generated free group. If F is a free group of rank k, then the rank of $\gamma_n F/\gamma_{n+1}F$ is

$$\frac{1}{n}\sum_{d|n}\mu(d)k^{n/d},$$

where $\mu(d)$ is the Möbius function defined as follows:

$$\mu(d) = \begin{cases} (-1)^r & \text{if } d \text{ is a product of } r \text{ different primes,} \\ 0 & \text{otherwise.} \end{cases}$$

We refer the reader to §11.4 of [7] for a proof.

According to Corollary 2.3, the elements of a nilpotent group G of class at most c must satisfy the identity

$$[g_1, \ldots, g_{c+1}] = 1 \tag{3.29}$$

for any elements $g_1, \ldots, g_{c+1} \in G$, along with the usual group axioms. Of course, the elements of G could satisfy other identities as well. For instance, they could all be torsion elements. If the elements of G satisfy *only* the group axioms, together with (3.29), then G is called a *free nilpotent group of class at most c*. This is made more precise in the next definition.

Definition 3.12 Let G be a nilpotent group of class at most c. Suppose that X is some nonempty set and $\varrho : X \to G$ is an injective set map. The group $G = (G, \varrho)$ is called *free nilpotent* on X if to each function $\mu : X \to H$, where H is any nilpotent group of class at most c, there exists a unique homomorphism $\beta : G \to H$ such that $\mu = \beta \circ \varrho$. If a nilpotent group of class at most c is free nilpotent on some set, then we call it a *free nilpotent group*. If $X \subset G$ and ϱ is the identity map, then G is *freely generated* by X.

It follows from Theorem 2.2 that every free nilpotent group of class c is isomorphic to $F/\gamma_{c+1}F$, where F is a free group. Furthermore, every nilpotent group is a quotient of a free nilpotent group.

Corollary 3.4 *If G is a finitely generated free nilpotent group of class c, freely generated by a finite set X, then the factors $\gamma_i G/\gamma_{i+1}G$ are free abelian groups, freely generated by the basic commutators on X of weight i, modulo $\gamma_{i+1}G$, for $i = 1, \ldots, c$. Furthermore, if $c_1, \ldots, c_t$ is a sequence of basic commutators of weights at most c, then every element of G can be uniquely written as $c_1^{e_1} \cdots c_t^{e_t}$, where $e_1, \ldots, e_t$ are integers and $gp(c_i)$ is infinite cyclic for each $i = 1, \ldots, t$.*

3.4 Proof of the Collection Formula

In this section, we prove Theorem 3.2. Our discussion is based on [7].

Suppose that $X = \{x_1, \ldots, x_r\}$ generates the group G, and consider the expression

$$(x_1 x_2 \cdots x_r)^n = \underbrace{(x_1 x_2 \cdots x_r) \cdots (x_1 x_2 \cdots x_r)}_{\text{n factors}}. \tag{3.30}$$

Let $x_i^{(1)}$, $x_i^{(2)}$, ..., $x_i^{(n)}$ be n labeled copies of x_i for $i = 1, 2, ..., r$, and rewrite (3.30) as

$$\left(x_1^{(1)} x_2^{(1)} \cdots x_r^{(1)}\right) \left(x_1^{(2)} x_2^{(2)} \cdots x_r^{(2)}\right) \cdots \left(x_1^{(n)} x_2^{(n)} \cdots x_r^{(n)}\right). \tag{3.31}$$

Our goal is to express (3.31) as a product of positive powers of basic commutators in x_1, ..., x_r. The idea is to collect basic commutators to the left just as we did in Section 3.1. There are several "stages" for this procedure. We consider the expression (3.31) to be stage zero. Stage one consists of moving $x_1^{(2)}$ right after $x_1^{(1)}$, then $x_1^{(3)}$ right after $x_1^{(2)}$, and so on. The placement of $x_1^{(n)}$ next to $x_1^{(n-1)}$ completes this stage of the collection process. Afterwards, we collect the x_2's, in order, immediately to the right of the x_1's. This completes the second stage of the process. The ith stage of the process consists of the collection of each appearance, in order, of the ith basic commutator. More specifically, we have

$$(x_1 x_2 \cdots x_r)^n = b_1^{e_1} b_2^{e_2} \cdots b_i^{e_i} R_1 \cdots R_s \tag{3.32}$$

at the end of the ith stage, where b_1, ..., b_i are the first i basic commutators, e_1, ..., e_i are positive integers, and R_1, ..., R_s are basic commutators arising after b_i. Of course, it must be shown that b_1, ..., b_i, R_1, ..., R_s are, indeed, basic commutators. This will be done in Lemma 3.8 below. First, we describe stage $(i + 1)$, assuming that stage i has been completed. Suppose that R_{j1}, ..., R_{jl} are the basic commutators equal to b_{i+1} for $1 \le j1 < \cdots < jl \le s$. We move R_{j1} immediately after $b_i^{e_i}$, then R_{j2} after R_{j1}, and so forth. At the end of stage $(i + 1)$, we find that (3.32) becomes

$$(x_1 x_2 \cdots x_r)^n = b_1^{e_1} b_2^{e_2} \cdots b_i^{e_i} b_{i+1}^{e_{i+1}} R_1^* \cdots R_k^*, \tag{3.33}$$

where $e_{i+1} = l$. Note that the sequence of R^*'s is different from the original sequence of R's since $R_{j1}, ..., R_{jl}$ (all equal to b_{i+1}) have been collected at this stage. In (3.32), we refer to $b_1^{e_1} b_2^{e_2} \cdots b_i^{e_i}$ as the *collected part* and to $R_1 \cdots R_s$ as the *uncollected part*.

Lemma 3.8 *At any given stage of the collection process described above, only basic commutators arise.*

Proof The proof is done by induction on the stage. At stage zero, we have (3.31). Only generators appear at this stage, and these are all basic commutators of weight one.

Assume that at the ith stage, R_1, R_2, ..., R_s are all basic commutators occurring after b_i. We claim that the same is true once stage $(i + 1)$ is completed; that is, after collecting all R's that are equal to b_{i+1}, only basic commutators are introduced. Indeed, each time we collect b_{i+1}, we introduce a commutator of the form

$$[b_j, b_{i+1}, ..., b_{i+1}], \text{ where } j > i + 1. \tag{3.34}$$

We claim that the commutator in (3.34) is basic. If we put $b_j = [b_u,\ b_v]$, then b_j resulted from collecting b_v during stage v. Hence, $v < i + 1$, so that $b_v < b_{i+1}$. This implies that $[b_j,\ b_{i+1}]$ is basic. By iterating this procedure we conclude that $[b_j,\ b_{i+1},\ \ldots,\ b_{i+1}]$ is, indeed, a basic commutator.   □

Our next task is to calculate $e_1,\ \ldots,\ e_{i+1}$. This is done by introducing a certain labeling system for the basic commutators in terms of their weights. In (3.31), we have labeled $x_1,\ x_2,\ \ldots,\ x_r$ with labels j as $x_1^{(j)},\ x_2^{(j)},\ \ldots,\ x_r^{(j)}$ for $j = 1,\ \ldots,\ n$. This describes the labeling system for the basic commutators of weight 1. Suppose that basic commutators b_i and b_j have weights w and u and labels $(\lambda_1,\ \ldots,\ \lambda_w)$ and $(\upsilon_1,\ \ldots,\ \upsilon_u)$ respectively. We define the label of the basic commutator $[b_i,\ b_j]$ to be $(\lambda_1,\ \ldots,\ \lambda_w,\ \upsilon_1,\ \ldots,\ \upsilon_u)$. In order to calculate $e_1,\ \ldots,\ e_{i+1}$, we need to determine the conditions for a basic commutator with a given label to

- exist in the uncollected part during stage i, and
- precede another in the uncollected part during stage i.

Note that $e_{i+1} = l$ is the amount of uncollected basic commutators equal to b_{i+1} at stage i.

Let E_k^i denote the condition that *the labeled commutator b_k exists at stage i*, and let P_{kt}^i be the condition that *the labeled commutator b_k precedes the labeled commutator b_t at stage i*. At stage zero, only basic commutators of weight 1 are present and $x_k^{(\lambda)}$ exists for any $1 \le k \le r$, and for any label $1 \le \lambda \le n$. Hence, the condition E_k^0 always holds regardless of the label assigned to x_k. In order to obtain the precedence conditions P_{kt}^0, we observe that

(i) when $k < t$, $x_k^{(\lambda)}$ precedes $x_t^{(\mu)}$ if $\lambda \le \mu$, and
(ii) when $k \ge t$, $x_k^{(\lambda)}$ precedes $x_t^{(\mu)}$ if $\lambda < \mu$.

Therefore, in terms of labels, P_{kt}^0 holds for $x_k^{(\lambda)}$ and $x_t^{(\mu)}$ when either $\lambda \le \mu$ or $\lambda < \mu$.

In what follows, we explore general conditions of this nature. Let $\lambda_1,\ \ldots,\ \lambda_m$ be a sequence of positive integers. Suppose that the sequence satisfies disjunctions or conjunctions of conditions of type $\lambda_v < \lambda_w$ or $\lambda_v \le \lambda_w$, where $1 \le v \le m$ and $1 \le w \le m$. We say that the sequence *satisfies a set of conditions $\mathscr{L}$ on $\lambda_1,\ \ldots,\ \lambda_m$*.

For example, consider a formal sequence $\lambda_1,\ \lambda_2,\ \lambda_3,\ \lambda_4$ and the set of conditions $\mathscr{L}$ given by $\lambda_1 \le \lambda_3$ or $\lambda_3 \le \lambda_4$. The sequence

$$\lambda_1 = 3,\ \lambda_2 = 2,\ \lambda_3 = 3,\ \lambda_4 = 5$$

satisfies the conditions, while the sequence

$$\lambda_1 = 3,\ \lambda_2 = 5,\ \lambda_3 = 2,\ \lambda_4 = 1$$

does not.

Lemma 3.9 *The conditions E_k^i that a commutator λ_k with label $(\lambda_1, \ldots, \lambda_m)$ exists at stage i are conditions $\mathscr{L}$ on $\lambda_1, \ldots, \lambda_m$, and the conditions P_{kt}^i that a commutator λ_k with label $(\lambda_1, \ldots, \lambda_m)$ precedes a commutator b_t with label $(\mu_1, \ldots, \mu_q)$ in the uncollected part of stage i are conditions $\mathscr{L}$ on $\lambda_1, \ldots, \lambda_m, \mu_1, \ldots, \mu_q$.*

Proof The proof is done by induction on the stage. We have already seen that at stage zero, existence and precedence conditions are conditions $\mathscr{L}$. Suppose that the lemma is true at stage i. At the end of stage $(i + 1)$, (3.33) resulted from (3.32) by collecting, in order, $R_{j1}, \ldots, R_{jl}$, where $R_{j1} = \cdots = R_{jl} = b_{i+1}$. Each step in this process involved a replacement of the form $SR = RS[S, R]$, where $R = b_{i+1}$ and $S = b_j$ for $j > i + 1$. Thus, all commutators existing during stage i which are different from b_{i+1} still exist at stage $(i+1)$ and are in the same order. Consequently,

$$E_k^{i+1} = E_k^i \quad \text{and} \quad P_{kt}^{i+1} = P_{kt}^i$$

for these commutators. Thus, it is enough to consider the existence of labeled commutators that arise at stage $(i + 1)$, and the precedence of pairs of labeled commutators where at least one of the commutators in the pair arises at stage $(i+1)$. A commutator arising at stage $(i + 1)$ has the form

$$b_k = \left[b_j, R_{u_1}, \ldots, R_{u_m}\right],$$

where $j > i + 1$ and $R_{u_1} = \cdots = R_{u_m} = b_{i+1}$. This commutator is obtained by moving R_{u_1} past b_j, then R_{u_2} past $\left[b_j, R_{u_1}\right]$, and so on, until R_{u_m} is moved past $\left[b_j, R_{u_1}, \ldots, R_{u_{m-1}}\right]$. Hence, for such a commutator, E_k^{i+1} is the conjunction of conditions for the existence of the labeled commutators $b_j, R_{u_1}, \ldots, R_{u_m}$ at stage i, together with the precedence conditions that these commutators are exactly in this order at stage i. This means that E_k^{i+1} is a condition $\mathscr{L}$ on the label of b_k.

Next, we show that P_{kt}^{i+1}, the condition that the labeled commutator b_k precedes the labeled commutator b_t at stage $(i+1)$, where b_k or b_t (or both) arose at this stage, is a condition $\mathscr{L}$ on the combined labels of b_k and b_t. Notice that if $j_1, j_2 > i + 1$ and

$$R_{u_1} = \cdots = R_{u_m} = R_{v_1} = \cdots = R_{v_w} = b_{i+1},$$

then

$$b_k = \begin{cases} b_{j_1} & \text{if } b_k \text{ exists at stage } i \\ \left[b_{j_1}, R_{u_1}, \ldots, R_{u_m}\right] & \text{if } b_k \text{ arises at stage } (i + 1), \end{cases}$$

and

$$b_t = \begin{cases} b_{j_2} & \text{if } b_t \text{ exists at stage } i \\ \left[b_{j_2}, R_{v_1}, \ldots, R_{v_w}\right] & \text{if } b_t \text{ arises at stage } (i + 1). \end{cases}$$

If $b_{j_1} \neq b_{j_2}$ as labeled commutators, then $P_{kt}^{i+1} = P_{j_1 j_2}^{i}$. By induction, P_{kt}^{i+1} is a condition $\mathscr{L}$ on the combined labels of b_k and b_t.

Suppose, on the other hand, that $b_{j_1} = b_{j_2}$ as labeled commutators. In this case, b_k precedes b_t if either of the following holds:

1. $b_k = b_{j_1}$, and thus $b_t = [b_k, R_{v_1}, \ldots, R_{v_w}]$. In this case, P_{kt}^{i+1} is the condition that the labeled commutators $b_k, R_{v_1}, \ldots, R_{v_w}$ exist precisely in this order at stage i.
2. There exists a largest positive integer e such that (as labeled commutators)

$$R_{u_1} = R_{v_1}, \ldots, R_{u_e} = R_{v_e}.$$

In this situation, one of the following holds:

(i) $e = m$, and thus $b_t = [b_{j_1}, R_{u_1}, \ldots, R_{u_m}, R_{v_{m+1}}, \ldots, R_{v_w}]$. Hence, P_{kt}^{i+1} is the condition that the labeled commutators

$$b_{j_1}, R_{u_1}, \ldots, R_{u_m}, R_{v_{m+1}}, \ldots, R_{v_w}$$

are precisely in this order at stage i.

(ii) $R_{v_{e+1}}$ precedes $R_{u_{e+1}}$ at stage i.

Thus, in every case, P_{kt}^{i+1} is a disjunction of precedence conditions, and by induction, a condition $\mathscr{L}$ on the combined labels of b_k and b_t.   □

Lemma 3.10 *The number of sequences of the form $\lambda_1, \ldots, \lambda_m$ with $1 \leq \lambda_i \leq n$ satisfying a given set of conditions $\mathscr{L}$ is*

$$a_1 n + a_2 \binom{n}{2} + \cdots + a_m \binom{n}{m},$$

where $a_1, \ldots, a_m$ are nonnegative integers which depend on the conditions $\mathscr{L}$, but not on n.

Proof Assume that $1 \leq t \leq n$, and let $\{S_1, \ldots, S_t\}$ be a partition of the set of indices $\{1, \ldots, m\}$. Choose numbers $v_1, \ldots, v_t \in \{1, \ldots, n\}$ such that $v_1 < \cdots < v_t$. For each $i \in S_j$, where $j = 1, \ldots, t$, set $\lambda_i = v_j$. The resulting sequence $\lambda_1, \ldots, \lambda_m$ satisfies a (generally non-strict) ordering determined by the partition $S_1, \ldots, S_t$.

For example, let $m = 5$ and consider the partition $S_1 = \{1, 5\}$, $S_2 = \{2, 3\}$, and $S_3 = \{4\}$. Choose $v_1 = 2$, $v_2 = 4$, and $v_3 = 5$. Then the sequence of λ's satisfies the ordering

$$\lambda_1 \leq \lambda_5 < \lambda_2 \leq \lambda_3 < \lambda_4.$$

For the rest of the proof, we identify each partition $\{S_1, \ldots, S_t\}$ with a specific ordering of the λ's. For each $t = 1, \ldots, n$, there may be several distinct orderings. However, if $t > n$, the procedure outlined above does not apply, and no ordering of the λ's can be produced.

For each $t = 1, \ldots, m$, let a_t be the number of orderings $S_1, \ldots, S_t$ satisfying the set of conditions $\mathscr{L}$. For each such ordering, there are $\binom{n}{t}$ choices for $v_1, \ldots, v_t$, and thus $\binom{n}{t}$ actual sequences of λ's satisfying this specific ordering. Notice that if $t > n$, then $a_t = 0$. Hence, the number of sequences $\lambda_1, \ldots, \lambda_m$ satisfying the given set of conditions $\mathscr{L}$ is

$$a_1 n + a_2 \binom{n}{2} + \cdots + a_m \binom{n}{m}.$$

Observe that each a_i depends on the set of conditions $\mathscr{L}$, but not on n. This completes the proof.   □

Recall that e_i in (3.32) is the number of commutators equal to b_i present in the uncollected part at stage $(i-1)$. Lemma 3.9 gives that e_i is, in fact, the number of sequences $\lambda_1, \ldots, \lambda_m$ satisfying certain conditions $\mathscr{L}$, where m is the weight of b_i. Theorem 3.2 follows from Lemma 3.10.

References

1. G. Baumslag, *Lecture Notes on Nilpotent Groups*. Regional Conference Series in Mathematics, No. 2 (American Mathematical Society, Providence, RI, 1971). MR0283082
2. B. Baumslag, B. Chandler, *Schaum's Outline of Theory and Problems of Group Theory* (McGraw-Hill, New York, 1968)
3. P. Hall, A contribution to the theory of groups of prime-power order. Proc. Lond. Math. Soc. S2–36(1), 29 (1934). MR1575964
4. M. Hall, Jr., A basis for free Lie rings and higher commutators in free groups. Proc. Am. Math. Soc. 1, 575–581 (1950). MR0038336
5. P. Hall, Some word-problems. J. Lond. Math. Soc. 33, 482–496 (1958). MR0102540
6. P. Hall, *The Edmonton Notes on Nilpotent Groups*. Queen Mary College Mathematics Notes. Mathematics Department (Queen Mary College, London, 1969). MR0283083
7. M. Hall, Jr., *The Theory of Groups* (Chelsea Publishing Co., New York, 1976). MR0414669. Reprinting of the 1968 edition
8. E.I. Khukhro, *Nilpotent Groups and Their Automorphisms*. De Gruyter Expositions in Mathematics, vol. 8 (Walter de Gruyter and Co., Berlin, 1993). MR1224233
9. J.C. Lennox, D.J.S. Robinson, *The Theory of Infinite Soluble Groups*. Oxford Mathematical Monographs (The Clarendon Press/Oxford University Press, Oxford, 2004). MR2093872
10. W. Magnus, Beziehungen zwischen Gruppen und Idealen in einem speziellen Ring. Math. Ann. 111(1), 259–280 (1935). MR1512992
11. W. Magnus, A. Karrass, D. Solitar, *Combinatorial Group Theory* (Dover Publications, Mineola, NY, 2004). MR2109550. Reprint of the 1976 second edition

Chapter 4
Normal Forms and Embeddings

This chapter deals with normal forms in finitely generated torsion-free nilpotent groups and embeddings of such groups into radicable nilpotent groups. In Section 4.1, we develop a way to expand a positive power of a product of elements as a product of a finite number of terms, each one lying in a specific lower central subgroup. These terms are powers of the so-called Hall-Petresco words. Section 4.2 pertains to the construction of a Mal'cev basis for a finitely generated torsion-free nilpotent group. If G is such a group, then one can use a Mal'cev basis to express an element of G in a normal form. In Section 4.3, we use a Mal'cev basis to embed a finitely generated torsion-free nilpotent group G in a nilpotent group that admits an action by a binomial ring. We apply this to prove a theorem of A. I. Mal'cev which states that G can be embedded in a radicable torsion-free nilpotent group G^* with the property that every element of G^* has a kth power in G for some positive integer k. The work in Section 4.3 invites a general study of nilpotent groups that admit an action by a binomial ring. These groups are termed nilpotent R-powered groups and are discussed in Section 4.4.

4.1 The Hall-Petresco Words

Every positive power of a product of group elements can be expressed as a product of basic commutators by Theorem 3.2. In this section, we present another way of expanding a positive power of a product. The key formula, due to P. Hall [5] and J. Petresco [15], uses a collection process. Our discussion is based on G. Baumslag's work [1].

4.1.1 m-Fold Commutators

We begin by defining an m-fold commutator.

Definition 4.1 Let G be a group. An *m-fold commutator* in G is defined inductively as follows:

1. A 1-fold commutator in G is just an element of G;
2. If a is an i-fold commutator and b is a j-fold commutator, then $[a,\ b]$ is an $(i+j)$-fold commutator.

Note that every weighted commutator of weight m of a group G relative to a generating set is an m-fold commutator.

Lemma 4.1 *Let G be any group. For any $n \in \mathbb{N}$, the subgroup $\gamma_n G$ contains every k-fold commutator in G for $k \geq n$.*

Proof The proof is done by induction on n. The result is clear for $n = 1$. Suppose that $n > 1$, and let c be an m-fold commutator in G where $m \geq n$. There exist m_1-fold and m_2-fold commutators c_1 and c_2 respectively, such that $c = [c_1,\ c_2]$ and $m_1 + m_2 = m$. By induction, $c_1 \in \gamma_{m_1} G$ and $c_2 \in \gamma_{m_2} G$. Hence,

$$[c_1,\ c_2] \in [\gamma_{m_1} G,\ \gamma_{m_2} G] \leq \gamma_{m_1+m_2} G = \gamma_m G \leq \gamma_n G$$

by Theorem 2.14. And so, $c \in \gamma_n G$. □

4.1.2 A Collection Process

Let G be a group, and suppose that $Y = \{y_1, y_2, \ldots, y_n\}$ is a subset of G. Let

$$p = y_1 y_2 \cdots y_n \tag{4.1}$$

be an element of G. Define the set $R = \{1, 2, \ldots, r\}$, where $r \leq n$, and let $\psi : Y \to R$ be a surjective map. We call the image $\psi(y_j)$ of y_j the *label* of y_j. Since ψ is surjective, each element of R is the label of some element of Y.

Choose a nonempty subset S of R. Let X_S denote the set of all m-fold commutators $c\ (m \geq |S|)$ such that the label of each component of c lies in S, and each element of S is the label of some component of c. We impose an ordering on the nonempty subsets of R, first by cardinality and then lexicographically. For example, if $R = \{1, 2, 3\}$, then

$$\{1\} < \{2\} < \{3\} < \{1, 2\} < \{1, 3\} < \{2, 3\} < \{1, 2, 3\}.$$

Lemma 4.2 *The element p given in (4.1) is expressible as*

$$p = \prod_{\emptyset \neq S \subseteq R} q_S, \qquad (4.2)$$

where q_S is a product of elements in X_S and the $2^r - 1$ factors q_S occur in the order imposed on them by the ordering of the nonempty subsets of R.

Proof Let y be a factor of p such that $\psi(y) = 1$ (such a y with this label exists because ψ is onto), and let y_l be the first y in p with this label. If $l = 1$, then the result is immediate. If $l > 1$, then we move y_l to the left of y_{l-1} by using the commutator identity

$$y_{l-1}y_l = y_l y_{l-1} [y_{l-1}, y_l].$$

Clearly, $[y_{l-1}, y_l] \in X_S$, where $S = \{1, \psi(y_{l-1})\}$ and $\psi(y_{l-1}) > 1$. We repeat this process, in order, for each remaining y whose label is 1. In the same way, we collect the y's whose labels are 2, 3, $\ldots$, r. We ultimately obtain the expression

$$p = q_{\{1\}} q_{\{2\}} \cdots q_{\{r\}} \bar{p},$$

where $\bar{p}$ is a product, each of whose factors belongs to some X_S, with S containing at least two labels. Following the prescribed ordering of the remaining nonempty subsets of R, we collect the factors in $\bar{p}$ in a similar way to finally arrive at the required expression for p. $\square$

Let S be a given nonempty subset of R, and define

$$p_S = y_{i_1} y_{i_2} \cdots y_{i_t} \quad (i_1 < i_2 < \cdots < i_t),$$

where $\{y_{i_1}, \ldots, y_{i_t}\}$ is the set of those y_j's occurring in p whose labels are in S. Observe that p_S can be obtained from p by setting $y_j = 1$ whenever $\psi(y_j) \notin S$. After making these substitutions in (4.2), we have that $q_T = 1$ whenever $T \nsubseteq S$ and q_T is unchanged if $T \subseteq S$. This is due to the fact that a commutator equals the identity whenever one or more of its components equals the identity. This proves the next lemma.

Lemma 4.3 *For each nonempty subset S of R,*

$$p_S = \prod_{\emptyset \neq T \subseteq S} q_T,$$

where the factors q_T occur in their prescribed order.

Lemma 4.3 enables us to express q_S in terms of the p_T by recurrence, where T ranges over the nonempty subsets of S. To illustrate this, let $\alpha,\ \beta \in R$ with $\alpha < \beta$, and put $S = \{\alpha,\ \beta\}$. Notice that $T = \{\{\alpha\},\ \{\beta\},\ \{\alpha,\ \beta\}\}$. We have

$$p_{\{\alpha\}} = q_{\{\alpha\}},\ p_{\{\beta\}} = q_{\{\beta\}},\ \text{and } p_{\{\alpha,\beta\}} = q_{\{\alpha\}}q_{\{\beta\}}q_{\{\alpha,\ \beta\}}.$$

And so,

$$q_{\{\alpha,\ \beta\}} = p_{\{\beta\}}^{-1}p_{\{\alpha\}}^{-1}p_{\{\alpha,\beta\}}.$$

4.1.3   The Hall-Petresco Words

We shall now derive the so-called Hall-Petresco words. Let $R = \{1,\ \ldots,\ r\}$ as before. Consider the product $p = y_1 y_2 \cdots y_{mr}$, where

$$y_1 = \cdots = y_r = x_1,$$
$$y_{r+1} = \cdots = y_{2r} = x_2,$$
$$\vdots$$
$$y_{(m-1)r+1} = \cdots = y_{mr} = x_m.$$

Thus, $p = x_1^r x_2^r \cdots x_m^r$. Put $Y = \{y_1,\ y_2,\ \ldots,\ y_{mr}\}$, and define the labeling map

$$\psi : Y \to R \text{ by } \psi(y_i) = j \text{ whenever } j \equiv i \bmod r.$$

For example, $\psi(y_2) = 2$, $\psi(y_{r+1}) = 1$, and $\psi\left(y_{(m-1)r+3}\right) = 3$. If S is a nonempty subset of R and $|S| = k$, then

$$p_S = x_1^k x_2^k \cdots x_m^k.$$

Note that p_S depends on k, but not on the actual elements of S. By the remark following Lemma 4.3, q_S also depends only on k. Therefore, we may write

$$q_S = \tau_k(x_1,\ x_2,\ \ldots,\ x_m) = \tau_k(\bar{x}).$$

Lemma 4.3 now gives

$$x_1^k x_2^k \cdots x_m^k = \tau_1(\bar{x})^k \tau_2(\bar{x})^{\binom{k}{2}} \cdots \tau_{k-1}(\bar{x})^{\binom{k}{k-1}} \tau_k(\bar{x}). \tag{4.3}$$

Definition 4.2 The elements $\tau_1(\overline{x})$, $\tau_2(\overline{x})$, $\cdots$, $\tau_k(\overline{x})$ in (4.3) are called the *Hall-Petresco words.*

Calculating these words is quite simple. To begin, set $k = 1$ in (4.3) and obtain

$$\tau_1(\overline{x}) = x_1 x_2 \cdots x_m. \qquad (4.4)$$

Next, put $k = 2$ in (4.3) and replace $\tau_1(\overline{x})$ by $x_1 x_2 \cdots x_m$ to get

$$x_1^2 x_2^2 \cdots x_m^2 = (x_1 x_2 \cdots x_m)^2 \tau_2(\overline{x}).$$

Thus,

$$\tau_2(\overline{x}) = (x_1 x_2 \cdots x_m)^{-2} x_1^2 x_2^2 \cdots x_m^2.$$

By continuing in this way, setting $k = 3$, 4, $\ldots$, we can find the rest of the Hall-Petresco words.

Recall that $\tau_s(\overline{x}) = q_S$ with $|S| = k$, and q_S is a product of m-fold commutators where $m \geq k$. This establishes the next theorem.

Theorem 4.1 *If G is any group, then*

$$\{\tau_k(x_1, x_2, \ldots, x_m) \mid x_1, x_2, \ldots, x_m \in G\} \subseteq \gamma_k G.$$

Remark 4.1 By rewriting (4.3) as

$$(x_1 x_2 \cdots x_m)^k = x_1^k x_2^k \cdots x_m^k \tau_k(\overline{x})^{-1} \tau_{k-1}(\overline{x})^{-\binom{k}{k-1}} \cdots \tau_2(\overline{x})^{-\binom{k}{2}}, \qquad (4.5)$$

we get an alternative way of expressing a positive power of a product. The point here is that $\tau_i(\overline{x})$ is contained in $\gamma_i G$ whenever $2 \leq i \leq k$. This will be useful in what follows.

Suppose that G is a nilpotent group of class c. By Theorems 2.2 and 4.1, we have that $\tau_k(\overline{g}) = 1$ for $k > c$. This leads to the next important result.

Corollary 4.1 *Let G be a nilpotent group of class c and g_1, $\ldots$, $g_n \in G$. For all $k \in \mathbb{N}$,*

$$g_1^k \cdots g_n^k = \tau_1(\overline{g})^k \tau_2(\overline{g})^{\binom{k}{2}} \cdots \tau_{c-1}(\overline{g})^{\binom{k}{c-1}} \tau_c(\overline{g})^{\binom{k}{c}},$$

where $\tau_i(\overline{g}) \in \gamma_i G$ for $1 \leq i \leq c$.

As usual, we set $\binom{k}{c} = 0$ whenever $c > k$.

Example 4.1 Suppose that G is nilpotent of class 3, and let g_1, $\ldots$, g_n be elements of G. Consider the product $g_1^k \cdots g_n^k$ for some $k \in \mathbb{N}$. We write this in terms of the

Hall-Petresco words. Since $\gamma_i G = 1$ for $i \geq 4$, $\tau_i(\bar{g}) = 1$ for $i \geq 4$ by Theorem 4.1. As discussed above, we find that

$$\tau_1(\bar{g}) = g_1 \cdots g_n, \quad \tau_2(\bar{g}) = (g_1 \cdots g_n)^{-2} g_1^2 \cdots g_n^2, \quad \text{and}$$

$$\tau_3(\bar{g}) = \left((g_1 \cdots g_n)^{-2} g_1^2 \cdots g_n^2 \right)^{-3} (g_1 \cdots g_n)^{-3} g_1^3 \cdots g_n^3.$$

By Corollary 4.1, we obtain

$$g_1^k \cdots g_n^k = (g_1 \cdots g_n)^k \left[(g_1 \cdots g_n)^{-2} g_1^2 \cdots g_n^2 \right]^{\binom{k}{2}} \cdot$$

$$\left[\left((g_1 \cdots g_n)^{-2} g_1^2 \cdots g_n^2 \right)^{-3} (g_1 \cdots g_n)^{-3} g_1^3 \cdots g_n^3 \right]^{\binom{k}{3}} \cdot$$

The next result is related to Lemma 3.2, but does not require the prime p to exceed the class c. The proof is based on [2].

Lemma 4.4 (N. Blackburn)   *Let G be a nilpotent group of class at most c, and let p be a fixed prime. There exists an integer $f(p, c)$ such that if $n \geq f(p, c)$, then every element of G^{p^n} is a $p^{n-f(p,\,c)}$th power.*

Proof If $p > c$, then the result is true by Lemma 3.2. In this case $f(p, c) = 0$, and in particular, $f(p, 1) = 0$ for abelian groups. We henceforth assume that $p \leq c$. The proof is done by induction on c, the basis of induction being $c = 2$.

Set $p = c = 2$, and let $n \in \mathbb{N}$. Let $g_1, \ldots, g_r \in G$, and consider the product

$$g_1^{2^n} \cdots g_r^{2^n}.$$

Since G has nilpotency class 2, we see from (3.14) that

$$(g_1 \cdots g_r)^{2^n} = g_1^{2^n} \cdots g_r^{2^n} \prod [g_i, g_j]^{\binom{2^n}{2}},$$

where the product is taken over all i and j satisfying the conditions $i > j$, $2 \leq i \leq r$, and $1 \leq j \leq r-1$. Observing that $\binom{2^n}{2} = 2^{n-1}(2^n - 1)$ and using the fact that each commutator is central, we obtain

$$g_1^{2^n} \cdots g_r^{2^n} = (g_1 \cdots g_r)^{2^n} \prod [g_j, g_i]^{2^{n-1}(2^n-1)}$$

$$= \left[(g_1 \cdots g_r)^2 \right]^{2^{n-1}} \left(\prod [g_j, g_i]^{2^n-1} \right)^{2^{n-1}}$$

$$= \left[(g_1 \cdots g_r)^2 \prod [g_j, g_i]^{2^n-1} \right]^{2^{n-1}}.$$

Therefore, $g_1^{2^n} \cdots g_r^{2^n}$ is a 2^{n-1}th power. We thus define $f(2, 2) = 1$, and this completes the basis of induction.

Now suppose that G is nilpotent of class $c > 2$ and $p \leq c$. Assume that the lemma holds for every nilpotent group of class at most $c - 1$. Let $g_1, \ldots, g_r \in G$, and put $\tau_i = \tau_i(g_1, \ldots, g_r)$. By Corollary 4.1, we have

$$g_1^{p^n} \cdots g_r^{p^n} = (g_1 \cdots g_r)^{p^n} \tau_2^{\binom{p^n}{2}} \cdots \tau_c^{\binom{p^n}{c}},$$

where $\tau_i \in \gamma_i G \leq \gamma_2 G$ for $2 \leq i \leq c$. Factor each i in the form $p^l u$, where u and p are relatively prime, and l is allowed to be zero. Since $p \leq c$, there exists a largest $s \in \mathbb{N}$ such that $p^s \leq c$. It follows that p^{n-l} divides $\binom{p^n}{p^l u}$ and $l \leq s$ (see Theorem 18 of [18]). Hence, p^{n-s} divides $\binom{p^n}{i}$ for each $i = 2, 3, \ldots, c$, and consequently, $\tau_i^{\binom{p^n}{i}}$ is always a p^{n-s}th power. Thus, we may write

$$g_1^{p^n} \cdots g_r^{p^n} = h_1^{p^{n-s}} \cdots h_c^{p^{n-s}},$$

where $h_1 \in G$ and $h_2, \ldots, h_c$ are contained in $\gamma_2 G$. By Lemma 2.9, $gp(h_1, \ldots, h_c)$ is of class less than c. By our induction hypothesis, there exists a number $f(p, c-1)$ such that if $n - s \geq f(p, c-1)$, then $g_1^{p^n} \cdots g_r^{p^n}$ is a $p^{n-s-f(p, c-1)}$th power. The result follows by setting $f(p, c) = f(p, c-1) + s$. ☐

Remark 4.2 In [12], A. I. Mal'cev proved a related result: if G is a nilpotent group of class c and $m \in \mathbb{N}$, then for any $g_1, \ldots, g_n \in G$, there exists $h \in G$ such that $g_1^{m^c} \cdots g_n^{m^c} = h^m$.

4.2   Normal Forms and Mal'cev Bases

In a finitely generated torsion-free nilpotent group, it is possible to obtain expressions for the multiplication and exponentiation of its elements by introducing a distinguished set of generators called a Mal'cev basis. This leads to a very useful normal form for such groups, bearing a certain resemblance to what occurs in vector spaces, free modules, and other algebraic structures where effective generating sets exist. In this section, we explain how to obtain such a basis and develop the corresponding normal form.

4.2.1   The Structure of a Finitely Generated Nilpotent Group

A finitely generated nilpotent group can be constructed by a finite sequence of cyclic extensions. To make sense of this, we define the notion of a polycyclic group.

Definition 4.3 A group G is called *polycyclic* if it has a subnormal series

$$1 = G_0 \trianglelefteq G_1 \trianglelefteq \cdots \trianglelefteq G_n = G \qquad (4.6)$$

such that G_{i+1}/G_i is cyclic for $0 \leq i \leq n-1$. The series (4.6) is called a *polycyclic series* for G. In addition, if each G_{i+1}/G_i is infinite, then G is called *poly-infinite cyclic* and the series (4.6) is called a *poly-infinite cyclic series*.

Clearly, every polycyclic group must be finitely generated.

Theorem 4.2 *Every subgroup and factor group of a polycyclic group is polycyclic.*

Proof Let G be a polycyclic group, and suppose that

$$1 = G_0 \trianglelefteq G_1 \trianglelefteq \cdots \trianglelefteq G_n = G$$

is a polycyclic series for G. Let $H \leq G$, and set $H_i = G_i \cap H$ for $i = 0, 1, \ldots, n$. Clearly, $H_i \trianglelefteq H_{i+1}$, and thus the series

$$1 = H_0 \trianglelefteq H_1 \trianglelefteq \cdots \trianglelefteq H_n = H$$

is subnormal. It is also polycyclic because, for each $i = 0, 1, \ldots, n-1$, there is a monomorphism from H_{i+1}/H_i to G_{i+1}/G_i whose image is cyclic. And so, H is polycyclic.

If $N \trianglelefteq G$, then $G_iN/N \trianglelefteq G_{i+1}N/N$ for $i = 0, 1, \ldots, n-1$. Thus, G/N has a subnormal series

$$1 = G_0N/N \trianglelefteq G_1N/N \trianglelefteq \cdots \trianglelefteq G_nN/N = G/N. \qquad (4.7)$$

Since $G_i \trianglelefteq G_{i+1}$, the Second and Third Isomorphism Theorems give

$$\frac{G_{i+1}N/N}{G_iN/N} \simeq \frac{G_{i+1}N}{G_iN} \simeq \frac{G_{i+1}}{G_{i+1} \cap G_iN}.$$

The factor group $G_{i+1}/(G_{i+1} \cap G_iN)$ must be cyclic because it is a quotient of G_{i+1}/G_i. Therefore, (4.7) is a polycyclic series. $\qquad \square$

Theorem 4.3 *Polycyclicity is preserved under extensions.*

Proof Let G be a group and $N \trianglelefteq G$, suppose that N and G/N are polycyclic. Let

$$1 = N_0 \trianglelefteq N_1 \trianglelefteq \cdots \trianglelefteq N_m = N$$

and

$$1 = G_0/N \trianglelefteq G_1/N \trianglelefteq \cdots \trianglelefteq G_n/N = G/N$$

be polycyclic series of N and G/N respectively. Then G has a subnormal series

$$1 = N_0 \trianglelefteq N_1 \trianglelefteq \cdots \trianglelefteq N_m \trianglelefteq G_1 \trianglelefteq \cdots \trianglelefteq G_n = G. \tag{4.8}$$

Since $G_{i+1}/G_i \cong (G_{i+1}/N)/(G_i/N)$ for $0 \leq i \leq n - 1$ by the Third Isomorphism Theorem, the series (4.8) is polycyclic. $\square$

Theorem 4.4 (R. Baer) *Every finitely generated nilpotent group has a polycyclic and central series.*

Proof Let G be a finitely generated nilpotent group of class c. Each upper central subgroup is finitely generated by Theorem 2.18. Hence, each factor group $\zeta_{i+1}G/\zeta_iG$ is finitely generated abelian for $0 \leq i \leq c-1$, and thus a direct product of cyclic groups. Consequently, the upper central series can be refined so that between each pair of upper central subgroups $\zeta_{i+1}G$ and ζ_iG, we obtain a normal series

$$\zeta_{i+1}G > U_{i+1,\,1} > U_{i+1,\,2} > \cdots > U_{i+1,\,k} = \zeta_iG. \tag{4.9}$$

Each factor group of consecutive terms of (4.9) is cyclic, and (4.9) forms part of a central series for G since, for $j = 1, \ldots, k - 1$,

$$\left[U_{i+1,\,j},\, G\right] \leq \left[\zeta_{i+1}G,\, G\right] \leq \zeta_iG \leq U_{i+1,\,j+1}.$$

This completes the proof. $\square$

An alterative proof can be obtained by refining the lower central series of G by using the fact that each factor group $\gamma_iG/\gamma_{i+1}G$ is finitely generated for $1 \leq i \leq c$ by Corollary 2.11.

Remark 4.3 Not every polycyclic group is finitely generated nilpotent. For instance, S_3 is a polycyclic group with polycyclic series $1 \triangleleft A_3 \triangleleft S_3$. However, S_3 is not nilpotent (refer to Remark 2.1).

In the case that G is torsion-free as well, each $\zeta_{i+1}G/\zeta_iG$ is torsion-free by Corollary 2.22. Thus, each $\zeta_{i+1}G/\zeta_iG$ is a direct product of infinite cyclic groups. This gives us the next result.

Theorem 4.5 *Every finitely generated torsion-free nilpotent group has a poly-infinite cyclic and central series.*

One important feature about polycyclic groups is contained in the next theorem.

Theorem 4.6 (K. A. Hirsch) *If G is a polycyclic group, then the number of infinite cyclic factors in any polycyclic series for G is an invariant of G.*

Proof Let

$$1 = G_0 \trianglelefteq G_1 \trianglelefteq \cdots \trianglelefteq G_k = G \tag{4.10}$$

be a polycyclic series for G. We claim that the number of infinite cyclic factors in the series (4.10) is the same as in any refinement of it.

Suppose that $G_i \neq G_{i+1}$ for some $0 \leq i \leq k - 1$ such that G_{i+1}/G_i is infinite cyclic, and let $G_i \lhd N \lhd G_{i+1}$ for some N. Since G_{i+1}/G_i is isomorphic to $\mathbb{Z}$, the subgroup N/G_i of G_{i+1}/G_i is isomorphic to $n\mathbb{Z}$ for some $n \in \mathbb{N}$. Thus, N/G_i is infinite cyclic. However, $G_{i+1}/N \cong (G_{i+1}/G_i)/(N/G_i)$ by the Third Isomorphism Theorem. Thus, G_{i+1}/N is isomorphic to the finite cyclic group $\mathbb{Z}/n\mathbb{Z}$. Hence, an insertion of N between G_i and G_{i+1} does not change the number of infinite cyclic factors between them. The result follows from Schreier's Refinement Theorem, which states that any two subnormal series of a group have isomorphic refinements.   □

Definition 4.4 The number of infinite cyclic factors in any polycyclic series for a polycyclic group G is called the *Hirsch length* or *torsion-free rank* of G. It is often denoted by $h(G)$.

The invariance of the Hirsch length of a polycyclic group is useful for proving results by induction on the Hirsch length. The next theorem is commonly used.

Theorem 4.7 *Let G be a polycyclic group. If $N \unlhd G$, then $h(G) = h(N) + h(G/N)$.*

Proof The result follows from Theorem 4.2.   □

4.2.2   Mal'cev Bases

Let G be a finitely generated torsion-free nilpotent group. By Theorems 4.5 and 4.6, we know that G has a (descending) poly-infinite cyclic and central series

$$G = G_1 \rhd G_2 \rhd \cdots \rhd G_{n+1} = 1 \tag{4.11}$$

of Hirsch length n. Since each factor of (4.11) is infinite cyclic, we may choose $u_i \in G_i$ so that $G_i = gp(G_{i+1}, u_i)$ for each $i = 1, \ldots, n$. Thus, every element of G can be uniquely expressed in the normal form

$$u_1^{\alpha_1} \cdots u_n^{\alpha_n} \quad (\alpha_1, \ldots, \alpha_n \in \mathbb{Z}).$$

Remark 4.4 By construction, we have that for each $j = 0, \ldots, n-1$, every element of G_{j+1} can be expressed in the normal form

$$u_{j+1}^{\alpha_{j+1}} \cdots u_n^{\alpha_n} \quad (\alpha_{j+1}, \ldots, \alpha_n \in \mathbb{Z}).$$

Moreover, since $[G_i, G_j] \leq G_k$ for some $k \geq 1 + max\{i, j\}$, where $1 \leq i, j \leq n$, we have

$$\left[u_i^{\alpha_i}, u_j^{\alpha_j} \right] = \prod_{m=k}^{n} u_m^{\alpha_m}.$$

Definition 4.5 The set $\bar{u} = \{u_1, \ldots, u_n\}$ associated with the series (4.11) is called a *Mal'cev basis* for G. The element $u_1^{\alpha_1} \cdots u_n^{\alpha_n}$ is said to have *(Mal'cev) coordinates* $\bar{\alpha} = (\alpha_1, \ldots, \alpha_n) \in \mathbb{Z}^n$ with respect to $\bar{u}$.

For simplicity, we will sometimes write $\bar{u}^{\bar{\alpha}}$ instead of $u_1^{\alpha_1} \cdots u_n^{\alpha_n}$.

Theorem 4.8 *Let G be a finitely generated free nilpotent group of class c, freely generated by a finite set X. Any basic sequence of basic commutators on X of weight at most c is a Mal'cev basis for G.*

Proof By Corollary 3.4, each $\gamma_i G/\gamma_{i+1} G$ is free abelian, freely generated by the basic commutators of weight i, modulo $\gamma_{i+1} G$. A refinement of the lower central series of G leads to a poly-infinite cyclic and central series

$$G = G_1 \rhd G_2 \rhd \cdots \rhd G_{t+1} = 1$$

such that G_j/G_{j+1} is generated by the jth basic commutator, modulo G_{j+1}, for each $j = 1, 2, \ldots, t$.   □

Example 4.2 Let G be a free nilpotent group of class 2, freely generated by the set $X = \{a, b\}$. A basic sequence of basic commutators on $\{a, b\}$ is

$$a, b, [b, a].$$

By Theorem 4.8, these basic commutators form a Mal'cev basis for G. Thus, every element of G can be written in the unique normal form

$$a^{n_1} b^{n_2} [b, a]^{n_3}$$

for some integers n_1, n_2, n_3.

Example 4.3 Let G be a free nilpotent group of class 3, freely generated by the set $X = \{a, b\}$. A basic sequence of basic commutators on X is

$$a, b, [b, a], [b, a, a], [b, a, b].$$

These basic commutators form a Mal'cev basis for G by Theorem 4.8. Every element of G can be uniquely written as

$$a^{n_1} b^{n_2} [b, a]^{n_3} [b, a, a]^{n_4} [b, a, b]^{n_5}$$

for some integers $n_1, \ldots, n_5$.

The next theorem illustrates how to multiply and exponentiate elements of a finitely generated torsion-free nilpotent group in normal form with respect to a given Mal'cev basis. We follow the proof given in [5].

Theorem 4.9 (P. Hall, A. I. Mal'cev) *Let G be a finitely generated torsion-free nilpotent group of class c, and let $\{u_1, \ldots, u_n\}$ be a Mal'cev basis of G with respect to the poly-infinite cyclic and central series*

$$G = G_1 \rhd G_2 \rhd \cdots \rhd G_{n+1} = 1. \tag{4.12}$$

If $x = u_1^{\alpha_1} \cdots u_n^{\alpha_n}$ and $y = u_1^{\beta_1} \cdots u_n^{\beta_n}$ for some $\alpha_i, \beta_i \in \mathbb{Z}$, and if $\lambda \in \mathbb{Z}$, then

$$xy = u_1^{f_1(\bar{\alpha}, \bar{\beta})} \cdots u_n^{f_n(\bar{\alpha}, \bar{\beta})} \quad \text{and} \quad x^\lambda = u_1^{g_1(\bar{\alpha}, \lambda)} \cdots u_n^{g_n(\bar{\alpha}, \lambda)},$$

where each f_i is a polynomial with rational coefficients in $2n$ variables, and each g_i is a polynomial with rational coefficients in $(n + 1)$ variables.

Proof The proof is done by induction on the Hirsch length n of G. If $n = 1$, then G is infinite cyclic and we have $f_1 = \alpha_1 + \beta_1$ and $g_1 = \lambda\alpha_1$. Let (M_k) and (E_k) denote, respectively, the hypotheses that $f_1, \ldots, f_k$ are polynomials with rational coefficients in $2k$ variables, and $g_1, \ldots, g_k$ are polynomials with rational coefficients in $(k + 1)$ variables for $k = 1, \ldots, n$.

Suppose that $n > 1$, and assume that (M_i) and (E_i) hold for all $i < n$. We first show that (M_n) is true. Since

$$\left(u_2^{\alpha_2} \cdots u_n^{\alpha_n}\right)\left(u_1^{\beta_1}\right) = u_1^{\beta_1} u_1^{-\beta_1} \left(\prod_{i=2}^{n} u_i^{\alpha_i}\right) u_1^{\beta_1}$$

$$= u_1^{\beta_1} \prod_{i=2}^{n} u_1^{-\beta_1} u_i^{\alpha_i} u_1^{\beta_1}$$

$$= u_1^{\beta_1} \prod_{i=2}^{n} \left(u_1^{-\beta_1} u_i^{-1} u_1^{\beta_1}\right)^{-\alpha_i},$$

we may write

$$xy = u_1^{\alpha_1 + \beta_1} \prod_{i=2}^{n} \left(u_1^{-\beta_1} u_i^{-1} u_1^{\beta_1}\right)^{-\alpha_i} u_2^{\beta_2} \cdots u_n^{\beta_n}. \tag{4.13}$$

Note that

$$u_1^{-\beta_1} u_i^{-1} u_1^{\beta_1} = u_1^{-\beta_1} \left(u_i^{-1} u_1 u_i\right)^{\beta_1} u_i^{-1}. \tag{4.14}$$

Since (4.12) is a central series for G and $u_i \in G_{i-1}$, $[u_1, u_i] \in G_i$. By Remark 4.4,

$$u_i^{-1} u_1 u_i = u_1 u_{i+1}^{c_{i,1}} \cdots u_n^{c_{i,n-i}}$$

for some constants $c_{i,j} \in \mathbb{Z}$.

The group $H_i = gp(u_1, u_{i+1}, \ldots, u_n)$ has Mal'cev basis $\{u_1, u_{i+1}, \ldots, u_n\}$ with $(n - i + 1)$ terms for $i > 1$. Thus, if $i > 1$, then

$$\left(u_i^{-1} u_1 u_i\right)^{\beta_1} = u_1^{\beta_1} u_{i+1}^{\varphi_{i,\,1}} \cdots u_n^{\varphi_{i,\,n-i}},$$

where, by (E_{n-i+1}), the $\varphi_{i,\,j}$ are polynomials in β_1 and the constants $c_{i,\,j}$. Next, we apply (M_{n-i+1}) to (4.14) and obtain

$$u_1^{-\beta_1} u_i^{-1} u_1^{\beta_1} = u_{i+1}^{\varphi_{i,\,1}} \cdots u_n^{\varphi_{i,\,n-i}} \cdot u_i^{-1}$$

$$= u_i^{-1} u_{i+1}^{\psi_{i,\,1}} \cdots u_n^{\psi_{i,\,n-i}},$$

where the $\psi_{i,\,j}$ are again polynomials in β_1 and the constants $c_{i,\,j}$. Hence,

$$\left(u_1^{-\beta_1} u_i^{-1} u_1^{\beta_1}\right)^{-\alpha_i} = u_i^{\alpha_i} u_{i+1}^{\theta_{i,\,1}} \cdots u_n^{\theta_{i,\,n-i}}$$

where, by (E_{n-i+1}) again, the $\theta_{i,\,j}$ are polynomials in β_1 and α_i. After substituting in (4.13) and using (M_{n-1}) repeatedly, we establish (M_n). Note that the polynomial that corresponds to u_1 in (4.13) is always $\alpha_1 + \beta_1$.

Next, we prove that (E_n) is true. Let $\left(E_n^+\right)$ denote (E_n) for $\lambda > 0$, and let

$$v_i = \tau_i\left(u_1^{\alpha_1}, \ldots, u_n^{\alpha_n}\right),$$

where the τ_i's are the Hall-Petresco words. Since each τ_i is a word in $u_1^{\alpha_1}, \ldots, u_n^{\alpha_n}$, repeated applications of (M_n) allows us to express v_i as

$$v_i = u_1^{\alpha_{i,\,1}} \cdots u_n^{\alpha_{i,\,n}},$$

where the $\alpha_{i,\,j}$ are polynomials in $\alpha_1, \ldots, \alpha_n$. By Theorem 4.1 and the fact that $\gamma_{c+1}G = 1$ (since G is of class c by assumption), we have $v_{c+1} = v_{c+2} = \cdots = 1$. Now,

$$v_1 = \tau_1\left(u_1^{\alpha_1}, \ldots, u_n^{\alpha_n}\right) = u_1^{\alpha_1} u_2^{\alpha_2} \cdots u_n^{\alpha_n} = x$$

by (4.4). For $\lambda > 0$, (4.3) gives

$$x^\lambda = u_1^{\lambda\alpha_1} u_2^{\lambda\alpha_2} \cdots u_n^{\lambda\alpha_n} v_c^{-\binom{\lambda}{c}} v_{c-1}^{-\binom{\lambda}{c-1}} \cdots v_2^{-\binom{\lambda}{2}}. \tag{4.15}$$

However, $v_k \in \gamma_2 G$ for $2 \leq k \leq c$ by Theorem 4.1, and $\gamma_2 G \leq G_1$ by Theorem 2.1. Consequently, by (E_{n-1}), each $v_i^{-\binom{\lambda}{i}}$ equals $u_2^{\eta_{i,\,2}} \cdots u_n^{\eta_{i,\,n}}$, where the $\eta_{i,\,j}$ are polynomials in λ and the α's. Substituting in (4.15) and applying (M_{n-1}) repeatedly gives $\left(E_n^+\right)$, with uniquely determined polynomials $g_i = g_i(\lambda, \alpha_1, \ldots, \alpha_n)$.

By (M_n), $x^{-1} = u_n^{-\alpha_n} \cdots u_1^{-\alpha_1} = u_1^{\delta_1} \cdots u_n^{\delta_n}$, where the δ_i are polynomials in the α's. If $\lambda > 0$, then (E_n^+) gives $x^{-\lambda} = u_1^{\varepsilon_1} \cdots u_n^{\varepsilon_n}$, where the ε_i are polynomials in the α's and λ. Hence, if $\lambda > 0$ and $\mu > 0$, then (M_n) and (E_n^+) give

$$x^{\mu - \lambda} = x^\mu x^{-\lambda} = u_1^{k_1} \cdots u_n^{k_n},$$

where $k_i = k_i(\lambda,\ \mu,\ \alpha_1,\ \ldots,\ \alpha_n)$ is a polynomial in λ, μ, and the α's. Given any integer μ, choose $\lambda > 0$ so that $\lambda + \mu > 0$. Then

$$x^\mu = x^{\lambda + \mu - \lambda} = u_1^{k_1(\lambda,\ \lambda + \mu,\ \alpha_1,\ \ldots,\ \alpha_n)} \cdots u_n^{k_n(\lambda,\ \lambda + \mu,\ \alpha_1,\ \ldots,\ \alpha_n)}$$

$$= u_1^{g_1(\mu,\ \alpha_1,\ \ldots,\ \alpha_n)} \cdots u_n^{g_n(\mu,\ \alpha_1,\ \ldots,\ \alpha_n)}$$

and (E_n) is proven.   □

Definition 4.6 The polynomials $f_i\left(\overline{\alpha},\ \overline{\beta}\right)$ and $g_i(\overline{\alpha},\ \lambda)$ in Theorem 4.9 are called the *multiplication* and *exponentiation polynomials* for G respectively with respect to $\overline{u}$. In vector form, we write

$$\overline{f}\left(\overline{\alpha},\ \overline{\beta}\right) = \left(f_1\left(\overline{\alpha},\ \overline{\beta}\right),\ \ldots,\ f_n\left(\overline{\alpha},\ \overline{\beta}\right)\right) \quad \text{and}$$

$$\overline{g}\left(\overline{\alpha},\ \lambda\right) = \left(g_1\left(\overline{\alpha},\ \lambda\right),\ \ldots,\ g_n\left(\overline{\alpha},\ \lambda\right)\right).$$

The multiplication and exponentiation polynomials can be obtained by collecting terms to the left, as illustrated in the next examples.

Example 4.4 For a finitely generated torsion-free abelian group with Mal'cev basis $\{u_1,\ \ldots,\ u_n\}$, we have

$$\left(u_1^{\alpha_1} \cdots u_n^{\alpha_n}\right)\left(u_1^{\beta_1} \cdots u_n^{\beta_n}\right) = u_1^{\alpha_1 + \beta_1} \cdots u_n^{\alpha_n + \beta_n} \quad \text{and}$$

$$\left(u_1^{\alpha_1} \cdots u_n^{\alpha_n}\right)^\lambda = u_1^{\alpha_1 \lambda} \cdots u_n^{\alpha_n \lambda}$$

for any $\alpha_i,\ \beta_i,\ \lambda \in \mathbb{Z}$. Therefore,

$$\overline{f}\left(\overline{\alpha},\ \overline{\beta}\right) = (\alpha_1 + \beta_1,\ \alpha_2 + \beta_2,\ \ldots,\ \alpha_n + \beta_n) \quad \text{and}$$

$$\overline{g}\left(\overline{\alpha},\ \lambda\right) = (\alpha_1 \lambda,\ \alpha_2 \lambda,\ \ldots,\ \alpha_n \lambda).$$

Example 4.5 Let G be a free nilpotent group of class 2, freely generated by $\{a,\ b\}$. Put $c = [a,\ b]$, and note that c is a central element of G. By Theorem 4.8, the set $\{a,\ b,\ c\}$ is a Mal'cev basis for G. Let $a^{\alpha_1} b^{\alpha_2} c^{\alpha_3}$ and $a^{\beta_1} b^{\beta_2} c^{\beta_3}$ be elements of G in normal form with respect to this basis. We use Lemma 1.13, together with the fact that $c \in Z(G)$, to obtain the multiplication polynomials:

$$\left(a^{\alpha_1} b^{\alpha_2} c^{\alpha_3} \right) \left(a^{\beta_1} b^{\beta_2} c^{\beta_3} \right) = a^{\alpha_1} \underline{b^{\alpha_2} a^{\beta_1}} \, b^{\beta_2} c^{\alpha_3 + \beta_3}$$

$$= a^{\alpha_1 + \beta_1} b^{\alpha_2} \left[b^{\alpha_2}, a^{\beta_1} \right] b^{\beta_2} c^{\alpha_3 + \beta_3}$$

$$= a^{\alpha_1 + \beta_1} b^{\alpha_2} [b, a]^{\alpha_2 \beta_1} b^{\beta_2} c^{\alpha_3 + \beta_3}$$

$$= a^{\alpha_1 + \beta_1} b^{\alpha_2} \left(c^{-1} \right)^{\alpha_2 \beta_1} b^{\beta_2} c^{\alpha_3 + \beta_3}$$

$$= a^{\alpha_1 + \beta_1} b^{\alpha_2 + \beta_2} c^{\alpha_3 + \beta_3 - \alpha_2 \beta_1}.$$

In the underlined product above, we used the identity $xy = yx[x, y]$. Thus,

$$\overline{f}\left(\overline{\alpha}, \overline{\beta} \right) = \left(\alpha_1 + \beta_1, \, \alpha_2 + \beta_2, \, \alpha_3 + \beta_3 - \alpha_2 \beta_1 \right). \qquad (4.16)$$

Next, we show that

$$(a^{\alpha_1} b^{\alpha_2} c^{\alpha_3})^\lambda = a^{\alpha_1 \lambda} b^{\alpha_2 \lambda} c^{\alpha_3 \lambda - \frac{\lambda(\lambda-1)}{2} \alpha_1 \alpha_2} \qquad (4.17)$$

for any $\lambda \in \mathbb{Z}$. It is obviously true for $\lambda = 0$ and $\lambda = 1$. We show that it holds for $\lambda > 1$ by induction. Assume that (4.17) is true for $\lambda - 1$. Using Lemma 1.13 as before, we get

$$(a^{\alpha_1} b^{\alpha_2} c^{\alpha_3})^\lambda = (a^{\alpha_1} b^{\alpha_2} c^{\alpha_3})^{\lambda-1} (a^{\alpha_1} b^{\alpha_2} c^{\alpha_3})$$

$$= a^{\alpha_1(\lambda-1)} b^{\alpha_2(\lambda-1)} c^{\alpha_3(\lambda-1) - \frac{(\lambda-1)(\lambda-2)}{2} \alpha_1 \alpha_2} \left(a^{\alpha_1} b^{\alpha_2} c^{\alpha_3} \right)$$

$$= a^{\alpha_1(\lambda-1)} \underline{b^{\alpha_2(\lambda-1)} a^{\alpha_1}} \, b^{\alpha_2} c^{\alpha_3(\lambda-1) - \frac{(\lambda-1)(\lambda-2)}{2} \alpha_1 \alpha_2 + \alpha_3}$$

$$= a^{\alpha_1(\lambda-1)} a^{\alpha_1} b^{\alpha_2(\lambda-1)} \left[b^{\alpha_2(\lambda-1)}, a^{\alpha_1} \right] b^{\alpha_2} c^{\alpha_3 \lambda - \frac{(\lambda-1)(\lambda-2)}{2} \alpha_1 \alpha_2}$$

$$= a^{\alpha_1 \lambda} b^{\alpha_2(\lambda-1)} [b, a]^{\alpha_1 \alpha_2(\lambda-1)} b^{\alpha_2} c^{\alpha_3 \lambda - \frac{(\lambda-1)(\lambda-2)}{2} \alpha_1 \alpha_2}$$

$$= a^{\alpha_1 \lambda} b^{\alpha_2(\lambda-1)} c^{-\alpha_1 \alpha_2(\lambda-1)} b^{\alpha_2} c^{\alpha_3 \lambda - \frac{(\lambda-1)(\lambda-2)}{2} \alpha_1 \alpha_2}$$

$$= a^{\alpha_1 \lambda} b^{\alpha_2(\lambda-1)} b^{\alpha_2} c^{-\alpha_1 \alpha_2(\lambda-1) + \alpha_3 \lambda - \frac{(\lambda-1)(\lambda-2)}{2} \alpha_1 \alpha_2}$$

$$= a^{\alpha_1 \lambda} b^{\alpha_2 \lambda} c^{\alpha_3 \lambda - \frac{\lambda(\lambda-1)}{2} \alpha_1 \alpha_2}.$$

Once again, the identity $xy = yx[x, y]$ has been used in the underlined products above and in what follows.

Next, we calculate $\left(a^{\alpha_1} b^{\alpha_2} c^{\alpha_3}\right)^{-1}$:

$$
\begin{aligned}
\left(a^{\alpha_1} b^{\alpha_2} c^{\alpha_3}\right)^{-1} &= c^{-\alpha_3} b^{-\alpha_2} a^{-\alpha_1} \\
&= \underline{b^{-\alpha_2} a^{-\alpha_1}} \, c^{-\alpha_3} \\
&= a^{-\alpha_1} b^{-\alpha_2} \left[b^{-\alpha_2}, \, a^{-\alpha_1}\right] c^{-\alpha_3} \\
&= a^{-\alpha_1} b^{-\alpha_2} [b, \, a]^{\alpha_1 \alpha_2} c^{-\alpha_3} \\
&= a^{-\alpha_1} b^{-\alpha_2} c^{-\alpha_1 \alpha_2 - \alpha_3},
\end{aligned}
$$

in agreement with (4.17) for $\lambda = -1$. Finally, if $\lambda < 0$, then

$$
\begin{aligned}
\left(a^{\alpha_1} b^{\alpha_2} c^{\alpha_3}\right)^{\lambda} &= \left(\left(a^{\alpha_1} b^{\alpha_2} c^{\alpha_3}\right)^{-1}\right)^{-\lambda} \\
&= \left(a^{-\alpha_1} b^{-\alpha_2} c^{-\alpha_1 \alpha_2 - \alpha_3}\right)^{-\lambda} \\
&= a^{\alpha_1 \lambda} b^{\alpha_2 \lambda} c^{\alpha_3 \lambda + \alpha_1 \alpha_2 \lambda - \frac{(-\lambda)(-\lambda - 1)}{2} \alpha_1 \alpha_2} \\
&= a^{\alpha_1 \lambda} b^{\alpha_2 \lambda} c^{\alpha_3 \lambda - \frac{\lambda(\lambda - 1)}{2} \alpha_1 \alpha_2}.
\end{aligned}
$$

Therefore,

$$
\bar{g}(\bar{\alpha}, \, \lambda) = \left(\alpha_1 \lambda, \, \alpha_2 \lambda, \, \alpha_3 \lambda - \frac{\lambda(\lambda - 1)}{2} \alpha_1 \alpha_2\right).
$$

We know from Examples 2.12 and 2.18 that $\mathcal{H}$ is a torsion-free nilpotent group of class 2. The next theorem shows that $\mathcal{H}$ is, in fact, free nilpotent of class 2 and rank 2. We use a computation similar to the one above.

Theorem 4.10 *If F is a free nilpotent group of class 2 and rank 2 with presentation*

$$
\langle a, \, b \mid [b, \, a, \, a] = 1, \, [b, \, a, \, b] = 1 \rangle,
$$

then F is isomorphic to $\mathcal{H}$.

Proof The set $\{a, \, b, \, [b, \, a]\}$ is a Mal'cev basis for F. Thus, every element of F can be uniquely written in normal form as $a^{\alpha} b^{\beta} [b, \, a]^{\gamma}$ for some integers α, β, and γ. Define the map $\varphi : F \to \mathcal{H}$ as

$$
a^{\alpha} b^{\beta} [b, \, a]^{\gamma} \mapsto \begin{pmatrix} 1 & \beta & \gamma \\ 0 & 1 & \alpha \\ 0 & 0 & 1 \end{pmatrix}.
$$

By repeating what was done in Example 4.5, we find that

$$\left(a^{\alpha_1} b^{\beta_1} [b, a]^{\gamma_1}\right) \left(a^{\alpha_2} b^{\beta_2} [b, a]^{\gamma_2}\right) = a^{\alpha_1 + \alpha_2} b^{\beta_1 + \beta_2} [b, a]^{\gamma_1 + \gamma_2 + \alpha_2 \beta_1}.$$

It follows that φ is a homomorphism. It is clearly a monomorphism since

$$a^{\alpha} b^{\beta} [b, a]^{\gamma} \mapsto \begin{pmatrix} 1 & 0 & 0 \\ 0 & 1 & 0 \\ 0 & 0 & 1 \end{pmatrix}$$

if and only if $\alpha = \beta = \gamma = 0$. It is also clear that φ is an epimorphism. $\qquad \square$

Example 4.6 Let G be a free nilpotent group of class 3 with a set of free generators $X = \{a, b\}$. By Example 4.3, the basic commutators

$$a, \ b, \ [b, a], \ [b, a, a], \ \text{and} \ [b, a, b]$$

form a Mal'cev basis for G. Suppose that

$$u = a^{\alpha_1} b^{\alpha_2} [b, a]^{\alpha_3} [b, a, a]^{\alpha_4} [b, a, b]^{\alpha_5} \ \text{and}$$
$$v = a^{\beta_1} b^{\beta_2} [b, a]^{\beta_3} [b, a, a]^{\beta_4} [b, a, b]^{\beta_5}$$

are elements of G for some $\alpha_i, \ \beta_i \in \mathbb{Z}$. If we put

$$uv = a^{f_1(\overline{\alpha}, \overline{\beta})} b^{f_2(\overline{\alpha}, \overline{\beta})} [b, a]^{f_3(\overline{\alpha}, \overline{\beta})} [b, a, a]^{f_4(\overline{\alpha}, \overline{\beta})} [b, a, b]^{f_5(\overline{\alpha}, \overline{\beta})}$$

and

$$u^{\lambda} = a^{g_1(\overline{\alpha}, \lambda)} b^{g_2(\overline{\alpha}, \lambda)} [b, a]^{g_3(\overline{\alpha}, \lambda)} [b, a, a]^{g_4(\overline{\alpha}, \lambda)} [b, a, b]^{g_5(\overline{\alpha}, \lambda)},$$

then the multiplication polynomials turn out to be

$$f_1\left(\overline{\alpha}, \overline{\beta}\right) = \alpha_1 + \beta_1,$$

$$f_2\left(\overline{\alpha}, \overline{\beta}\right) = \alpha_2 + \beta_2,$$

$$f_3\left(\overline{\alpha}, \overline{\beta}\right) = \alpha_3 + \beta_3 + \alpha_2 \beta_1,$$

$$f_4\left(\overline{\alpha}, \overline{\beta}\right) = \alpha_4 + \beta_4 + \alpha_3 \beta_1 + \frac{\alpha_2 \beta_1 (\beta_1 - 1)}{2},$$

$$f_5\left(\overline{\alpha}, \overline{\beta}\right) = \alpha_5 + \beta_5 + \alpha_3 \beta_2 + \frac{\alpha_2 \beta_1 (\alpha_2 - 1)}{2} + \alpha_2 \beta_1 \beta_2,$$

and the exponentiation polynomials are

$$g_1(\bar{\alpha}, \lambda) = \lambda\alpha_1,$$

$$g_2(\bar{\alpha}, \lambda) = \lambda\alpha_2,$$

$$g_3(\bar{\alpha}, \lambda) = \lambda\alpha_3 + \frac{\lambda(\lambda-1)\alpha_2\alpha_1}{2},$$

$$g_4(\bar{\alpha}, \lambda) = \lambda\alpha_4 + \frac{\lambda(\lambda-1)\alpha_3\alpha_1}{2} - \frac{\lambda(\lambda-1)\alpha_2\alpha_1}{4} + \frac{\lambda(\lambda-1)(2\lambda-1)\alpha_2\alpha_1^2}{12},$$

$$g_5(\bar{\alpha}, \lambda) = \lambda\alpha_5 + \frac{\lambda(\lambda-1)\alpha_3\alpha_2}{2} - \frac{\lambda(\lambda-1)\alpha_2\alpha_1}{4} + \frac{\lambda(\lambda-1)(4\lambda+1)\alpha_1\alpha_2^2}{12}.$$

These formulas can be found in [14].

4.3 The R-Completion of a Finitely Generated Torsion-Free Nilpotent Group

We have shown that every finitely generated torsion-free nilpotent group has a Mal'cev basis. If the group happens to be free nilpotent, then such a basis may be chosen to consist of basic commutators according to Theorem 4.8. In either case, it is possible to embed the group into another nilpotent group which admits an action by a binomial ring. In this section, we discuss such an embedding. The material discussed here is based on [4, 5, 8], and [12].

Definition 4.7 A ring R is called a *binomial ring* if it is a commutative integral domain of characteristic zero with unity such that for any $r \in R$ and $k \in \mathbb{N}$, the element

$$\binom{r}{k} = \frac{r(r-1)\cdots(r-k+1)}{k!}$$

is well defined in R.

For example, $\mathbb{Z}$, $\mathbb{Q}$, any field F of characteristic zero, the polynomial ring $F[x]$, and the ring of p-adic integers for any prime p are binomial rings.

4.3.1 R-Completions

Let G be a finitely generated torsion-free nilpotent group with a specified Mal'cev basis $\bar{u} = \{u_1, \ldots, u_n\}$, and let R be a binomial ring. Assume that $\bar{f}(\bar{\alpha}, \bar{\beta})$ and

$\bar{g}(\bar{\alpha}, \lambda)$ are the multiplication and exponentiation polynomials for G respectively with respect to $\bar{u}$. Consider the set of formal products

$$G^R = \left\{ u_1^{\alpha_1} \cdots u_n^{\alpha_n} \,\middle|\, \alpha_i \in R \right\}, \tag{4.18}$$

and define multiplication and *R*-exponentiation in G^R by means of the polynomials $\bar{f}_i\left(\bar{\alpha}, \bar{\beta}\right)$ and $g_i\left(\bar{\alpha}, \lambda\right)$, where the arguments for each polynomial are elements of R. More precisely,

$$\bar{u}^{\bar{\alpha}}\,\bar{u}^{\bar{\beta}} = \bar{u}^{\bar{f}(\bar{\alpha},\,\bar{\beta})} \text{ and } \left(\bar{u}^{\bar{\alpha}}\right)^{\lambda} = \bar{u}^{\bar{g}(\bar{\alpha},\,\lambda)}, \tag{4.19}$$

where $\bar{\alpha}, \bar{\beta} \in R^n$ and $\lambda \in R$.

Theorem 4.11 (P. Hall) *The set G^R defined in (4.18), together with the multiplication defined in (4.19), is a group.*

The proof relies on the following:

Lemma 4.5 *Let $f(x_1, \ldots, x_k)$ be a polynomial of degree n over a field F of characteristic zero. If $f(a_1, \ldots, a_k) = 0$ for all possible choices of elements $a_i \in \mathbb{Z}$, then $f(x_1, \ldots, x_k)$ is the zero polynomial.*

Proof The proof is done by induction on k. If $k = 1$, then $f(x_1)$ must be the zero polynomial. Otherwise, it would have at most n roots, contradicting the hypothesis that every integer is a root.

Suppose that the lemma is true for all polynomials with $(k - 1)$ variables. Write $f(x_1, \ldots, x_k)$ as a polynomial in the variable x_k with coefficients in $F[x_1, \ldots, x_{k-1}]$ as such:

$$f(x_1, \ldots, x_k) = f_n(x_1, \ldots, x_{k-1})x_k^n + f_{n-1}(x_1, \ldots, x_{k-1})x_k^{n-1}$$
$$+ \cdots + f_0(x_1, \ldots, x_{k-1}).$$

For each $(a_1, \ldots, a_{k-1}) \in \mathbb{Z}^{k-1}$, the polynomial

$$f(a_1, \ldots, a_{k-1}, x_k) = f_n(a_1, \ldots, a_{k-1})x_k^n + f_{n-1}(a_1, \ldots, a_{k-1})x_k^{n-1}$$
$$+ \cdots + f_0(a_1, \ldots, a_{k-1})$$

is contained in $F[x_k]$ and has degree n or less. Since each integer is a root, it must be that $f(a_1, \ldots, a_{k-1}, x_k)$ is the zero polynomial. And so, $f_i(a_1, \ldots, a_{k-1}) = 0$ for all $(a_1, \ldots, a_{k-1}) \in \mathbb{Z}^{k-1}$ and $i = 0, 1, \ldots, n$. By induction, $f_i(x_1, \ldots, x_{k-1})$ is the zero polynomial for $i = 0, 1, \ldots, n$. It follows that $f(x_1, \ldots, x_k)$ is the zero polynomial. $\square$

We now prove Theorem 4.11 by verifying the group axioms.

- Associativity: We want to show that

$$\left(\bar{u}^{\bar{\alpha}} \, \bar{u}^{\bar{\beta}} \right) \bar{u}^{\bar{\gamma}} = \bar{u}^{\bar{\alpha}} \left(\bar{u}^{\bar{\beta}} \, \bar{u}^{\bar{\gamma}} \right)$$

for any $\bar{\alpha}, \bar{\beta}, \bar{\gamma} \in R^n$. Observe that

$$\left(\bar{u}^{\bar{\alpha}} \, \bar{u}^{\bar{\beta}} \right) \bar{u}^{\bar{\gamma}} = \bar{u}^{\bar{f}(\bar{\alpha}, \bar{\beta})} \, \bar{u}^{\bar{\gamma}} = \bar{u}^{\bar{f}(\bar{f}(\bar{\alpha}, \bar{\beta}), \bar{\gamma})}$$

and

$$\bar{u}^{\bar{\alpha}} \left(\bar{u}^{\bar{\beta}} \, \bar{u}^{\bar{\gamma}} \right) = \bar{u}^{\bar{\alpha}} \, \bar{u}^{\bar{f}(\bar{\beta}, \bar{\gamma})} = \bar{u}^{\bar{f}(\bar{\alpha}, \bar{f}(\bar{\beta}, \bar{\gamma}))}.$$

We claim that

$$\bar{f} \left(\bar{f} \left(\bar{\alpha}, \bar{\beta} \right), \bar{\gamma} \right) = \bar{f} \left(\bar{\alpha}, \bar{f} \left(\bar{\beta}, \bar{\gamma} \right) \right),$$

or equivalently, that

$$f_i \left(\bar{f} \left(\bar{\alpha}, \bar{\beta} \right), \bar{\gamma} \right) = f_i \left(\bar{\alpha}, \bar{f} \left(\bar{\beta}, \bar{\gamma} \right) \right)$$

for all $\bar{\alpha}, \bar{\beta}, \bar{\gamma} \in R^n$ and $1 \le i \le n$. Consider the function

$$h_i \left(\bar{x}, \bar{y}, \bar{z} \right) = f_i \left(\bar{f} \left(\bar{x}, \bar{y} \right), \bar{z} \right) - f_i \left(\bar{x}, \bar{f} \left(\bar{y}, \bar{z} \right) \right),$$

where $\bar{x}, \bar{y}, \bar{z} \in R^n$. Since elements in G are multiplied using the Mal'cev basis $\bar{u}$ and the corresponding polynomials, $h_i \left(\bar{x}, \bar{y}, \bar{z} \right) = 0$ whenever $\bar{x}, \bar{y}$, and $\bar{z}$ are elements of $\mathbb{Z}^n$. By Lemma 4.5, $h_i \equiv 0$ and the claim is proven.

- Identity: We claim that $\bar{u}^{\bar{0}} = u_1^0 \cdots u_n^0$ is the identity element in G^R. Clearly, $\bar{u}^{\bar{0}}$ is the identity element in G^R if and only if $\bar{u}^{\bar{\alpha}} \, \bar{u}^{\bar{0}} = \bar{u}^{\bar{\alpha}}$ for any $\bar{u}^{\bar{\alpha}} \in G^R$. Since $\bar{u}^{\bar{\alpha}} \, \bar{u}^{\bar{0}} = \bar{u}^{\bar{f}(\bar{\alpha}, \bar{0})}$, we need to show that $\bar{f} \left(\bar{\alpha}, \bar{0} \right) = \bar{\alpha}$ for all $\bar{\alpha} \in R^n$. Consider the function

$$h \left(\bar{x} \right) = \bar{f} \left(\bar{x}, \bar{0} \right) - \bar{x}$$

for $\bar{x}, \bar{0} \in R^n$. By a similar reasoning as before, it is clear that $h \left(\bar{x} \right) = \bar{0}$ for all $\bar{x} \in \mathbb{Z}^n$. By Lemma 4.5, $h \equiv 0$, and thus $\bar{f} \left(\bar{\alpha}, \bar{0} \right) = \bar{\alpha}$ for all $\bar{\alpha} \in R^n$.

- Inverses: Let $\bar{u}^{\bar{\alpha}} \in G^R$. We claim that $\bar{u}^{\bar{\alpha}}$ has an inverse. Assuming this to be the case, there exists $\left(\bar{u}^{\bar{\alpha}}\right)^{-1} \in G^R$ such that $\bar{u}^{\bar{\alpha}} \left(\bar{u}^{\bar{\alpha}}\right)^{-1} = \bar{u}^{\bar{0}}$. Notice that $\bar{u}^{\bar{\alpha}} \left(\bar{u}^{\bar{\alpha}}\right)^{-1} = \bar{u}^{\bar{0}}$ implies that $\bar{u}^{\bar{\alpha}} \, \bar{u}^{\bar{g}(\bar{\alpha}, -1)} = \bar{u}^{\bar{0}}$. Thus $\bar{u}^{\bar{f}(\bar{\alpha}, \, \bar{g}(\bar{\alpha}, -1))} = \bar{u}^{\bar{0}}$. Once again, Lemma 4.5 gives

$$\bar{f}\left(\bar{\alpha}, \, \bar{g}\left(\bar{\alpha}, -1\right)\right) = \bar{0}$$

for all $\bar{\alpha} \in R^n$. The element $\bar{u}^{\bar{g}(\bar{\alpha}, -1)}$ is, indeed, the inverse of $\bar{u}^{\bar{\alpha}}$. □

Definition 4.8 The group G^R described above is called the *R-completion* of *G* with respect to the Mal'cev basis $\bar{u}$.

Remark 4.5 Since the map $(\alpha_1, \ldots, \alpha_n) \mapsto (1 \cdot \alpha_1, \ldots, 1 \cdot \alpha_n)$ is an embedding of the ring $\mathbb{Z}^n$ into R^n, *G* embeds as a subgroup in G^R.

Theorem 4.12 *If G is a finitely generated torsion-free nilpotent group of class c, then the R-completion of G with respect to any Mal'cev basis is also nilpotent of class c.*

Proof Let $\{u_1, \ldots, u_n\}$ be a Mal'cev basis for *G*. Suppose that

$$u_1^{\alpha_{1, 1}} \cdots u_n^{\alpha_{1, n}}, \ldots, u_1^{\alpha_{c+1, 1}} \cdots u_n^{\alpha_{c+1, n}}$$

are any $(c + 1)$ elements of G^R, where $\alpha_{j, k} \in R$. We claim that

$$\left[u_1^{\alpha_{1, 1}} \cdots u_n^{\alpha_{1, n}}, \ldots, u_1^{\alpha_{c+1, 1}} \cdots u_n^{\alpha_{c+1, n}}\right] = 1. \tag{4.20}$$

Using the operations in G^R, we can express the left-hand side of (4.20) in the form

$$u_1^{P_1(\bar{\alpha})} \cdots u_n^{P_n(\bar{\alpha})},$$

where $P_i(\bar{\alpha})$ is a polynomial in the $\alpha_{j, k}$ for $1 \leq i \leq n$. We show that $P_i(\bar{\alpha}) = 0$ for each $i = 1, \ldots, n$. This equality certainly holds whenever each $\alpha_{j, k}$ is an integer because *G* has nilpotency class *c*. Therefore, Lemma 4.5 implies that $P_i(\bar{\alpha}) = 0$ for all $\alpha_{j, k} \in R$. Thus, (4.20) holds and G^R is nilpotent of class at most *c* by Corollary 2.3. However, the class of G^R must be at least *c* because *G* embeds into G^R. Therefore, G^R is of class exactly *c*. □

Remark 4.6 Even though an *R*-completion of a finitely generated torsion-free nilpotent group *G* depends on the chosen Mal'cev basis, all *R*-completions of *G* are isomorphic (see Theorem 4.23). Thus, G^R is unique up to isomorphism. Henceforth, we omit any specific Mal'cev basis for G^R unless needed.

Lemma 4.5 can be used to derive other identities in G^R which hold in G. In particular,

$$g^\alpha g^\beta = g^{\alpha+\beta}, \quad (g^\alpha)^\beta = g^{\alpha\beta}, \quad (h^{-1}gh)^\alpha = h^{-1}g^\alpha h, \quad \text{and}$$

$$g_1^\alpha \cdots g_n^\alpha = \tau_1(\overline{g})^\alpha \tau_2(\overline{g})^{\binom{\alpha}{2}} \cdots \tau_{k-1}(\overline{g})^{\binom{\alpha}{k-1}} \tau_k(\overline{g})^{\binom{\alpha}{k}}$$

for every α, $\beta \in R$ and g, h, g_1, $\ldots$, $g_n \in G$, where k is the class of the nilpotent group generated by $\{g_1, g_2, \ldots, g_n\}$ and $\overline{g} = (g_1, \ldots, g_n) \in G \times \cdots \times G$.

4.3.2 Mal'cev Completions

Theorem 4.12 leads naturally to the study of nilpotent groups allowing an action by $R = \mathbb{Q}$, which amounts to the investigation of extraction of roots in nilpotent groups. Recall from the paragraph preceding Theorem 2.7 that an element g of a group G is said to have an nth root in G for some integer $n > 1$ if there is an element $h \in G$ such that $h^n = g$.

Definition 4.9 A group G is called a $\mathbb{Q}$-*powered group* (or a $\mathscr{D}$-*group*) if each element of G has a unique nth root for every natural number $n > 1$.

If $g \in G$ and $h^n = g$ for some unique element $h \in G$, then it is quite natural to write $h = g^{1/n}$. This admits the equalities

$$\left(g^{1/n}\right)^n = g \quad \text{and} \quad \left(g^n\right)^{1/n} = g.$$

Thus, a $\mathbb{Q}$-powered group G may be viewed as an *algebraic system* with the additional unary operations of taking nth roots for every $n \in \mathbb{N}$ (see Section 1.3 of [7]). More generally, we may define

$$g^{m/n} = \left(g^{1/n}\right)^m \quad (m \in \mathbb{Z}, \, n \in \mathbb{N}).$$

This means that elements in G have rational powers. For example, $g^{2/3}$ and $g^{-4/9}$ are elements of G. The usual group laws are satisfied in $\mathbb{Q}$-powered groups. In particular,

$$g^m g^n = g^{m+n} \quad \text{and} \quad (g^m)^n = g^{mn} \quad (m, \, n \in \mathbb{Q}).$$

Definition 4.10 Let G be a torsion-free locally nilpotent group. A *Mal'cev completion* of G is a locally nilpotent $\mathbb{Q}$-powered group G^* in which G embeds as a subgroup under some map ϑ in such a way that for every $g \in G^*$, there exists $k \in \mathbb{N}$ such that $g^k \in \vartheta(G)$.

If G happens to be a subgroup of G^*, then we take ϑ to be the inclusion map. For the remainder of this section, we focus on finitely generated torsion-free nilpotent groups.

Theorem 4.13 *If G is a finitely generated torsion-free nilpotent group, then $G^{\mathbb{Q}}$ is a Mal'cev completion of G of the same nilpotency class as G.*

Proof By Theorem 4.12 and Remark 4.5, $G^{\mathbb{Q}}$ is a nilpotent group containing G as a subgroup and having the same nilpotency class as G. We show that $G^{\mathbb{Q}}$ is a $\mathbb{Q}$-powered group. If $g \in G^{\mathbb{Q}}$ and $g^m = 1$ for some integer $m > 1$, then

$$g = (g^m)^{1/m} = 1^{1/m} = 1.$$

Hence, $G^{\mathbb{Q}}$ is torsion-free. By Theorem 2.7, every element of G must have either no nth roots or one nth root for every integer $n > 1$. To see that the latter holds, notice that if $r > 1$ and $h \in G$, then

$$\left(h^{1/r}\right)^r = h.$$

Hence, every element $h \in G$ has a unique nth root. And so, $G^{\mathbb{Q}}$ is a $\mathbb{Q}$-powered group.

We need to prove that every element of $G^{\mathbb{Q}}$ has a positive power contained in G. Let $\{u_1, \ldots, u_n\}$ be a Mal'cev basis for G. Set $G^{\mathbb{Q}} = G_1^{\mathbb{Q}}$, and define

$$G_{i+1}^{\mathbb{Q}} = \left\{ u_1^{\alpha_1} \cdots u_n^{\alpha_n} \,\middle|\, \alpha_1 = \alpha_2 = \cdots = \alpha_i = 0 \right\}$$

for $1 \leq i \leq n$. These subgroups form a (descending) central series

$$G^{\mathbb{Q}} = G_1^{\mathbb{Q}} > G_2^{\mathbb{Q}} > \cdots > G_{n+1}^{\mathbb{Q}} = 1.$$

We prove by induction on $(n - i + 1)$ that if $g \in G_i^{\mathbb{Q}}$, then there exists $k \in \mathbb{N}$ such that $g^k \in G$.

- If $n - i + 1 = 0$, then $i = n + 1$ and $g \in G_{n+1}^{\mathbb{Q}} = 1$. The result is trivial in this case.
- If $n - i + 1 = 1$, then $i = n$ and $g = u_n^{\alpha_n} \in G_n^{\mathbb{Q}}$ for some $\alpha_n \in \mathbb{Q}$. If $\alpha_n \neq 0$ and we take k to be the denominator of α_n, then $g^k \in G$.
- Assume that the result holds for $n - i + 1 = \ell$ where $1 \leq \ell \leq n - 1$. Thus, every element of $G_{n-\ell+1}^{\mathbb{Q}}$ has some positive integral power in G. Let $g = u_{n-\ell}^{\alpha_{n-\ell}} \cdots u_n^{\alpha_n}$ be an element of $G_{n-\ell}^{\mathbb{Q}}$. For any $s \in \mathbb{N}$, we have

$$g^s = \left(u_{n-\ell}^{\alpha_{n-\ell}} \cdot u_{n-\ell+1}^{\alpha_{n-\ell+1}} \cdots u_n^{\alpha_n} \right)^s$$

$$= u_{n-\ell}^{\alpha_{n-\ell}s} \left(u_{n-\ell+1}^{\alpha_{n-\ell+1}} \cdots u_n^{\alpha_n} \right)^s \tau_c^{-\binom{s}{c}} \cdots \tau_2^{-\binom{s}{2}},$$

where τ_m is the Hall-Petresco word $\tau_m\left(u_{n-\ell}^{\alpha_{n-\ell}},\ u_{n-\ell+1}^{\alpha_{n-\ell+1}}\cdots u_n^{\alpha_n}\right)$ for $2\leq m\leq c$ and c is the nilpotency class of $H=gp\left(u_{n-\ell}^{\alpha_{n-\ell}},\ u_{n-\ell+1}^{\alpha_{n-\ell+1}}\cdots u_n^{\alpha_n}\right)$. By Theorem 4.1, $\tau_m\in\gamma_m H\leq\gamma_2 H$. Since $H\leq G_{n-\ell}^{\mathbb{Q}}$, we have

$$\tau_m\in\gamma_2 G_{n-\ell}^{\mathbb{Q}}=\left[G_{n-\ell}^{\mathbb{Q}},\ G_{n-\ell}^{\mathbb{Q}}\right]\leq\left[G_{n-\ell}^{\mathbb{Q}},\ G^{\mathbb{Q}}\right]\leq G_{n-\ell+1}^{\mathbb{Q}}$$

for $2\leq m\leq c$. By induction, $u_{n-\ell+1}^{\alpha_{n-\ell+1}}\cdots u_n^{\alpha_n}$, as well as each τ_m, has a positive integral power in G. Since $u_{n-\ell}^{\alpha_{n-\ell}}$ also has a positive integral power in G, we can construct an s such that $g^s\in G$. Let $s_0,\ s_1,\ \ldots,\ s_c$ be such that $u_{n-\ell}^{\alpha_{n-\ell}s_0},\ \left(u_{n-\ell+1}^{\alpha_{n-\ell+1}}\cdots u_n^{\alpha_n}\right)^{s_1}$ and $\left(\tau_m^{-1}\right)^{s_m}$ are all elements of G for $2\leq m\leq c$. Set

$$s=c!s_0 s_1\cdots s_c.$$

We claim that $g^s\in G$. Clearly, $u_{n-\ell}^{\alpha_{n-\ell}s}$ and $\left(u_{n-\ell+1}^{\alpha_{n-\ell+1}}\cdots u_n^{\alpha_n}\right)^s$ are both contained in G. To verify that each $\left(\tau_m^{-1}\right)^{\binom{s}{m}}$ also lies in G for $2\leq m\leq c$, observe that $m!$ always divides $c!$, so that s_m is a factor of $\binom{s}{m}$. $\qquad\square$

Theorem 4.14 *Given any two Mal'cev completions of a finitely generated torsion-free nilpotent group G, there exists a unique isomorphism between them which extends the identity automorphism of G.*

It is in this sense that the Mal'cev completion of a finitely generated torsion-free nilpotent group G is unique up to isomorphism.

Proof It suffices to show that there is an isomorphism from $G^{\mathbb{Q}}$ to any Mal'cev completion that restricts to the identity on G. To this end, choose a Mal'cev basis $\{u_1,\ \ldots,\ u_n\}$ for G, and suppose that the exponentiation polynomials for G with respect to this basis are $g_1(\overline{\alpha},\ \lambda),\ldots,g_n(\overline{\alpha},\ \lambda)$. By Theorem 4.13, $G^{\mathbb{Q}}$ is a Mal'cev completion of G. Let G^* be another Mal'cev completion of G, and let ϑ be an embedding of G into G^*. Define a map

$$\Psi:G^{\mathbb{Q}}\longrightarrow G^*\ \text{ by }\ \Psi\left(u_1^{\alpha_1}\cdots u_n^{\alpha_n}\right)=\vartheta(u_1)^{\alpha_1}\cdots\vartheta(u_n)^{\alpha_n}.$$

We claim that Ψ is an isomorphism. Using the polynomials, it is easy to show that Ψ is a homomorphism.

- $\underline{\Psi\text{ is one-to-one}}$: Suppose that $g=u_1^{\beta_1}\cdots u_n^{\beta_n}\in ker\ \Psi$. Since $G^{\mathbb{Q}}$ is a Mal'cev completion of G, there exists $k\in\mathbb{N}$ such that $g^k\in G$. Hence,

$$g^k=\left(u_1^{\beta_1}\cdots u_n^{\beta_n}\right)^k=u_1^{g_1(\overline{\beta},\ k)}\cdots u_n^{g_n(\overline{\beta},\ k)},$$

where each $g_i\left(\bar{\beta},\ k\right)$ is integral valued. Therefore,

$$\Psi\left(g^k\right) = \Psi\left(u_1^{g_1\left(\bar{\beta},\ k\right)}\cdots u_n^{g_n\left(\bar{\beta},\ k\right)}\right)$$

$$= \vartheta(u_1)^{g_1\left(\bar{\beta},\ k\right)}\cdots \vartheta(u_n)^{g_n\left(\bar{\beta},\ k\right)}$$

$$= \vartheta\left(u_1^{g_1\left(\bar{\beta},\ k\right)}\cdots u_n^{g_n\left(\bar{\beta},\ k\right)}\right).$$

Now, $g \in ker\ \Psi$ implies $g^k \in ker\ \Psi$. Hence,

$$\vartheta\left(u_1^{g_1\left(\bar{\beta},\ k\right)}\cdots u_n^{g_n\left(\bar{\beta},\ k\right)}\right) = 1.$$

This means that $u_1^{g_1\left(\bar{\beta},\ k\right)}\cdots u_n^{g_n\left(\bar{\beta},\ k\right)} = 1$ because ϑ is one-to-one. Thus, $g^k = 1$. Since G is torsion-free, $g = 1$.

- Ψ is onto: Suppose that $g^* \in G^*$. Since G^* is a Mal'cev completion of G, there exists $m \in \mathbb{N}$ such that $(g^*)^m \in \vartheta(G)$. Hence, $(g^*)^m = \vartheta(g)$ for some $g \in G$. If we write g in the normal form $u_1^{\alpha_1}\cdots u_n^{\alpha_n}$, where $\alpha_i \in \mathbb{Z}$, then

$$\Psi(g) = \Psi\left(u_1^{\alpha_1}\cdots u_n^{\alpha_n}\right)$$

$$= \vartheta(u_1)^{\alpha_1}\cdots \vartheta(u_n)^{\alpha_n}$$

$$= \vartheta\left(u_1^{\alpha_1}\cdots u_n^{\alpha_n}\right)$$

$$= \vartheta(g).$$

And so, $(g^*)^m = \Psi(g)$. Now, $\Psi(g)$ has a unique mth root since G^* is a $\mathbb{Q}$-powered group. Moreover, $\Psi\left(g^{1/m}\right) = (\Psi(g))^{1/m}$ since Ψ is a homomorphism between $\mathbb{Q}$-powered groups. Hence,

$$(g^*)^m = \left[(\Psi(g))^{1/m}\right]^m = \left[\Psi\left(g^{1/m}\right)\right]^m.$$

Therefore, $g^* = \Psi\left(g^{1/m}\right)$ by Theorem 2.7. Thus, Ψ is onto.   □

Combining Theorems 4.13 and 4.14 gives:

Theorem 4.15 (A. I. Mal'cev) *If G is a finitely generated torsion-free nilpotent group of class c, then G can be embedded in a Mal'cev completion G^* of class c. Furthermore, G^* is unique up to isomorphism.*

Remark 4.7 A. I. Mal'cev actually proved using Lie group theory that every torsion-free locally nilpotent group can be embedded in a Mal'cev completion. A proof of this which involves inverse limits can be found in Section 2.1 of [8].

4.4 Nilpotent R-Powered Groups

According to Theorem 4.11, the R-completion G^R of a finitely generated torsion-free nilpotent group G with respect to a given Mal'cev basis is a well-defined group for any binomial ring R. The group G^R is an example of a *nilpotent R-powered group*. In this section, we discuss some of the theory of nilpotent R-powered groups.

4.4.1 Definition of a Nilpotent R-Powered Group

The axioms which define a nilpotent R-powered group were given by P. Hall [4].

Definition 4.11 Let G be a (locally) nilpotent group, and let R be a binomial ring. Suppose that G comes equipped with an action by R :

$$G \times R \longrightarrow G \text{ defined by } (g, \alpha) \mapsto g^\alpha.$$

We say that G is a *nilpotent R-powered group* if the following axioms are satisfied for all $g, h \in G$ and $\alpha, \beta \in R$:

(i) $g^1 = g$, $g^\alpha g^\beta = g^{\alpha+\beta}$, and $(g^\alpha)^\beta = g^{\alpha\beta}$;

(ii) $(h^{-1}gh)^\alpha = h^{-1}g^\alpha h$;

(iii) *The Hall-Petresco axiom*:

$$g_1^\alpha \cdots g_n^\alpha = \tau_1(\overline{g})^\alpha \tau_2(\overline{g})^{\binom{\alpha}{2}} \cdots \tau_{k-1}(\overline{g})^{\binom{\alpha}{k-1}} \tau_k(\overline{g})^{\binom{\alpha}{k}}$$

for all $g_i \in G$, where k is the class of $gp(g_1, \ldots, g_n)$ and $\overline{g} = (g_1, \ldots, g_n)$ is contained in $G \times \cdots \times G$.

For the rest of this chapter, R will always be a binomial ring unless otherwise told.

Lemma 4.6 *If G is a nilpotent R-powered group, then $(g^\alpha)^{-1} = g^{-\alpha}$, $g^0 = 1$, and $1^\alpha = 1$ for all $g \in G$ and $\alpha \in R$.*

Proof The result follows immediately from the axioms.   □

4.4.2 Examples of Nilpotent R-Powered Groups

The next examples illustrate how naturally nilpotent R-powered groups arise [5].

Example 4.7 Let G be an abelian R-powered group, and suppose that $g_1, g_2 \in G$ and $\alpha \in R$. By Theorem 4.1, $\tau_i(g_1, g_2) = 1$ for $i \geq 2$ because $\gamma_i G = 1$ whenever $i \geq 2$. The Hall-Petresco axiom gives

$$g_1^\alpha g_2^\alpha = (g_1 g_2)^\alpha.$$

Therefore, G can be viewed as an R-module by interpreting the group multiplication and R-exponentiation operations of G as the R-module operations of addition and scalar multiplication respectively.

Example 4.8 If G is a finitely generated torsion-free nilpotent group, then G^R is a nilpotent R-powered group. In particular,

- $G^{\mathbb{Z}}$ is just G;
- $G^{\mathbb{Q}}$ is the Mal'cev completion of G by Theorem 4.13;
- $G^{\mathbb{R}}$ is a real connected torsion-free nilpotent Lie group (see [1] and [13]);
- if R is the ring of p-adic integers for a given prime p, then G^R is the p-adic completion of G (see [3] and [16]).

Example 4.9 Let $\widetilde{R}$ be any ring with unity 1 which contains R in its ring center. Let I be a nilpotent ideal of $\widetilde{R}$ with $I^{n+1} = \{0\}$. It was shown in Section 2.2 that the set $G = \{1 + a \mid a \in I\}$ is a subgroup of the group of units of $\widetilde{R}$ and is nilpotent of class at most n. For $\lambda \in R$ and $a \in I$, define R-exponentiation in G by

$$(1 + a)^{\lambda} = 1 + \lambda a + \binom{\lambda}{2} a^2 + \cdots + \binom{\lambda}{n} a^n.$$

This operation turns G into a nilpotent R-powered group. In particular, $UT_n(R)$ is a nilpotent R-powered group.

4.4.3 R-Subgroups and Factor R-Groups

Definition 4.12 Let G be a nilpotent R-powered group. A subgroup H of G is called an R-*subgroup* of G if $g^{\alpha} \in H$ whenever $g \in H$ and $\alpha \in R$.

We write $H \leq_R G$ whenever H is an R-subgroup of G. If H is a normal subgroup of G, then we write $H \trianglelefteq_R G$.

The intersection of a collection of R-subgroups of a nilpotent R-powered group is clearly an R-subgroup. If G is a nilpotent R-powered group and $S \subseteq G$, then we denote the intersection of all R-subgroups containing S by $gp_R(S)$. Thus, $gp_R(S)$ is the smallest R-subgroup of G containing S.

Definition 4.13 We call S a set of R-*generators* of $gp_R(S)$. If $|S| < \infty$, then $gp_R(S)$ is *finitely R-generated* by S.

One can construct $gp_R(S)$ in the following way: if we set $S_0 = gp(S)$ and recursively define $S_{n+1} = gp\left(g_n^{\alpha_n} \mid g_n \in S_n, \ \alpha_n \in R\right)$, then

$$gp_R(S) = \bigcup_{n=0}^{\infty} S_n.$$

Definition 4.14 If H_1 and H_2 are R-subgroups of a nilpotent R-powered group G, then

$$[H_1, H_2]_R = gp_R([h_1, h_2] \mid h_1 \in H_1, h_2 \in H_2)$$

is called the *commutator R-subgroup of H_1 and H_2*. For a collection of R-subgroups $H_1, \ldots, H_i$ of G, we recursively define, for $i > 2$,

$$[H_1, \ldots, H_i]_R = \big[[H_1, \ldots, H_{i-1}]_R, H_i\big]_R.$$

Theorem 4.16 *Let G be a nilpotent R-powered group, and let N be a normal R-subgroup of G. The R-action on G induces an R-action on G/N defined by*

$$(gN)^\alpha = g^\alpha N \quad (g \in G, \ \alpha \in R)$$

which turns G/N into a factor R-group *of G.*

Proof Let $g, h \in G$ and $\alpha \in R$. We need to prove that if $gN = hN$, then $g^\alpha N = h^\alpha N$. Suppose that $gN = hN$. There exists $n \in N$ such that $gn = h$. Thus, $(gn)^\alpha = h^\alpha$. Set $\tau_i = \tau_i(g, n)$. By the Hall-Petresco axiom,

$$g^\alpha n^\alpha = (gn)^\alpha \tau_2^{\binom{\alpha}{2}} \cdots \tau_{k-1}^{\binom{\alpha}{k-1}} \tau_k^{\binom{\alpha}{k}},$$

where k is the class of $gp(g, n)$. Equivalently,

$$(gn)^{-\alpha} g^\alpha n^\alpha = \tau_2^{\binom{\alpha}{2}} \cdots \tau_{k-1}^{\binom{\alpha}{k-1}} \tau_k^{\binom{\alpha}{k}}. \tag{4.21}$$

We assert that $\tau_i \in N$ for $i = 2, 3, \ldots, k$. The proof is done by induction on i. If $i = 2$, then setting $\alpha = 2$ in (4.21) yields $(gn)^{-2} g^2 n^2 = \tau_2$. Re-expressing the left-hand side gives

$$(gn)^{-2} g^2 n^2 = n^{-1}\big(g^{-1} n^{-1} g\big) n^2.$$

Since $N \trianglelefteq_R G$, $g^{-1} n^{-1} g \in N$, and thus $\tau_2 \in N$. This gives the basis for induction.

Before continuing, we show that if $(gn)^{-i} g^i n^i \in N$ for $i = 2, 3, \ldots, k-1$, then $(gn)^{-(i+1)} g^{i+1} n^{i+1} \in N$. We proceed as follows:

$$\begin{aligned}
(gn)^{-(i+1)} g^{i+1} n^{i+1} &= (gn)^{-i}(gn)^{-1} g^{i+1} n^{i+1} \\
&= (gn)^{-i} n^{-1} g^{-1} g g^i n^{i+1} \\
&= (gn)^{-i} n^{-1} g^i n^{i+1} \\
&= \big((gn)^{-i} g^i n^i\big) n^{-i}\big(g^{-i} n^{-1} g^i\big) n^{i+1} \in N.
\end{aligned}$$

Now, assume that $\tau_j \in N$ for $j = 2, 3, \ldots, i$, where $i < k$. Putting $\alpha = i$ in (4.21) leads to $(gn)^{-i}g^i n^i \in N$. And so, $(gn)^{-(i+1)}g^{i+1}n^{i+1} \in N$. For $\alpha = i + 1$, (4.21) becomes

$$(gn)^{-(i+1)}g^{i+1}n^{i+1} = \tau_2^{\binom{i+1}{2}} \cdots \tau_i^{\binom{i+1}{i}} \tau_{i+1}.$$

By induction,

$$\tau_{i+1} = \bar{n}(gn)^{-(i+1)}g^{i+1}n^{i+1} \in N$$

for some $\bar{n} \in N$. This proves the assertion. Hence, (4.21) reduces to $g^\alpha n^\alpha = (gn)^\alpha n_0$ for some $n_0 \in N$. Therefore, $(gn)^\alpha = g^\alpha n_1$ for some $n_1 \in N$. Thus, $(gn)^\alpha N = g^\alpha N$. Since $(gn)^\alpha = h^\alpha$, we have $g^\alpha N = h^\alpha N$ as claimed. The rest of the proof requires a verification of the axioms. $\qquad\square$

4.4.4 R-Morphisms

Definition 4.15 Let G and H be nilpotent R-powered groups. A homomorphism $\varphi : G \to H$ is called an R-*homomorphism* if

$$\varphi\big(g^\alpha\big) = (\varphi(g))^\alpha$$

for all $g \in G$ and $\alpha \in R$.

It is not hard to check that $ker\ \varphi \trianglelefteq_R G$ and $im\ \varphi \leq_R H$. We say that an R-homomorphism is an R-*monomorphism (R-epimorphism, R-isomorphism)* if it is a monomorphism (epimorphism, isomorphism). If G and H are R-isomorphic, then we write $G \cong_R H$.

The usual isomorphism theorems for groups carry over to nilpotent R-powered groups. We merely state them here.

Theorem 4.17 *Let G be a nilpotent R-powered group.*

(i) *If H is a nilpotent R-powered group and $\varphi : G \to H$ is an R-homomorphism, then*

$$G/ker\ \varphi \cong_R im\ \varphi.$$

(ii) *If $H \leq_R G$ and $N \trianglelefteq_R G$, then*

$$HN/N \cong_R H/(H \cap N).$$

(iii) *If $K \leq_R H \trianglelefteq_R G$ and $K \trianglelefteq_R G$, then*

$$G/H \cong_R (G/K)/(H/K).$$

The proof of Theorem 4.17 (ii) makes use of the next lemma.

Lemma 4.7 *Let G be a nilpotent R-powered group. If $N \trianglelefteq_R G$ and $H \leq_R G$, then $HN = gp_R(H, N) \leq_R G$ and $H \cap N \trianglelefteq_R H$. If $H \trianglelefteq_R G$ as well, then $HN \trianglelefteq_R G$.*

Proof We claim that HN is an R-subgroup of G. Let $hn \in HN$ and $\beta \in R$. By repeating what was done in the proof of Theorem 4.16, we find that $h^\beta n^\beta = (hn)^\beta n_0$ for some $n_0 \in N$. Thus,

$$(hn)^\beta = h^\beta n^\beta n_0^{-1} \in HN.$$

Hence, HN is closed under R-exponentiation. The rest is straightforward.   □

4.4.5   Direct Products

Let $\{G_i \mid i \in I\}$ be a collection of nilpotent R-powered groups of bounded nilpotency class, indexed by a nonempty countable set I. The *unrestricted direct product* or *Cartesian product* of the G_i's is the nilpotent R-powered group

$$\overline{G} = \prod_{i \in I} G_i = \left\{ f : I \to \bigcup_{i \in I} G_i \,\middle|\, f(i) \in G_i \text{ for all } i \in I \right\}$$

with multiplication and R-exponentiation defined by

- $(f_1 f_2)(i) = f_1(i)f_2(i)$ for all f_1, $f_2 \in \overline{G}$ and $i \in I$, and
- $(f^\alpha)(i) = (f(i))^\alpha$ for all $f \in \overline{G}$, $i \in I$, and $\alpha \in R$.

As usual, the elements of $\overline{G}$ can be viewed as vectors $(g_1, \ldots, g_i, \ldots)$ whose i^{th} coordinate is $g_i = f(i) \in G_i$ for all $i \in I$. By viewing them in this way, we get

- $(g_1, \ldots, g_i, \ldots)(h_1, \ldots, h_i, \ldots) = (g_1 h_1, \ldots, g_i h_i, \ldots)$ for all g_i, $h_i \in G_i$, and
- $(g_1, \ldots, g_i, \ldots)^\alpha = \left(g_1^\alpha, \ldots, g_i^\alpha, \ldots \right)$ for all $g_i \in G_i$ and $\alpha \in R$.

Definition 4.16 The *external direct product* of the G_i's is the subset of $\prod_{i \in I} G_i$ consisting of those functions f for which $f(i) = 1$ except for finitely many $i \in I$.

Using vector notation, $(g_1, \ldots, g_i, \ldots) \in \overline{G}$ is contained in the external direct product if all but finitely many of the g_i equals the identity.

Definition 4.17 Suppose that G is a nilpotent R-powered group, and let $\{G_i \mid i \in I\}$ be a family of R-subgroups of G indexed by a nonempty countable set I. Then G is the *internal direct product* of the G_i's, denoted by $G = \prod_{i \in I} G_i$, if the following conditions are met:

(i) $G_i \trianglelefteq_R G$ for each $i \in I$;
(ii) $G = gp_R(G_i \mid i \in I)$;
(iii) $G_i \cap gp_R(G_j \mid i, j \in I, j \neq i) = 1$.

Every element of G can be written uniquely as a product of g_i's, where $g_i \in G_i$. Furthermore, the elements of G_i commute with the elements of G_j whenever $i \neq j$. These can be proven in the same way as for ordinary groups.

4.4.6 Abelian R-Groups

According to Example 4.7, abelian R-groups are just R-modules. Thus, the structure of such groups depends on the ring structure of R just as in the case of R-modules. For instance:

- If R is Euclidean, then every R-submodule of a cyclic R-module is cyclic. Similarly, if $G = gp_R(g)$ for some $g \in G$ (hence, G is a *cyclic R-group*) and R is Euclidean, then every R-subgroup of G is a cyclic R-group.
- If R is a noetherian ring, then every R-submodule of a finitely generated R-module is finitely generated. Consequently, if R is a noetherian binomial ring and G is a finitely R-generated abelian R-group with $H \leq_R G$, then H is also a finitely R-generated abelian R-group.
- Every finitely generated R-module is a direct sum of cyclic R-modules whenever R is a PID. Thus, every finitely R-generated abelian R-group is a direct product of cyclic R-groups whenever R is a PID.

These results about R-modules can be found in [17].

4.4.7 Upper and Lower Central Series

The various types of series we have encountered can be formed for nilpotent R-powered groups. One simply has to modify the usual definitions in the right way. For instance, an *R-series* of a nilpotent R-powered group G is a series

$$1 = G_0 \leq G_1 \leq \cdots \leq G_n = G, \tag{4.22}$$

where $G_i \leq_R G$ for $i = 0, 1, \ldots, n$. If $G_i \trianglelefteq_R G_{i+1}$ for $i = 0, 1, \ldots, n-1$, then (4.22) is called a *subnormal R-series*. The notions of a *normal R-series*, a *central R-series*, and so on, should now be apparent.

The upper and lower central subgroups of a nilpotent R-powered group are normal R-subgroups (see [5] or [19]). The proof of this relies on the next lemma.

Lemma 4.8 *Let G be a nilpotent R-powered group. If g, $h \in G$ and $[g, h] \in Z(G)$, then*

$$[g, h^\alpha] = [g^\alpha, h] = [g, h]^\alpha$$

for any $\alpha \in R$.

Proof The result is true when $\alpha \in \mathbb{Z}$ by Lemma 1.13. Suppose that α is an arbitrary element of R. By the Hall-Petresco axiom,

$$
\begin{aligned}
[g, h^\alpha] &= \left(g^{-1}h^{-1}g\right)^\alpha h^\alpha \\
&= [g, h]^\alpha \tau_2\left(g^{-1}h^{-1}g, h\right)^{\binom{\alpha}{2}} \cdots \tau_k\left(g^{-1}h^{-1}g, h\right)^{\binom{\alpha}{k}},
\end{aligned}
$$

where k is the class of $gp\left(g^{-1}h^{-1}g, h\right)$. By replacing α by $2, 3, \ldots, k$ and using the fact that $[g, h^\alpha] = [g, h]^\alpha$ for such α, it follows that $\tau_i\left(g^{-1}h^{-1}g, h\right) = 1$ for $2 \leq i \leq k$. Thus, $[g, h^\alpha] = [g, h]^\alpha$ for any $\alpha \in R$. Similarly, $[g^\alpha, h] = [g, h]^\alpha$. □

Theorem 4.18 *The upper and lower central subgroups of a nilpotent R-powered group are normal R-subgroups.*

Consequently, the upper and lower central series of a nilpotent R-powered group are central R-series.

Proof Let G be a nilpotent R-powered group of class c.

1. First, we prove that $\zeta_i G \trianglelefteq_R G$ for $i = 1, 2, \ldots, c$ by induction on i.

   - If $z \in Z(G)$ and $g \in G$, then $g^{-1}zg = z$ implies $\left(g^{-1}zg\right)^\alpha = z^\alpha$ for any $\alpha \in R$. Thus, $g^{-1}z^\alpha g = z^\alpha$, and consequently, $z^\alpha \in Z(G)$. Therefore, $Z(G) \trianglelefteq_R G$ and the case $i = 1$ is established.
   - Assume that $\zeta_{k-1} G \trianglelefteq_R G$ for $1 < k \leq c$. By Theorem 4.16, $G/\zeta_{k-1}G$ is a nilpotent R-powered group. Thus,

$$\zeta_k G/\zeta_{k-1}G = Z(G/\zeta_{k-1}G) \trianglelefteq_R G/\zeta_{k-1}G.$$

   Hence, $\zeta_k G \trianglelefteq_R G$ as claimed.

2. We prove by induction on c that the lower central subgroups of a nilpotent R-powered group are normal R-subgroups. If $c = 1$, then the result is immediate. Assume the assertion is true for nilpotent R-powered groups of class less than c.

- We first show that $\gamma_c G \trianglelefteq_R G$. By definition,

$$\gamma_c G = [\gamma_{c-1}G, \ G] = gp([x, \ y] \mid x \in \gamma_{c-1}G, \ y \in G).$$

  It is enough to show that if $[g, \ h] \in \gamma_c G$ for any $g \in \gamma_{c-1}G$ and $h \in G$, then $[g, \ h]^\alpha \in \gamma_c G$ for any $\alpha \in R$. Well, $\gamma_c G \leq Z(G)$ implies that $[g, \ h]^\alpha = [g, \ h^\alpha]$ by Lemma 4.8. Since $g \in \gamma_{c-1}G$, $[g, \ h^\alpha] \in \gamma_c G$. Thus, $[g, \ h]^\alpha \in \gamma_c G$ as claimed.
- Next, we show that $\gamma_i G \trianglelefteq_R G$ for $1 \leq i < c$. Since $G/\gamma_c G$ is a nilpotent R-powered group of class $c - 1$, $\gamma_i(G/\gamma_c G) \trianglelefteq_R G/\gamma_c G$ for $1 \leq i \leq c - 1$ by induction. However, $\gamma_i(G/\gamma_c G) = \gamma_i G/\gamma_c G$ by Corollary 2.1. And so, each $\gamma_i G$ is a normal R-subgroup of G.   □

The next result is an immediate consequence of Theorem 4.18.

Corollary 4.2 *If G is a nilpotent R-powered group, then each factor R-group $\gamma_i G/\gamma_{i+1}G$ and $\zeta_{i+1}G/\zeta_i G$ is an R-module.*

Lemma 4.9 *If G is a nilpotent R-powered group, then*

$$\gamma_n G = gp([g_1, \ \ldots, \ g_n] \mid g_i \in G) = gp_R([g_1, \ \ldots, \ g_n] \mid g_i \in G).$$

This follows from an application of the commutator calculus and the axioms of a nilpotent R-powered group. Compare Lemma 4.9 with Lemma 2.6.

4.4.8 Tensor Product of the Abelianization

In Chapter 2, we saw that the abelianization of a nilpotent group can transfer certain properties to the group itself. The same is true for nilpotent R-powered groups.

Lemma 4.10 (The Three R-Subgroup Lemma) *Let G be a nilpotent R-powered group, and suppose that H, K, and L are R-subgroups of G. If any two of the R-subgroups $[H, \ K, \ L]_R$, $[K, \ L, \ H]_R$, and $[L, \ H, \ K]_R$ are contained in a normal R-subgroup of G, then so is the third.*

The proof is the same as that of Lemma 2.18.

Theorem 4.19 *If G is a nilpotent R-powered group and i, j ≥ 0 are integers, then*

$$[\gamma_i G,\ \gamma_j G]_R \leq_R \gamma_{i+j} G.$$

Proof The result follows from Lemma 4.10 (see Theorem 2.14 (i)). □

The next theorem can be found in [19].

Theorem 4.20 *Let G be a nilpotent R-powered group. For every integer n > 1, there exists a well-defined R-module epimorphism*

$$\Psi_n : \gamma_{n-1}G/\gamma_n G \bigotimes_R Ab(G) \to \gamma_n G/\gamma_{n+1}G$$

given by

$$\Psi_n(g\gamma_n G \otimes h\gamma_2 G) = [g,\ h]\gamma_{n+1}G.$$

Proof We need to show that Ψ_n respects R-exponentiation; that is,

$$[g,\ h^\alpha]\gamma_{n+1}G = [g^\alpha,\ h]\gamma_{n+1}G = [g,\ h]^\alpha \gamma_{n+1}G$$

for any $g \in \gamma_{n-1}G$, $h \in G$, and $\alpha \in R$. Set $\overline{\tau}_i = \tau_i\left(g^{-1}h^{-1}g,\ h\right)$, and suppose that $gp\left(g^{-1}h^{-1}g,\ h\right)$ is of class k. By the Hall-Petresco axiom,

$$[g,\ h^\alpha] = \left(g^{-1}h^{-1}g\right)^\alpha h^\alpha = [g,\ h]^\alpha \overline{\tau}_2^{\binom{\alpha}{2}} \cdots \overline{\tau}_k^{\binom{\alpha}{k}}. \qquad (4.23)$$

Using the commutator calculus, it is not difficult to show that $\overline{\tau}_i \in \gamma_{n+1}G$ for each $i = 2,\ \ldots,\ k$. Consequently, (4.23) reduces to $[g,\ h^\alpha] = [g,\ h]^\alpha k$ for all $\alpha \in R$, where $k \in \gamma_{n+1}G$. Therefore,

$$[g,\ h^\alpha]\gamma_{n+1}G = [g,\ h]^\alpha \gamma_{n+1}G.$$

A similar argument shows that

$$[g^\alpha,\ h]\gamma_{n+1}G = [g,\ h]^\alpha \gamma_{n+1}G.$$

The rest of the proof is identical to the proof of Theorem 2.15. □

The next two results are proven in a similar way as that of Corollaries 2.10 and 2.11.

Corollary 4.3 *Suppose that G is a nilpotent R-powered group. For each $n \in \mathbb{N}$, the mapping from $\bigotimes_R^n Ab(G)$ to $\gamma_n G/\gamma_{n+1}G$ defined by*

$$g_1\gamma_2 G \otimes \cdots \otimes g_n\gamma_2 G \mapsto [g_1,\ \ldots,\ g_n]\gamma_{n+1}G$$

is an R-module epimorphism.

Corollary 4.4 *Let G be a finitely R-generated nilpotent R-powered group with R-generating set $X = \{x_1, \ldots, x_k\}$. For each $n \in \mathbb{N}$, the factor R-group $\gamma_n G / \gamma_{n+1} G$ is finitely R-generated, modulo $\gamma_{n+1}G$, by the simple commutators of weight n of the form $[x_{i_1}, \ldots, x_{i_n}]$, where the x_{i_j}'s vary over all elements of X and are not necessarily distinct.*

4.4.9 Condition Max-R

Definition 4.18 A nilpotent R-powered group G is said to satisfy *Max-R* (the *maximal condition on R-subgroups*) if every R-subgroup of G is finitely R-generated.

A finitely R-generated nilpotent R-powered group satisfies Max-R whenever R is a noetherian ring [6]. This is highlighted in the next theorem.

Theorem 4.21 *If R is noetherian and G is a finitely R-generated nilpotent R-powered group, then G satisfies Max-R.*

In particular, a finitely R-generated nilpotent R-powered group satisfies Max-R whenever R is a PID.

Proof We mimic the proof of Theorem 2.18. Suppose that G is of class c, and let H be an R-subgroup of G. By Corollary 4.4, each factor R-group $\gamma_i G / \gamma_{i+1} G$ is a finitely R-generated R-module for $1 \le i \le c$. If $H_i = H \cap \gamma_i G$ for $1 \le i \le c + 1$, then

$$H = H_1 \ge_R H_2 \ge_R \cdots \ge_R H_c \ge_R H_{c+1} = 1$$

is a central R-series for H and

$$\frac{H_i}{H_{i+1}} = \frac{H \cap \gamma_i G}{H \cap \gamma_{i+1} G} \cong_R \frac{\gamma_{i+1} G (H \cap \gamma_i G)}{\gamma_{i+1} G}$$

by Theorem 4.17 (ii). Therefore, H_i / H_{i+1} is R-isomorphic to an R-submodule of $\gamma_i G / \gamma_{i+1} G$ which is finitely generated because R is noetherian. Thus, each H_i / H_{i+1} is finitely R-generated. In particular, $H_c = H_c / H_{c+1}$ is finitely R-generated and the result follows. □

Corollary 4.5 *Let G be a nilpotent R-powered group, where R is noetherian. If $N \trianglelefteq_R G$, and both G/N and N satisfy Max-R, then G also satisfies Max-R.*

Proof Since G/N and N satisfy Max-R, both are finitely R-generated. By repeating the proof of Lemma 2.19, we find that G is finitely R-generated. The result follows from Theorem 4.21. □

The next result is an analogue of Corollary 2.12. It is a consequence of Corollaries 4.4 and 4.5, Theorem 4.21, and the fact that the tensor product of finitely many finitely generated R-modules is finitely generated.

Corollary 4.6 *Suppose that G is a nilpotent R-powered group of class c and R is noetherian. If $Ab(G)$ is finitely R-generated, then G satisfies Max-R. Hence, G is finitely R-generated.*

4.4.10   An R-Series of a Finitely R-Generated Nilpotent R-Powered Group

Every finitely generated nilpotent group has a polycyclic and central series by Theorem 4.4. A similar property is enjoyed by finitely R-generated nilpotent R-powered groups [19].

Definition 4.19 Let $\mathscr{Q}$ be a property of nilpotent R-powered groups. A subnormal R-series

$$1 = G_0 \trianglelefteq_R G_1 \trianglelefteq_R \cdots \trianglelefteq_R G_n = G$$

of a nilpotent R-powered group G is called a *poly-R $\mathscr{Q}$ series* if each G_{i+1}/G_i is a $\mathscr{Q}$ R-group for $i = 0, 1, \ldots, n-1$.

Let G be a finitely R-generated nilpotent R-powered group. By Corollary 4.4, each factor group $\gamma_i G/\gamma_{i+1} G$ is a finitely R-generated abelian R-group. By refining the lower central series of G (see the comment preceding Theorem 4.4), we obtain:

Theorem 4.22 *A nilpotent R-powered group is finitely R-generated if and only if it has a poly-R-cyclic and central R-series.*

Definition 4.20 The minimal length of all poly-R cyclic and central R-series for a finitely R-generated nilpotent R-powered group is called its *Hirsch R-length*.

Any two R-completions of a finitely generated torsion-free nilpotent group are R-isomorphic. This is a consequence of the next theorem due to P. Hall [5].

Theorem 4.23 *Let G be a finitely generated torsion-free nilpotent group, and let H be any nilpotent R-powered group. If $\varphi : G \to H$ is a homomorphism, then there exists a unique R-homomorphism $\Phi : G^R \to H$ which extends φ.*

We give the idea behind the proof. Let $\{u_1, \ldots, u_n\}$ be a Mal'cev basis for G and let $\{f_1, \ldots, f_n\}$ and $\{g_1, \ldots, g_n\}$ be the multiplication and exponentiation polynomials respectively with respect to this basis. Define the map

$$\Phi : G^R \to H \quad \text{by} \quad \Phi\left(u_1^{\varrho_1} \cdots u_n^{\varrho_n}\right) = \varphi(u_1)^{\varrho_1} \cdots \varphi(u_n)^{\varrho_n} \qquad (\varrho_i \in R).$$

The assertion is that Φ is an R-homomorphism which extends φ. The fact that Φ extends φ is clear. The bulk of the work relies on showing that if $u_1^{\alpha_1} \cdots u_n^{\alpha_n}$ and $u_1^{\beta_1} \cdots u_n^{\beta_n}$ are elements of G^R for some $\alpha_i, \beta_i \in R$, then

$$\left(\varphi(u_1)^{\alpha_1} \cdots \varphi(u_n)^{\alpha_n}\right)\left(\varphi(u_1)^{\beta_1} \cdots \varphi(u_n)^{\beta_n}\right) = \varphi(u_1)^{f_1\left(\bar{\alpha}, \bar{\beta}\right)} \cdots \varphi(u_n)^{f_n\left(\bar{\alpha}, \bar{\beta}\right)}$$

and

$$\left(\varphi(u_1)^{\alpha_1}\cdots\varphi(u_n)^{\alpha_n}\right)^{\lambda} = \varphi(u_1)^{g_1(\overline{\alpha},\,\lambda)}\cdots\varphi(u_n)^{g_n(\overline{\alpha},\,\lambda)},$$

where $\lambda \in R$. This can be proven by induction on the Hirsch R-length of G^R using the technique given in the proof of Theorem 4.9.

Corollary 4.7 *Suppose that G is a finitely generated torsion-free nilpotent group. Let $G_1 = (G, \mathscr{B}_1)$ and $G_2 = (G, \mathscr{B}_2)$ denote G with respect to distinct Mal'cev bases $\mathscr{B}_1$ and $\mathscr{B}_2$ respectively. Then $G_1^R \cong_R G_2^R$.*

4.4.11 Free Nilpotent R-Powered Groups

Definition 4.21 Let G be a nilpotent R-powered group of class c. A nonempty subset of generators X of G is said to *freely R-generate G* if for every function $\mu : X \to H$, where H is any nilpotent R-powered group of class at most c, there is a unique R-homomorphism $\beta : G \to H$ which coincides with μ on X.

A nilpotent R-powered group is called a *free nilpotent R-powered group* if it is freely R-generated by some subset. Such groups are just nilpotent R-powered groups which satisfy only the usual group axioms and those given in Definition 4.11. They arise as the R-completion of finitely generated free nilpotent groups [4].

Lemma 4.11 *Let F be a free nilpotent group, freely generated by $X = \{x_1, \ldots, x_n\}$. The R-completion F^R of F is a free nilpotent R-powered group, freely R-generated by X.*

The proof follows from what was done for Theorem 4.23, where the Mal'cev basis for F is chosen to consist of basic commutators on X (see Corollary 3.4).

4.4.12 ω-Torsion and R-Torsion

Much is known about P-torsion and P-torsion-free nilpotent groups. In [11], S. Majewicz and M. Zyman studied ω-torsion and ω-torsion-free nilpotent R-powered groups, where ω is a set of non-associate primes in a unique factorization domain R. Some of this work had already been done by P. Hall [5] and R.B. Warfield [19] when R is any binomial ring and ω is the set of all primes in R.

Definition 4.22 Let R be a unique factorization domain (UFD), and let ω be a set of non-associate primes in R. A nonzero element $\alpha \in R$ is called an *ω-member* if it is a non-unit and all its prime divisors (up to units) are in ω.

From this point on, ω will be a set of non-associate primes in R. We require R to be a UFD in Definition 4.22 so that there is no distinction between irreducible and prime elements. In this situation, the product of ω-members is again an ω-member.

For example, if $R = \mathbb{Z}$ and $\omega = \{2, -3, 7\}$, then 6 and -14 are both ω-members. If $R = \mathbb{Q}[x]$ and $\omega = \left\{ x, \frac{1}{5}x^2 + \frac{1}{5}, -3x^2 - 3x - 3 \right\}$, then

$$-7x\left(x^2 + 1\right), \; \frac{2}{9}\left(x^2 + 1\right)^4\left(x^2 + x + 1\right)^3, \text{ and } \frac{10}{11}x^3\left(x^2 + x + 1\right)^2$$

are ω-members.

Definition 4.23 Let R be a UFD, and suppose that G is a nilpotent R-powered group. An element $g \in G$ is called an *ω-torsion element* if $g^\alpha = 1$ for some ω-member α. If every element of G is ω-torsion, then G is an *ω-torsion group*. We say that G is *ω-torsion-free* if the only ω-torsion element of G is the identity.

The set of ω-torsion elements of G is denoted by $\tau_\omega(G)$. If $g^\alpha = 1$ for some ω-member α, then it is clear that $g^\beta = 1$ for any associate β of α. If π is a prime in R and $\omega = \{\pi\}$, then we use the terms π-torsion and π-torsion-free.

If ω is the set of all of the primes in R (up to units), then every nonzero element of R is an ω-member. In this case, the product of any two ω-members is an ω-member whether R is a UFD or not.

Definition 4.24 Let R be any binomial ring, and let G be a nilpotent R-powered group. An element $g \in G$ is called an *R-torsion element* if there exists $0 \neq \alpha \in R$ such that $g^\alpha = 1$. If every element of G is R-torsion, then G is called an *R-torsion group*. If the only R-torsion element of G is the identity, then G is termed *R-torsion-free*.

We write $\tau(G)$ for the set of R-torsion elements of G.

Theorem 4.24 *Let G be a nilpotent R-powered group.*

(i) If R is a UFD, then $\tau_\omega(G) \trianglelefteq_R G$.
(ii) If R is any binomial ring, then $\tau(G) \trianglelefteq_R G$.

Proof We prove (i) by induction on the class c of G. The proof of (ii) is similar. We may assume that G is R-generated by the elements of $\tau_\omega(G)$. The claim is that every element of G is an ω-torsion element.

Suppose that $c = 1$. If $g, h \in \tau_\omega(G)$, then there exist ω-members α and β such that $g^\alpha = h^\beta = 1$. Since G is an abelian R-group, we have

$$(gh)^{\alpha\beta} = \left(g^{\alpha\beta}\right)\left(h^{\alpha\beta}\right) = \left(g^\alpha\right)^\beta\left(h^\beta\right)^\alpha = 1.$$

Clearly, $\alpha\beta$ is an ω-member, and thus gh is an ω-torsion element. It follows from the axioms that every element of G is an ω-torsion element as claimed.

Next, suppose that (i) holds for all nilpotent R-powered groups of class less than $c > 1$. First, we prove that every element of the abelian R-group $\gamma_c G$ is an ω-torsion element. To this end, let $[g_1, \ldots, g_c]$ be any commutator of weight c in $\gamma_c G$, where $g_1, \ldots, g_c \in G$. Since $G/\gamma_2 G$ is abelian and R-generated by the elements of $\tau_\omega(G)\gamma_2 G/\gamma_2 G \leq_R \tau_\omega(G/\gamma_2 G)$, it follows by induction that there exist ω-members

$\alpha_1, \ldots, \alpha_c$ such that $g_i^{\alpha_i} \in \gamma_2 G$ for $1 \leq i \leq c$. By Corollary 4.3, there exists a R-epimorphism from $\bigotimes_R^c Ab(G)$ to $\gamma_c G / \gamma_{c+1} G = \gamma_c G$ given by

$$x_1 \gamma_2 G \otimes \cdots \otimes x_c \gamma_2 G \mapsto [x_1, \ldots, x_c] \gamma_{c+1} G = [x_1, \ldots, x_c].$$

Thus,

$$[g_1, \ldots, g_c]^{\alpha_1 \cdots \alpha_c} = [g_1^{\alpha_1}, \ldots, g_c^{\alpha_c}] \in [\gamma_2 G, \ldots, \gamma_2 G] = 1$$

because $[\gamma_2 G, \ldots, \gamma_2 G] \leq \gamma_{2c} G \leq \gamma_{c+1} G = 1$ by Theorem 4.19. Since $\alpha_1 \cdots \alpha_c$ is an ω-member, $[g_1, \ldots, g_c]$ is an ω-torsion element of $\gamma_c G$. It follows from Lemma 4.9 and the fact that $\gamma_c G$ is abelian that every element of $\gamma_c G$ is an ω-torsion element.

Choose an element $g \in G$. Since $G / \gamma_c G$ is of class $c - 1$ and is R-generated by the elements of $\tau_\omega(G) \gamma_c G / \gamma_c G \leq_R \tau_\omega(G / \gamma_c G)$, it follows by induction that there exists an ω-member μ such that $(g \gamma_c G)^\mu = \gamma_c G$ in $G / \gamma_c G$. Thus, $g^\mu = h$ for some $h \in \gamma_c G$. As we deduced above, there is an ω-member σ such that $h^\sigma = 1$. Therefore,

$$g^{\mu \sigma} = (g^\mu)^\sigma = h^\sigma = 1.$$

Since $\mu \sigma$ is an ω-member, $g \in \tau_\omega(G)$.   □

We shall refer to $\tau_\omega(G)$ as the *ω-torsion subgroup* of G and $\tau(G)$ as the *R-torsion subgroup* of G.

Corollary 4.8 *Let G be a nilpotent R-powered group.*

(i) If R is a UFD, then $G / \tau_\omega(G)$ is ω-torsion-free.
(ii) If R is any binomial ring, then $G / \tau(G)$ is R-torsion-free.

Proof We prove (i). Let $g \tau_\omega(G) \in G / \tau_\omega(G)$, and suppose that $(g \tau_\omega(G))^\alpha = \tau_\omega(G)$ for some ω-member α. Then $g^\alpha \in \tau_\omega(G)$, and thus there is an ω-member β for which $g^{\alpha \beta} = (g^\alpha)^\beta = 1$. Since $\alpha \beta$ is an ω-member, $g \in \tau_\omega(G)$.   □

Lemma 4.12 *Suppose that G is a nilpotent R-powered group and $N \trianglelefteq_R G$.*

(i) If R is a UFD and both N and G/N are ω-torsion groups, then G is an ω-torsion group.
(ii) If R is any binomial ring and both N and G/N are R-torsion groups, then G is an R-torsion group.

Proof See Lemma 2.13.   □

According to Theorem 2.26, every nilpotent group is the direct product of its p-torsion subgroups for various primes p. If G happens to be finite, then these p-torsion subgroups are just the Sylow subgroups of G by Theorem 2.13. Our aim is to obtain a similar result for nilpotent R-powered groups. This result will depend on the ring structure of R. We begin with a definition which resembles the notion of a Sylow p-subgroup in the finite case.

Definition 4.25 Let π be a prime in R. If G is a nilpotent R-powered group, then the π-*primary component* of G, denoted by G_π, is

$$G_\pi = \left\{ g \in G \ \middle| \ g^{\pi^k} = 1 \ \text{for some} \ k \in \mathbb{N} \right\}.$$

We say that G is π-*primary* if $G = G_\pi$.

If π_1 and π_2 are primes in R, then it is clear that $G_{\pi_1} = G_{\pi_2}$ whenever π_1 and π_2 are associates. Thus, a π-primary component is unique up to associates.

Definition 4.26 If G is a nilpotent R-powered group and $1 \neq g \in G$, then the set

$$ann(g) = \left\{ \alpha \in R \ \middle| \ g^\alpha = 1 \right\}$$

is an ideal of R, referred to as the *annihilator* (or *order ideal*) of g.

The following is the analogue of Theorem 2.26 and can be found in [9] for the case $R = \mathbb{Q}[x]$.

Theorem 4.25 (S. Majewicz) *Suppose that R is a PID, and let $\mathbb{P}$ be the set of all primes in R (up to associates). If G is an R-torsion nilpotent R-powered group, then*

$$G = \prod_{\pi \in \mathbb{P}} G_\pi.$$

Proof By Theorem 4.24 (ii), $G_\pi \trianglelefteq_R G$ for any prime $\pi \in R$. We claim that

$$G = gp_R(G_\pi \mid \pi \in \mathbb{P}).$$

Choose any element $g \neq 1$ in G. Since R is a PID and G is R-torsion, there exists $\delta \in R$ such that $ann(g) = \langle \delta \rangle \neq \{0\}$, where $\langle \delta \rangle$ denotes the ideal of R generated by δ. Write δ as

$$\delta = \pi_1^{m_1} \cdots \pi_n^{m_n}$$

for some $m_i \in \mathbb{N}$ and primes $\pi_i \in R$, where no pair of primes are associates. Set $\alpha_i = \delta / \pi_i^{m_i}$ for each i, and observe that $g^{\alpha_i} \in G_{\pi_i}$. Now, the greatest common divisor of the α_i's is 1 since they are pairwise relatively prime. Thus, there are elements $\sigma_i \in R$ such that $\sum_{i=1}^n \alpha_i \sigma_i = 1$. Consequently,

$$g = g^{\alpha_1 \sigma_1 + \cdots + \alpha_n \sigma_n}$$

$$= \left(g^{\alpha_1} \right)^{\sigma_1} \cdots \left(g^{\alpha_n} \right)^{\sigma_n},$$

which is an element of $G_{\pi_1} G_{\pi_2} \cdots G_{\pi_n}$ as claimed.

Next, we show that

$$G_{\pi_i} \cap gp_R\left(G_{\pi_j} \mid \pi_j \in \mathbb{P}, \, j \neq i\right) = 1$$

for primes π_i and π_j. Let $S = \{\pi_1, \ldots, \pi_n, \alpha\}$ be a set of primes in R, and suppose that no two elements in S are associates. Suppose that $g \in G_\alpha \cap gp_R(G_{\pi_1}, \ldots, G_{\pi_n})$. We claim that $g = 1$. By definition, $g \in G_\alpha$ implies that there exists $q \in \mathbb{N}$ such that $g^{\alpha^q} = 1$. Furthermore, $g \in G_{\pi_1} \cdots G_{\pi_n}$ implies that there exists $\mu \in R$ such that $g^\mu = 1$ and $\mu = \pi_1^{\mu_1} \cdots \pi_n^{\mu_n}$ for some $\mu_i \in \mathbb{N}$. Since α and μ are relatively prime, there exist elements $\sigma_1, \sigma_2 \in R$ with $\sigma_1 \alpha^q + \sigma_2 \mu = 1$. Therefore,

$$g = g^{\sigma_1 \alpha^q + \sigma_2 \mu} = \left(g^{\alpha^q}\right)^{\sigma_1} \left(g^\mu\right)^{\sigma_2} = 1$$

as required. $\qquad\square$

The next theorem appears in [19].

Theorem 4.26 *Let G be a nilpotent R-powered group.*

(i) *Suppose that R is a UFD. Then G is ω-torsion-free if and only if the following holds:*

> *If g, $h \in G$ and $g^\alpha = h^\alpha$ for some ω-member α, then $g = h$.*

(ii) *Let R be any binomial ring. Then G is R-torsion-free if and only if the following holds:*

> *If g, $h \in G$ and $g^\alpha = h^\alpha$ for some $0 \neq \alpha \in R$, then $g = h$.*

Proof Repeat the proof of Theorem 2.7. $\qquad\square$

Just as in the case of ordinary nilpotent groups, the center of a nilpotent R-powered group has an influence on the structure of its upper central factors (see Corollaries 2.20 and 2.21).

Theorem 4.27 *Let G be a nilpotent R-powered group.*

(i) *Suppose that R is a UFD. If $Z(G)$ is ω-torsion-free, then G and $\zeta_{i+1}G/\zeta_i G$ are ω-torsion-free for $i \geq 0$.*

(ii) *If R is any binomial ring and $Z(G)$ is R-torsion-free, then G and $\zeta_{i+1}G/\zeta_i G$ are R-torsion-free for $i \geq 0$.*

Proof The result follows from Lemma 4.8 and induction on i. $\qquad\square$

Corollary 4.9 *Let G be a nilpotent R-powered group.*

(i) *If R is a UFD and G is ω-torsion-free, then $G/Z(G)$ is ω-torsion-free.*

(ii) *If R is any binomial ring and G is R-torsion-free, then $G/Z(G)$ is R-torsion-free.*

Theorem 4.28 *Let G be a nilpotent R-powered group.*

(i) If R is a UFD and Ab(G) is an ω-torsion group, then G is an ω-torsion group.
(ii) If R is any binomial ring and Ab(G) is an R-torsion group, then G is an R-torsion group.

Proof We mimic the proof of Theorem 2.19, using Corollary 4.3 and the fact that the tensor product of ω-torsion R-modules is an ω-torsion R-module whenever R is a UFD.   □

4.4.13   Finite ω-Type and Finite Type

Every finitely generated torsion nilpotent group is finite by Theorem 2.25. This result motivated S. Majewicz and M. Zyman to study nilpotent R-powered groups which are finitely R-generated and ω-torsion (see [9, 10], and [11]).

Definition 4.27 Let R be a UFD. A finitely R-generated ω-torsion group G is said to be of *finite ω-type*.

This is a natural analogue of a finite P-group. If $\omega = \{\pi\}$ for a prime $\pi \in R$, then G is of *finite π-type*.

Definition 4.28 Let R be *any* binomial ring. A nilpotent R-powered group is of *finite type* if it is finitely R-generated and R-torsion.

The next lemma follows from Theorem 4.21.

Lemma 4.13 *Let G be a nilpotent R-powered group, and let $H \leq_R G$.*

1. Let R be a UFD, and suppose that G is of finite ω-type.

(i) If R is noetherian, then H is of finite ω-type.
(ii) If $H \trianglelefteq_R G$, then G/H is of finite ω-type.

2. Suppose that G is of finite type.

(i) If R is noetherian, then H is of finite type.
(ii) If $H \trianglelefteq_R G$, then G/H is of finite type.

Lemma 4.14 *Let G be a nilpotent R-powered group.*

(i) Suppose that R is a UFD. If $H \trianglelefteq_R G$ and both H and G/H are of finite ω-type, then G is of finite ω-type.
(ii) If $H \trianglelefteq_R G$ and both H and G/H are of finite type, then G is of finite type.

Proof Clearly, G is finitely R-generated since H and G/H are finitely R-generated. To prove that G is ω-torsion (R-torsion), repeat the proof of Lemma 2.13.   □

Putting together Lemmas 4.13 and 4.14 gives:

Theorem 4.29 *Let R be a noetherian ring, and suppose that G is a nilpotent R-powered group with $H \trianglelefteq_R G$.*

(i) If R is a UFD, then G is of finite ω-type if and only if H and G/H are both of finite ω-type.

(ii) G is of finite type if and only if H and G/H are both of finite type.

Lemma 4.15 *Let R be noetherian, and let G be a nilpotent R-powered group.*

(i) If R is a UFD and $Ab(G)$ is of finite ω-type, then G is of finite ω-type.

(ii) If $Ab(G)$ is of finite type, then G is of finite type.

Proof The result follows from Corollary 4.6 and Theorem 4.28. □

Theorem 4.30 *Suppose that G is a finitely R-generated nilpotent R-powered group with finite R-generating set $X = \{x_1, \ldots, x_k\}$.*

(i) If R is a UFD and each element of X is an ω-torsion element, then G is of finite ω-type.

(ii) If each element of X is an R-torsion element, then G is of finite type.

Proof We prove (i) by induction on the class c of G. The proof of (ii) is the same. If $c = 1$, then G is abelian. In this case, every element in G can be expressed in the form $x_1^{\alpha_1} \cdots x_k^{\alpha_k}$ for some $\alpha_1, \ldots, \alpha_k \in R$. If $x_i^{\beta_i} = 1$ for some ω-members β_i, then it follows that $g^{\beta_1 \cdots \beta_k} = 1$ for any $g \in G$. Since $\beta_1 \cdots \beta_k$ is an ω-member, G is an ω-torsion group.

Let $c > 1$, and assume that the result is true for nilpotent R-powered groups of class less than c. By Corollary 4.4, the abelian R-group $\gamma_c G$ is R-generated by simple commutators of the form $[x_{j_1}, x_{j_2}, \ldots, x_{j_c}]$, where the x_{j_k}'s vary over all elements of X and are not necessarily distinct. Since each element of X is an ω-torsion element, $\gamma_c G$ is of finite ω-type by induction. Moreover, $G/\gamma_c G$ is of class $c-1$ with R-generators $\{x_1 \gamma_c G, \ldots, x_k \gamma_c G\}$, each of which is an ω-torsion element. By induction, $G/\gamma_c G$ is of finite ω-type. An application of Lemma 4.14 completes the proof. □

4.4.14 Order, Exponent, and Power-Commutativity

In [10], S. Majewicz and M. Zyman introduced the notions of "order" and "exponent" for nilpotent R-powered groups in order to study the structure of power-commutative nilpotent R-powered groups (see Definition 4.30 below).

At this time, the theory of nilpotent R-powered groups which revolves around these notions is still in its very early stages. It is likely that many of the known results on finite (nilpotent) groups have analogues for nilpotent R-powered groups, where the ring structure of R plays a key role.

Let R be a PID, and suppose that G is a nilpotent R-powered group. If $g \in \tau(G)$, then $ann(g) = \langle \mu \rangle \neq 0$ for some $\mu \in R$. Any other generator of $ann(g)$ is an associate of μ in R, and the relation

$$\mu_1 \sim \mu_2 \text{ if and only if } \mu_1 \text{ and } \mu_2 \text{ are associates}$$

is an equivalence relation on R. We call μ the *order* of g and write $|g| = \mu$. Observe that it is well defined up to associates. Thus, $|g| = \mu$ implies that $g^{\mu_0} = 1$ for any $\mu_0 \in R$ which is an associate of μ. Clearly, $|g|$ divides α whenever $g^\alpha = 1$ for some $\alpha \in R$.

Lemma 4.16 *If R is a PID, then the elements of coprime order in an R-torsion group commute.*

Proof The result follows from Theorem 4.25. $\qquad\qquad\qquad\qquad\qquad\qquad\square$

Next, suppose that R is a PID and G is a nilpotent R-powered group of finite type. The proof of Theorem 4.30 (i) illustrates that the orders of the elements of a finite R-generating set of G determine the order of each element of G, and each such order is a product of finitely many primes in R. This implies the existence of an element $\alpha \in R$ that annihilates every element of G.

Definition 4.29 Let G be a nilpotent R-powered group of finite type, where R is a PID. We say that G has *exponent* $\pi_1\pi_2\cdots\pi_n$, where the π_j's are primes (not necessarily distinct) in R, if

$$G^{\pi_1\pi_2\cdots\pi_n} = 1 \text{ but } G^{\pi_1\pi_2\cdots\widehat{\pi_i}\cdots\pi_n} \neq 1 \text{ for each } 1 \leq i \leq n.$$

As usual, $\pi_1\pi_2\cdots\widehat{\pi_i}\cdots\pi_n$ means $\pi_1\pi_2\cdots\pi_{i-1}\pi_{i+1}\cdots\pi_n$, and

$$G^\beta = gp_R\left(g^\beta \mid g \in G\right).$$

Observe that the exponent of an R-subgroup and a factor R-group of G divides the exponent of G.

Theorem 4.31 *Let R be a PID, and suppose that G is a finitely R-generated nilpotent R-powered group. Then $Z(G)$ is of finite type if and only if G is of finite type.*

The theorem is also true if "finite type" is replaced by "finite ω-type."

Proof By Lemma 4.13, $Z(G)$ is of finite type whenever G is of finite type. We prove the converse by induction on the class c of G. Suppose that $Z(G)$ is of finite type. The result is obvious if $c = 1$. Let $c > 1$, and suppose that the theorem is true for every finitely R-generated nilpotent R-powered group of class less than c. Suppose that

$Z(G)$ has exponent α, and let $a \in \zeta_2 G$. If g is any element of G, then $[g, a] \in Z(G)$. Now, $[g, a]^\alpha = 1$ implies that $[g, a^\alpha] = 1$ by Lemma 4.8. Therefore, $a^\alpha \in Z(G)$, and thus $\zeta_2 G / Z(G)$ is an R-torsion group. Furthermore, $\zeta_2 G / Z(G)$ is finitely R-generated by Theorem 4.21 because

$$\zeta_2 G / Z(G) = Z(G/Z(G)) \trianglelefteq_R G/Z(G)$$

and $G/Z(G)$ is finitely R-generated. Thus, $\zeta_2 G / Z(G)$ is of finite type. By induction, $G/Z(G)$ is also of finite type because it is of class $c - 1$. Since $Z(G)$ is of finite type by hypothesis, the result follows from Theorem 4.29. $\qquad \square$

We now turn our attention to power-commutative nilpotent R-powered groups. The study of these groups was motivated by Wu [20]. Here we provide only two results from [10].

Definition 4.30 A nilpotent R-powered group G is called *power-commutative* if $[g^\alpha, h] = 1$ implies $[g, h] = 1$ for all $g, h \in G$ and $\alpha \in R$ whenever $g^\alpha \neq 1$.

Clearly, every abelian R-group is power-commutative and every R-subgroup of a power-commutative group is also power-commutative.

Lemma 4.17 *Every R-torsion-free group is power-commutative.*

Proof Let G be an R-torsion-free group. If $[g^\alpha, h] = 1$ for $g, h \in G$ and $0 \neq \alpha \in R$, then $(h^{-1}gh)^\alpha = g^\alpha$. Apply Theorem 4.26 (ii). $\qquad \square$

Lemma 4.18 *Suppose that R is a PID. If G is a nilpotent R-powered group of finite type and has exponent π for some prime $\pi \in R$, then it is power-commutative.*

Proof Suppose that $[g^\alpha, h] = 1$, where $g, h \in G$, $\alpha \in R$, and $g^\alpha \neq 1$. Since $G^\pi = 1$, π and α are relatively prime. Thus, there exists $a, b \in R$ such that $a\alpha + b\pi = 1$. Therefore, the equalities $[g^\alpha, h] = 1$ and $g^\pi = 1$ yield

$$\left[g, h\right] = \left[g^{a\alpha + b\pi}, h\right] = \left[(g^\alpha)^a (g^\pi)^b, h\right] = \left[(g^\alpha)^a, h\right] = 1.$$

This completes the proof. $\qquad \square$

Theorem 4.32 *Suppose that R is a PID, and let G be a nilpotent R-powered group of finite type which is not of finite π-type for any prime $\pi \in R$. If G is power-commutative, then it must be abelian.*

Proof By Theorem 4.25, it suffices to show that each π-primary component of G is abelian. Suppose that $\pi_1, \ldots, \pi_n, \alpha$ are primes in R, no two being associates. Assume that $g \neq 1$ and $h \neq 1$ are elements of G_α and $a \in G_{\pi_1} \times \cdots \times G_{\pi_n}$, and suppose that $|g| = \alpha^r$ and $|h| = \alpha^s$ for some $r, s \in \mathbb{N}$. We will show that $[g, h] = 1$. Clearly, a and g are of coprime order. By Lemma 4.16, $[a, g] = 1$. By the axioms, we get

$$\left((ag)^{\alpha^r}\right)^{|a|} = (ag)^{|a|\alpha^r} = a^{|a|\alpha^r} g^{|a|\alpha^r} = \left(a^{|a|}\right)^{\alpha^r} (g^{\alpha^r})^{|a|} = 1.$$

This means that $(ag)^{\alpha^r}$ has order dividing $|a|$ which is relatively prime to α^r. However,

$$(ag)^{\alpha^r} = a^{\alpha^r} g^{\alpha^r} = a^{\alpha^r},$$

implying that a^{α^r} has order dividing $|a|$. Since $|h|$ and $|a|$ are relatively prime,

$$\left[(ag)^{\alpha^r},\ h\right] = \left[a^{\alpha^r},\ h\right] = 1. \tag{4.24}$$

Since G is power-commutative, (4.24) gives $[ag,\ h] = 1$. By Lemma 1.4 and the fact that h and a commute, we get

$$[ag,\ h] = [a,\ h]^g [g,\ h] = [g,\ h].$$

And so, $[g,\ h] = 1$. □

References

1. G. Baumslag, *Lecture Notes on Nilpotent Groups*. Regional Conference Series in Mathematics, No. 2 (American Mathematical Society, Providence, RI, 1971). MR0283082
2. N. Blackburn, Conjugacy in nilpotent groups. Proc. Am. Math. Soc. **16**, 143–148 (1965). MR0172925
3. J.D. Dixon, et al., *Analytic Pro-p-Groups*. London Mathematical Society Lecture Note Series, vol. 157 (Cambridge University Press, Cambridge, 1991). MR1152800
4. P. Hall, Some word-problems. J. Lond. Math. Soc. **33**, 482–496 (1958). MR0102540
5. P. Hall, *The Edmonton Notes on Nilpotent Groups*. Queen Mary College Mathematics Notes, Mathematics Department (Queen Mary College, London, 1969). MR0283083
6. M.I. Kargapolov, V.N. Remeslennikov, N.S. Romanovskii, V.A. Roman'kov, V.A. Churkin, Algorithmic problems for σ-power groups. J. Algebra Logic (Translated) **8**(6), 364–373 (1969)
7. E.I. Khukhro, *p-Automorphisms of Finite p-Groups*. London Mathematical Society Lecture Note Series, vol. 246 (Cambridge University Press, Cambridge, 1998). MR1615819
8. J.C. Lennox, D.J.S. Robinson, *The Theory of Infinite Soluble Groups*. Oxford Mathematical Monographs (The Clarendon Press/Oxford University Press, Oxford, 2004). MR2093872
9. S. Majewicz, Nilpotent R-powered groups and $\mathbb{Z}[x]$-groups, Ph.D. Dissertation, City University of New York, 2004
10. S. Majewicz, M. Zyman, Power-commutative nilpotent R-powered groups. Groups Complex. Cryptol. **1**(2), 297–309 (2009). MR2598996
11. S. Majewicz, M. Zyman, On the extraction of roots in exponential A-groups II. Commun. Algebra **40**(1), 64–86 (2012). MR2876289
12. A.I. Mal'cev, Nilpotent torsion-free groups. Izv. Akad. Nauk. SSSR. Ser. Mat. **13**, 201–212 (1949). MR0028843
13. A.I. Mal'cev, On a class of homogeneous spaces. Izv. Akad. Nauk. SSSR. Ser. Mat. **13**, 9–32 (1949). MR0028842
14. A.W. Mostowski, Computational algorithms for deciding some problems for nilpotent groups. Fund. Math. **59**, 137–152 (1966). MR0224694
15. J. Petresco, Sur les commutateurs. Math. Z. **61**, 348–356 (1954). MR0066380
16. P.F. Pickel, Finitely generated nilpotent groups with isomorphic finite quotients. Trans. Am. Math. Soc. **160**, 327–341 (1971). MR0291287

17. J.J. Rotman, *Advanced Modern Algebra* (Prentice Hall, Upper Saddle River, NJ, 2002). MR2043445
18. D. Singmaster, Divisibility of binomial and multinomial coefficients by primes and prime powers, http://www.fq.math.ca/Books/Collection/singmaster.pdf
19. R.B. Warfield, Jr., *Nilpotent Groups*. Lecture Notes in Mathematics, vol. 513 (Springer, Berlin, 1976). MR0409661
20. Y.-F. Wu, On locally finite power-commutative groups. J. Group Theory **3**(1), 57–65 (2000). MR1736517

Chapter 5
Isolators, Extraction of Roots, and P-Localization

When considering rings acting on groups, one may specialize to subrings of $\mathbb{Q}$. A group action by a subring of $\mathbb{Q}$ may be interpreted as root extraction in the said group, which is the underlying subject of this chapter. In Section 5.1, we discuss the theory of isolators and isolated subgroups. This topic is related to the study of root extraction in nilpotent groups, which is the theme of Section 5.2. We discuss some of the theory of root extraction in groups, emphasizing the main results for nilpotent groups. Residual properties of nilpotent groups are also discussed in Section 5.2, as well as embeddings of nilpotent groups into nilpotent groups with roots. Section 5.3 contains an introduction to the theory of localization of nilpotent groups. A group is said to be P-local, for a set of primes P, if every group element has a unique nth root whenever n is relatively prime to the elements of P. The main result in this section is: Given a nilpotent group G, there exists a nilpotent group G_P of class at most the class of G which is the best approximation to G among all P-local nilpotent groups.

In Sections 5.1 and 5.2, P will always denote a nonempty set of primes and P' will be the set of all primes not in P. We further declare that if p is a single prime, then p' is the set of all primes excluding p.

5.1 The Theory of Isolators

Isolators are closely related to the extraction of roots in (locally) nilpotent groups. In this section, we present some of the theory of isolators. Our discussion is based on the work of P. Hall [10].

5.1.1 Basic Properties of P-Isolated Subgroups

We begin with the definition of a P-isolated subgroup.

© Springer International Publishing AG 2017
A.E. Clement et al., *The Theory of Nilpotent Groups*,
DOI 10.1007/978-3-319-66213-8_5

Definition 5.1 Let G be any group. A subgroup H of G is called *P-isolated* in G if $g \in G$ and $g^n \in H$ for some P-number n imply $g \in H$.

If P is the set of all primes, then H is *isolated* in G. If $P = \{p\}$, then we say that H is *p-isolated* in G.

One important aspect of P-isolated normal subgroups is that they give rise to P-torsion-free quotients.

Lemma 5.1 *Let G be any group. If $N \trianglelefteq G$, then G/N is P-torsion-free if and only if N is P-isolated in G.*

Proof If N is P-isolated in G and $gN \in G/N$ such that $(gN)^m = N$ for some P-number m, then $g^m \in N$. Hence, $g \in N$, implying that G/N is P-torsion-free.

Next, suppose that G/N is P-torsion-free. If m is a P-number, then $(gN)^m = N$ implies $gN = N$; that is, $g^m \in N$ implies $g \in N$. Thus, N is P-isolated in G. □

Lemma 5.2 *Let G be a group, and suppose that H is P-isolated in G.*

(i) If K is P-isolated in H, then K is P-isolated in G.
(ii) If $K \leq G$, then $H \cap K$ is P-isolated in K.

Proof

(i) If $g \in G$ such that $g^n \in K$ for some P-number n, then $g^n \in H$. Thus, $g \in H$ because H is P-isolated in G. Since K is P-isolated in H, $g \in K$.
(ii) If $k^n \in H \cap K$ for some P-number n and $k \in K$, then $k \in H$ because H is P-isolated in G. Therefore, $k \in H \cap K$. □

There is a natural correspondence between P-isolated subgroups and their quotients.

Theorem 5.1 *Let G be a group with a P-isolated normal subgroup H. If $K \leq G$ and $H \trianglelefteq K$, then K is P-isolated in G if and only if K/H is P-isolated in G/H.*

Proof Assume that K is P-isolated in G, and suppose that $(gH)^n \in K/H$ for some $g \in G$ and P-number n. Since $g^n \in K$ and K is P-isolated in G, we have $g \in K$. Therefore, $gH \in K/H$.

Conversely, suppose that K/H is P-isolated in G/H. If $g^n = k$ for some $g \in G$, $k \in K$, and P-number n, then $(gH)^n = kH$ in K/H. Thus, $gH \in K/H$ and $g \in K$. □

The automorphic image of a P-isolated subgroup is always P-isolated.

Lemma 5.3 *Let G be a group, and suppose that H is a P-isolated subgroup of G. If $\psi \in Aut(G)$, then $\psi(H)$ is also P-isolated in G.*

Proof Let $g \in G$, and assume that $g^n \in \psi(H)$ for some P-number n. There exists an element $h \in H$ such that $g^n = \psi(h)$. Thus,

$$\left(\psi^{-1}(g)\right)^n = \psi^{-1}(g^n) = h$$

is contained in H. Since H is P-isolated, $\psi^{-1}(g) \in H$. And so, $g \in \psi(H)$. □

Under certain conditions, the intersection or union of P-isolated subgroups is P-isolated.

Lemma 5.4 *Let G be any group, and suppose that $\{G_i \mid i \in I\}$ is a family of P-isolated subgroups of G.*

(i) If I is any index set, then $\bigcap_{i \in I} G_i$ is a P-isolated subgroup of G.
(ii) If I is a well-ordered index set and $G_i \leq G_j$ whenever $i \leq j$, then $\bigcup_{i \in I} G_i$ is a P-isolated subgroup of G.

Proof (i) Suppose that $g^n \in \bigcap_{i \in I} G_i$ for some P-number n and $g \in G$. Then $g^n \in G_i$ for each $i \in I$. Since each G_i is P-isolated in G, g is contained in each G_i. Thus, $g \in \bigcap_{i \in I} G_i$. The proof of (ii) is similar.   □

5.1.2 The P-Isolator

Lemma 5.4 (i) suggests the following definition:

Definition 5.2 Let G be a group and $S \subset G$. The *P-isolator* of S in G, denoted by $I_P(S, G)$, is the intersection of all P-isolated subgroups of G that contain S. If P is the set of all primes, then we write $I(S, G)$ and call it the *isolator* of S in G.

Clearly, $I_P(S, G)$ is the unique minimal P-isolated subgroup of G containing S. It is constructed in the following way: Put $S_1 = S$ and $G_1 = gp(S)$, and recursively define $G_{i+1} = gp(S_{i+1})$, where $S_{i+1} = \{g \in G \mid g^n \in G_i \text{ for some } P\text{-number } n\}$. Then

$$I_P(S, G) = \bigcup_{i=1}^{\infty} G_i.$$

To see why this is true, let $H = \bigcup_{i=1}^{\infty} G_i$. Suppose that $g \in G$ and n is a P-number such that $g^n \in H$. Then $g^n \in G_i$ for some $i \in \mathbb{N}$, and consequently, $g \in S_{i+1}$. Hence, $g \in G_{i+1} \leq H$, and consequently, H is a P-isolated subgroup of G containing S. Thus, $I_P(S, G) \leq H$.

Next, we prove that $I_P(S, G) \geq H$. It suffices to show that $I_P(S, G) \geq G_k$ for all $k \in \mathbb{N}$. This is clearly true when $k = 1$, giving the basis for induction on k. Suppose that $I_P(S, G) \geq G_m$ for $m > 1$. If $g \in S_{m+1}$, then $g^n \in G_m$ for some P-number n, and thus $g^n \in I_P(S, G)$. Since $I_P(S, G)$ is P-isolated in G, we have $g \in I_P(S, G)$. We have shown that $I_P(S, G)$ contains S_{m+1}, and thus $I_P(S, G) \geq G_{m+1}$ as asserted.

Lemma 5.5 *Let G be any group. If $H \trianglelefteq G$ and $I_P(H, G) \leq G$, then $I_P(H, G) \trianglelefteq G$.*

Proof Let $g \in G$. If $h \in I_P(H, G)$, then there exists a P-number n such that $h^n \in H$. Since $H \trianglelefteq G$, we have $\left(g^{-1}hg\right)^n = g^{-1}h^n g \in H$. And so, $g^{-1}hg \in I_P(H, G)$.   □

We give an explicit description of the P-isolator of a subgroup of a nilpotent group in the next theorem.

Theorem 5.2 *If G is a nilpotent group and $H \leq G$, then the set*

$$T = \{g \in G \mid g^n \in H \text{ for some } P\text{-number } n\}$$

is a subgroup of G and equals $I_P(H, G)$. If G is P-torsion-free, then the nilpotency classes of $I_P(H, G)$ and H are equal.

Proof We first prove that T is a subgroup of G by induction on the class c of G. Clearly, $g^{-1} \in T$ whenever $g \in T$. We need to show that T is closed under multiplication. If $c = 1$ and $g, h \in T$, then there exist P-numbers m and n such that $g^m \in H$ and $h^n \in H$. Since $H \leq G$ and G is abelian,

$$(gh)^{mn} = (g^m)^n (h^n)^m \in H.$$

Hence, $gh \in T$ because mn is a P-number.

We proceed by induction on c. Suppose that $c > 1$, and assume that the result is true for all nilpotent groups of class less than c. Without loss of generality, we may assume that $G = gp(T)$. We assert that $G = T$. First, we show that the abelian group $\gamma_c G$ is contained in T. Let $[g_1, \ldots, g_c]$ be any commutator of weight c in $\gamma_c G$, where $g_1, \ldots, g_c \in G$. Then $g_1 \gamma_2 G, \ldots, g_c \gamma_2 G$ are contained in $Ab(G)$. By the induction hypothesis, there exist P-numbers $m_1, \ldots, m_c$ such that $(g_i \gamma_2 G)^{m_i} \in H\gamma_2 G/\gamma_2 G$ for $1 \leq i \leq c$. Hence, there exist elements $a_1, \ldots, a_c$ in $\gamma_2 G$ such that $g_i^{m_i} a_i \in H$ for $1 \leq i \leq c$. By Corollary 2.10, the mapping

$$\bigotimes_{\mathbb{Z}}^{c} Ab(G) \to \gamma_c G/\gamma_{c+1} G = \gamma_c G$$

given by

$$x_1 \gamma_2 G \otimes \cdots \otimes x_c \gamma_2 G \mapsto [x_1, \ldots, x_c]\gamma_{c+1}G = [x_1, \ldots, x_c]$$

is a $\mathbb{Z}$-module epimorphism and multiplicative in each variable. Thus,

$$[g_1^{m_1} a_1, \ldots, g_c^{m_c} a_c] = [g_1, \ldots, g_c]^{m_1 \cdots m_c}[a_1, \ldots, a_c].$$

By Theorem 2.14 (i), $[a_1, \ldots, a_c] \in \gamma_{2c} G = 1$. Hence, $[g_1, \ldots, g_c]^{m_1 \cdots m_c} \in H$. Since $m_1 \cdots m_c$ is a P-number, $[g_1, \ldots, g_c] \in T$, and thus $\gamma_c G$ is contained in T.

Choose an arbitrary element g of G. Since $G/\gamma_c G$ is of class less than c, there exists a P-number m and an element x of H such that $(g\gamma_c G)^m = x\gamma_c G$; that is, $g^m = xy$ for some $y \in \gamma_c G$. Furthermore, there exists a P-number n satisfying $y^n \in H$ because T contains $\gamma_c G$. Therefore,

$$g^{mn} = (xy)^n = x^n y^n \in H,$$

and thus $g \in T$ and $G = T$ as asserted.

Next, we show that $T = I_P(H, G)$. Suppose that $g \in G$ and $g^r \in T$ for some P-number r. There exists a P-number s such that $(g^r)^s = g^{rs} \in H$, and thus $g \in T$. Since T is a subgroup of G, T must be a P-isolated subgroup of G which contains H; that is, $T \geq I_P(H, G)$. The reverse inclusion follows from the construction of the P-isolator.

Now, suppose that G is P-torsion-free, and let H and $I_P(H, G)$ have nilpotency classes c and k respectively. Clearly, $c \leq k$ since $H \leq I_P(H, G)$. We claim that $c \geq k$. Since H is of class c, each $(c + 1)$-fold commutator of elements of H equals the identity. Let $g_1, \ldots, g_{c+1} \in I_P(H, G)$, and consider the $(c+1)$-fold commutator $[g_1, \ldots, g_{c+1}]$ of $I_P(H, G)$. We need to show that $[g_1, \ldots, g_{c+1}] = 1$. For each $i = 1, 2, \ldots, c + 1$, there exists a P-number n_i such that $g_i^{n_i} \in H$. Thus,

$$\left[g_1^{n_1}, \ldots, g_{c+1}^{n_{c+1}} \right] = 1 \tag{5.1}$$

in H. Arguing as before, we can re-express (5.1) as

$$[g_1, \ldots, g_{c+1}]^{n_1 \cdots n_{c+1}} = 1.$$

Since H is P-torsion-free and $n_1 \cdots n_{c+1}$ is a P-number, $[g_1, \ldots, g_{c+1}] = 1$. Therefore, $k \leq c$. □

Note that we obtain Theorem 2.26 by setting $H = 1$ in Theorem 5.2.

Corollary 5.1 *If G is a nilpotent group and $H \trianglelefteq G$, then $I_P(H, G) \trianglelefteq G$.*

Proof Apply Theorem 5.2 and Lemma 5.5. □

Theorem 5.2 also holds for locally nilpotent groups.

Theorem 5.3 *If G is a locally nilpotent group and $H \leq G$, then the set*

$$T = \{g \in G \mid g^n \in H \text{ for some } P\text{-number } n\}$$

is a subgroup of G and equals $I_P(H, G)$.

Proof Let $g, h \in T$. Put $K = gp(g, h)$, and realize that K is nilpotent since it is a finitely generated subgroup of the locally nilpotent group G. There exist P-numbers r and s such that $g^r \in H \cap K$ and $h^s \in H \cap K$. Since $H \cap K \leq K$, we may apply Theorem 5.2 to K and conclude that the set

$$\{x \in K \mid x^n \in H \cap K \text{ for some } P\text{-number } n\}$$

is a subgroup of K. Thus, there is a P-number t such that $(gh)^t \in H \cap K \leq H$ for some P-number t. And so, $gh \in T$. The rest of the proof is similar to the one given for Theorem 5.2. □

It is worth mentioning that Theorem 2.23 can also be used to prove Theorems 5.2 and 5.3. Moreover, Theorem 2.27 follows from Theorem 5.3 if we set $H = 1$.

Corollary 5.2 *If G is a torsion-free locally nilpotent group contained in a locally nilpotent $\mathbb{Q}$-powered group $\overline{G}$, then $I(G, \overline{G})$ is a Mal'cev completion of G.*

Proof Let $g \in I(G, \overline{G})$ with $g \notin G$. By Theorem 5.3, $I(G, \overline{G})$ is locally nilpotent, and there exists $k \in \mathbb{N}$ such that $g^k \in G$. We need to prove that g has a unique nth root in $I(G, \overline{G})$ for any $n > 1$. Clearly, g has a unique nth root in $\overline{G}$ because $g \in \overline{G}$ and $\overline{G}$ is a $\mathbb{Q}$-powered group. Since

$$g = \left(g^{1/n}\right)^n \in I(G, \overline{G}),$$

there exists $m > 1$ such that $\left(g^{1/n}\right)^{mn} \in G$. Thus $g^{1/n} \in I(G, \overline{G})$, as required.   □

Remark 5.1 In general, the P-isolator of a subgroup H of an arbitrary group G does not have to be equal to $\{g \in G \mid g^n \in H \text{ for some } P\text{-number } n\}$. For example, the set of torsion elements of the infinite dihedral group D_∞ does not form a subgroup (see the comment before Theorem 2.26).

Remark 5.2 In the literature, the P-isolator of a subgroup H of G is sometimes defined as the *set*

$$I_P(H, G) = \{g \in G \mid g^n \in H \text{ for some } P\text{-number } n\}.$$

If $I_P(H, G)$ happens to be a subgroup of G for all subgroups $H \leq G$, then G is regarded as having the *P-isolator property*.

5.1.3   *P-Equivalency*

Let H and K be subgroups of a group G. If for all $h \in H$ and $k \in K$, there are P-numbers m and n such that $h^m \in K$ and $k^n \in H$, then H and K are said to be *P-equivalent* (see Definition 5.5).

In [10], P. Hall laid out the properties of P-equivalent subgroups. Our intention is to discuss some of these results. We begin with a brief discussion of verbal subgroups.

Definition 5.3 Let F be a free group on the set $X = \{x_1, x_2, \ldots\}$, and suppose that $\emptyset \neq V \subset F$. Let G be any group, and let

$$\vartheta = x_{i_1}^{k_1} \cdots x_{i_n}^{k_n} \in V \quad (x_{i_j} \in X, \ k_j \in \mathbb{Z}).$$

(i) If $g_1, \ldots, g_n \in G$, then the *value* of ϑ at the n-tuple $(g_1, \ldots, g_n)$ is

$$\vartheta(g_1, \ldots, g_n) = g_1^{k_1} \cdots g_n^{k_n}.$$

(ii) The word ϑ is termed a *law* in G if $\vartheta(g_1, \ldots, g_n) = 1$ for all n-tuples of elements of G.

(iii) The *verbal subgroup* of G determined by V, denoted by $V(G)$, is the subgroup of G generated by all values in G of words in V. Hence,

$$V(G) = gp(\vartheta(g_1, g_2, \ldots) \mid g_i \in G, \ \vartheta \in V).$$

(iv) The class of all groups G satisfying $V(G) = 1$ is called the *variety* relative to V.

Example 5.1 If $V = \{[x_1, x_2]\}$ and G is any group, then

$$V(G) = gp([g, h] \mid g, h \in G) = [G, G].$$

If G is abelian, then the word $\vartheta = [x_1, x_2]$ is a law in G. The class of abelian groups forms a variety.

Example 5.2 Suppose that $V = \{[x_1, \ldots, x_{n+1}]\}$ and G is a group. The verbal subgroup of G is

$$V(G) = gp([g_1, \ldots, g_{n+1}] \mid g_i \in G) = \gamma_n G$$

by Lemma 2.6. By Corollary 2.3, the word $\vartheta = [x_1, \ldots, x_{n+1}]$ is a law in any nilpotent group of class at most n. Hence, the class of nilpotent groups of class at most n is a variety.

Remark 5.3 It follows from Definition 5.3 that any subgroup or homomorphic image of a group in a variety is again in that variety. This result can be compared to Theorem 2.4.

Definition 5.4 Let $\vartheta = \vartheta(x_1, \ldots, x_n)$ be a word in the n variables $x_1, \ldots, x_n$. Let G be any group, and let $H_1, \ldots, H_n$ be subgroups of G. The *generalized verbal subgroup* $\vartheta(H_1, \ldots, H_n)$ is defined as

$$\vartheta(H_1, \ldots, H_n) = gp(\vartheta(h_1, \ldots, h_n) \mid h_i \in H_i \text{ for } i = 1, \ldots, n).$$

The next theorem on generalized verbal subgroups will be useful in determining certain facts about P-isolators.

Theorem 5.4 *Let G be a finitely generated nilpotent group. Suppose that H_i and K_i are subgroups of G such that $[H_i : K_i] = m_i$ for $i = 1, \ldots, n$ and $m_i \in \mathbb{N}$. Let $H = \vartheta(H_1, \ldots, H_n)$ and $K = \vartheta(K_1, \ldots, K_n)$ be the corresponding generalized verbal subgroups for some n-variable word ϑ. Then $[H : K]$ is finite and divides some power of $m = m_1 \cdots m_n$.*

The proof relies on the next lemma which gives a criterion for a finitely generated nilpotent group to be a finite p-group.

Lemma 5.6 *Let G be a finitely generated nilpotent group, and let p be a fixed prime. Suppose that H and K are subgroups of G with $K < H$ such that $[H : K]$ is either infinite or finite and divisible by p. Suppose in addition, that $[NH : NK]$ is a finite p'-number for every nontrivial normal subgroup N of G, then G is a finite p-group.*

Proof We begin by observing that $N \nleq K$ and $N \cap H \neq 1$ for any normal subgroup N of G. For if either $N \leq K$ or $N \cap H = 1$, then $[H : K] = [NH : NK]$, contradicting the hypothesis.

In light of Lemma 2.26, it suffices to show that $Z(G)$ is a p-group. First, we prove that $Z(G)$ is a torsion group. Let $1 \neq z \in Z(G)$, and put $N = gp(z)$. Clearly, $1 \neq N \trianglelefteq G$. Assume, on the contrary, that z is of infinite order. Since $N \cap H \neq 1$, we may assume that $z \in N \cap H$. Thus, $N \leq H$. Now, $N \cap K \leq K$ and $N \cap K \leq H$ implies that

$$[(N \cap K)H : (N \cap K)K] = [H : K].$$

However, $[H : K]$ is either infinite or divisible by p, yet $[(N \cap K)H : (N \cap K)K]$ is a finite p'-number because $N \cap K \lhd G$. We conclude that $N \cap K = 1$.

Let $M = gp(z^p)$, and note that $1 \neq M \trianglelefteq G$ and $[N : M] = p$. By hypothesis, $[MH : MK]$ is a p'-number. On the other hand, $M \leq H$ implies that

$$[MH : MK] = [H : MK] = [H : NK][NK : MK] = [H : NK]p,$$

which is not a p'-number. This contradiction confirms that z must be of finite order. Thus, N is finite and $Z(G)$ is a torsion group.

Next, we show that $Z(G)$ is a p-group. Put $N = gp(z)$ as before, and suppose that N has prime order. We may assume once again that $N \leq H$ because $N \cap H \neq 1$. We have

$$[H : K] = [H : NK][NK : K] = [NH : NK]|N|.$$

By hypothesis, $[NH : NK]$ is a finite p'-number. Therefore, $[H : K]$ is finite, and the hypothesis implies that it is divisible by p. This proves that $|N| = p$, and consequently, $Z(G)$ is a p-group as claimed.   □

We now prove Theorem 5.4. Assume on the contrary, that $[H : K]$ is either infinite or finite with a prime divisor p not dividing m. By Theorem 2.18, G satisfies Max. Thus, there exists a nontrivial normal subgroup M of G which is maximal subject to the condition that $[MH : MK]$ is either infinite or finite with a prime divisor p that does not divide m.

Set $G^* = G/M$, and let $H_i^* = MH_i/M$ and $K_i^* = MK_i/M$ for each i. By Lemma 5.6, G^* is a finite p-group. Hence, H_i^* and K_i^* are also finite p-groups. This means that $[H_i^* : K_i^*]$ is a pth power. Now,

$$[H_i^* : K_i^*] = [MH_i/M : MK_i/M] = [H_i : K_i(M \cap H_i)],$$

and thus

$$[H_i : K_i] = [H_i : K_i(M \cap H_i)][K_i(M \cap H_i) : K_i]$$
$$= [H_i^* : K_i^*][K_i(M \cap H_i) : K_i].$$

Since $[H_i : K_i] = m_i$, it must be that $[H_i^* : K_i^*]$ divides m_i. However, $[H_i^* : K_i^*]$ is a pth power and p is relatively prime to m. We conclude that $[H_i^* : K_i^*] = 1$; that is, $H_i^* = K_i^*$ for each i. Therefore,

$$H^* = \vartheta(H_1^*, \ldots, H_n^*) = \vartheta(K_1^*, \ldots, K_n^*) = K^*.$$

It follows that $MH = MK$, a contradiction. This completes the proof of Theorem 5.4.

Definition 5.5 Let G be any group with subgroups H and K. Then H is *P-equivalent* to K if for every pair of elements $h \in H$ and $k \in K$, there exist *P*-numbers m and n such that $h^m \in K$ and $k^n \in H$. We write $H \underset{P}{\sim} K$.

For the rest of this section, the *P*-isolator of a subgroup H of a group G will be written as $\overline{H}$.

Remark 5.4 Let G be a locally nilpotent group.

1. If $H \leq G$, then $H \underset{P}{\sim} \overline{H}$.
2. If H and K are subgroups of G, then $H \underset{P}{\sim} K$ if and only if $H \leq \overline{K}$ and $K \leq \overline{H}$.

Theorem 5.5 *Let G be a locally nilpotent group. For each $i = 1, \ldots, n$, suppose that $H_i \underset{P}{\sim} K_i$ for subgroups H_i and K_i of G. If $\vartheta(x_1, \ldots, x_n)$ is a word in the variables $x_1, \ldots, x_n$, then*

$$\vartheta(H_1, \ldots, H_n) \underset{P}{\sim} \vartheta(K_1, \ldots, K_n).$$

Proof Set $H = \vartheta(H_1, \ldots, H_n)$ and $K = \vartheta(K_1, \ldots, K_n)$. By Remark 5.4, it suffices to prove that $H \leq \overline{K}$. Let $w = \vartheta(h_1, \ldots, h_n)$, where $h_i \in H_i$ for each i. We claim that there exists a *P*-number m such that $w^m \in K$; that is, $w \in \overline{K}$.

Let $M = gp(h_1, \ldots, h_n)$. By hypothesis, M is nilpotent because it is a finitely generated subgroup of G. Since $H_i \underset{P}{\sim} K_i$ for each i, there exists a *P*-number m_i such that $h_i^{m_i} \in K_i$. For each i, define

$$M_i = gp(h_i) \leq H_i \quad \text{and} \quad N_i = gp(h_i^{m_i}) \leq K_i.$$

Clearly, $N_i \leq M_i$, and thus $\vartheta(N_1, \ldots, N_n) \leq \vartheta(M_1, \ldots, M_n)$. Furthermore, each M_i and N_i is nilpotent because each H_i and K_i is locally nilpotent according to Lemma 2.20 (ii). Thus, both $\vartheta(N_1, \ldots, N_n)$ and $\vartheta(M_1, \ldots, M_n)$ are nilpotent. By Theorem 5.4,

$$[\vartheta(M_1, \ldots, M_n) : \vartheta(N_1, \ldots, N_n)]$$

is a P-number because $[M_i : N_i] = m_i$. Furthermore, $w \in \vartheta(M_1, \ldots, M_n)$. By Corollary 2.6, there is a P-number m such that $w^m \in \vartheta(N_1, \ldots, N_n) \leq K$.   □

Let $\vartheta(x_1, x_2) = [x_1, x_2]$, and let G be a locally nilpotent group. Suppose that $H_1 \underset{P}{\sim} K_1$ and $H_2 \underset{P}{\sim} K_2$ for subgroups H_1, H_2, K_1, and K_2 of G. By Theorem 5.5,

$$[H_1, H_2] \underset{P}{\sim} [K_1, K_2].$$

The result above, together with Remark 5.4, give the following:

Corollary 5.3 *Let G be a locally nilpotent group. If H and K are subgroups of G, then $\left[\overline{H}, \overline{K}\right] \leq \overline{[H, K]}$.*

Several properties of the isolator carry over to the centralizer of a factor group in a group. To make sense of this, we need a definition.

Definition 5.6 Let G be any group with $N \trianglelefteq G$ and $N \leq H \leq G$. The subgroup K of G which satisfies the conditions $N \leq K$ and $K/N = C_{G/N}(H/N)$ is termed the *centralizer* of H/N in G, written as $K = C_G(H/N)$.

Thus, $C_G(H/N) = \{g \in G \mid [g, h] \in N \text{ for all } h \in H\}$.

Lemma 5.7 *Let G be a locally nilpotent group with $S \trianglelefteq G$. Suppose that R is a subgroup of G such that $S \trianglelefteq R$, and put $C = C_G(R/S)$. The following hold:*

(i) *$\overline{C}$ is a subgroup of $C_G\left(\overline{R}/\overline{S}\right)$; that is, $\left[\overline{C}, \overline{R}\right] \leq \overline{S}$;*
(ii) *If $S = \overline{S}$, then $C = \overline{C}$;*
(iii) *If $\overline{S} \geq R$, then C is P'-isolated in G;*
(iv) *If $R \trianglelefteq G$ and R/S is a finite P-torsion group, then G/C is a finite P-torsion group.*

Proof

(i) By Corollary 5.1, $\overline{S} \trianglelefteq G$. Hence, $\overline{S} \trianglelefteq \overline{R}$. Clearly, $[C, R] \leq S$, and thus $\overline{[C, R]} \leq \overline{S}$. By Corollary 5.3,

$$\left[\overline{C}, \overline{R}\right] \leq \overline{[C, R]}$$

and the result follows.
(ii) By (i), $\left[\overline{C}, \overline{R}\right] \leq \overline{S}$. Since $S = \overline{S}$, $\left[\overline{C}, \overline{R}\right] \leq S$. Hence $\left[\overline{C}, R\right] \leq S$, and thus $\overline{C} = C$.
(iii) Suppose that $x^m \in C$ for some P'-number m. Put $X = gp(x)$ and $Y = gp(x^m)$. Clearly, $X \underset{P'}{\sim} Y$. By Theorem 5.5, $[X, R] \underset{P'}{\sim} [Y, R]$. We claim that

$$[X, R] \leq \overline{S} \cap I_{P'}(S, G).$$

We show first that $[X, R] \leq I_{P'}(S, G)$. Let $w \in [X, R]$. Since $[X, R] \underset{P'}{\sim} [Y, R]$, there exists a P'-number l such that $w^l \in [Y, R]$. Now, $Y \leq C$ implies that $[Y, R] \leq S$, and thus $w^l \in S \leq I_{P'}(S, G)$. Since $I_{P'}(S, G)$ is P'-isolated, $w \in I_{P'}(S, G)$. Secondly, $R \underset{P}{\sim} S$ since $\overline{S} \geq R$ by hypothesis. Invoking Theorem 5.5 again gives $[X, R] \underset{P}{\sim} [X, S]$. Since $S \trianglelefteq G$, we have $[X, S] \leq S$. Hence, $\overline{[X, S]} \leq \overline{S}$, and consequently, $[X, R] \leq \overline{S}$. This proves the claim. We therefore have that $[X, R] \leq S$ and $x \in C$, as required.

(iv) Since $R \trianglelefteq G$, we have that $C \trianglelefteq G$. By the Third Isomorphism Theorem,

$$G/C \cong \frac{G/S}{C/S} = \frac{N_{G/S}(R/S)}{C_{G/S}(R/S)}.$$

Thus, G/C is isomorphic to a subgroup of $Aut(R/S)$ by Theorem 1.3, and consequently, G/C is finite because R/S is assumed to be finite. Furthermore, $R \leq \overline{S}$. By (iii), C is P'-isolated in G. Therefore, G/C is finite but P'-torsion-free; that is, G/C is a finite P-torsion group. $\qquad\square$

The next theorem is a major result on isolators.

Theorem 5.6 (V. M. Gluškov) *If G is a locally nilpotent group and $H \leq G$, then*

$$I_P(N_G(H), G) \leq N_G(I_P(H, G)).$$

Equality holds if G is finitely generated, and thus nilpotent.

Proof Set $N = N_G(H)$. By Proposition 1.1 (ii), $[N, H] \leq H$ because $H \trianglelefteq N$. By Corollary 5.3,

$$[\overline{N}, \overline{H}] \leq \overline{[N, H]} \leq \overline{H}.$$

This implies that $\overline{N} \leq N_G(\overline{H})$ again by Proposition 1.1 (ii).

Suppose that G is a finitely generated nilpotent group, and let $x \in N_G(\overline{H})$. We must show that $x \in \overline{N}$. Since G is finitely generated, we may assume that $G = gp(x, \overline{N})$. By the previous paragraph, $\overline{N} \leq N_G(\overline{H})$. Thus, $G = N_G(\overline{H})$, and consequently, $\overline{H} \trianglelefteq G$. Moreover, $[\overline{H} : H] = m$ is a P-number by Theorems 2.24 and 5.3. By Corollary 2.6, we have $\overline{H}^m \leq H$. In fact, $\overline{H}^m \trianglelefteq G$ because $\overline{H} \trianglelefteq G$.

Let C denote the centralizer of $\overline{H}/\overline{H}^m$ in G. Observe that $\overline{H}/\overline{H}^m$ is a finite P-torsion group. By Lemma 5.7 (iv), G/C is a finite P-torsion group. Thus, for every $gC \in G/C$, there exists a P-number l such that $(gC)^l = C$; that is, $g^l \in C$. We conclude that $G = \overline{C}$. However, C normalizes H. For if $c \in C$ and $h \in H$, then

$$[c, h] \in \overline{H}^m \leq H.$$

And so, $c^{-1}h^{-1}c \in H$. Hence, $C \leq N$, and thus $\overline{N} = G$. Since $G = gp(x, \overline{N})$, we conclude that $x \in \overline{N}$. $\qquad\square$

Corollary 5.4 *Let G be a locally nilpotent group. If H is a P-isolated subgroup of G, then so is $N_G(H)$.*

Proof Let $g \in G$, and suppose that there is a P-number n such that $g^n \in N_G(H)$. Then $g \in N_G(\overline{H})$. By Theorem 5.6, $g \in N_G(\overline{H})$. Since H is P-isolated in G, we have $\overline{H} = H$. And so, $g \in N_G(H)$.   □

5.1.4   P-Isolators and Transfinite Ordinals

In the definitions that follow, certain group theoretic notions are extended to the transfinite case.

Definition 5.7 Let H be a subgroup of a group G. We define the *sequence of successive normalizers* of H as follows:

1. $H_0 = H$.
2. If α is a successor ordinal, then $H_\alpha = N_G(H_{\alpha-1})$.
3. If α is a limit ordinal, then $H_\alpha = \cup_{\beta<\alpha} H_\beta$.

We call H_α the α*th normalizer* of H.

Definition 5.8 Let G be a group with subgroups H and K. If $H \leq K$, then we say that H is an *ascendant subgroup* of K if there is a series of subgroups $\{H_\alpha\}_{\alpha \leq \beta}$ satisfying:

1. $H_0 = H$ and $H_\beta = K$;
2. For each successor ordinal $\alpha < \beta$, $H_{\alpha-1} \triangleleft H_\alpha$;
3. If β is a limit ordinal, then $H_\beta = \cup_{\alpha<\beta} H_\alpha$.

Note that if β is a finite ordinal, then H is subnormal in K.

Lemma 5.8 *Let G be a locally nilpotent group with subgroups H and K such that $H \leq K$. For each ordinal α, let H_α be the αth normalizer of H.*

(i) If $H = \overline{H}$, then $H_\alpha = \overline{H}_\alpha$.
(ii) If H is an ascendant subgroup of K, then $\overline{H}$ is an ascendant subgroup of $\overline{K}$.

Proof

(i) The proof is done by transfinite induction on α. Suppose that $H = \overline{H}$. If $\alpha = 0$, then $H_0 = \overline{H}_0$ by hypothesis. Assume that the result holds for all ordinals $\mu < \alpha$.

   • If α is a successor ordinal and $H_{\alpha-1} = \overline{H}_{\alpha-1}$, then

$$\overline{H}_\alpha = \overline{N_G(H_{\alpha-1})} \leq N_G(\overline{H}_{\alpha-1}) = N_G(H_{\alpha-1}) = H_\alpha$$

   by Theorem 5.6. Thus, $H_\alpha = \overline{H}_\alpha$.

- Suppose that α is a limit ordinal, and assume that $H_\mu = \overline{H}_\mu$ for all $\mu < \alpha$. We claim that $H_\alpha = \overline{H}_\alpha$. If $x \in \overline{H}_\alpha$, then there exists a P-number m such that $x^m \in H_\alpha$. Thus, $x^m \in H_\mu$ for some $\mu < \alpha$. Since $H_\mu = \overline{H}_\mu$, $x \in H_\mu$. Therefore, $x \in H_\alpha$ and $H_\alpha = \overline{H}_\alpha$ as claimed.

(ii) Since H is an ascendant subgroup of K, there exists a series $\{H_\alpha\}_{\alpha<\beta}$ satisfying the conditions of Definition 5.8. We claim that the series $\{\overline{H}_\alpha\}_{\alpha<\beta}$ satisfies Definition 5.8 for $\overline{H}$ and $\overline{K}$. Clearly, $\overline{H}_0 = \overline{H}$, $\overline{H}_\beta = \overline{K}$, and $\overline{H}_\alpha \leq \overline{H}_{\alpha+1}$ for ordinals β and α. By Theorem 5.6 and the fact that $H_\alpha \lhd H_{\alpha+1}$, we have

$$N_G\left(\overline{H}_\alpha\right) \geq \overline{N_G(H_\alpha)} \geq \overline{H}_{\alpha+1}.$$

Thus, $\overline{H}_\alpha \lhd \overline{H}_{\alpha+1}$.

For a limit ordinal μ, we let $x \in \overline{H}_\mu$. There exists a P-number m such that $x^m \in H_\mu$. Thus, $x^m \in H_\alpha$ for some $\alpha < \mu$. Hence, $x \in \overline{H}_\alpha$ and $\overline{H}_\mu = \cup_{\alpha<\mu}\overline{H}_\alpha$. Consequently, $\overline{H}$ is an ascendant subgroup of $\overline{K}$. $\qquad\square$

The definition of the upper central series may be generalized to the transfinite case.

Definition 5.9 The *transfinite upper central series* of a group G is the ascending series of subgroups $\{\zeta_\alpha G\}$ defined by

1. $\zeta_0 G = 1$,
2. $\zeta_\alpha G/\zeta_{\alpha-1}G = Z\left(G/\zeta_{\alpha-1}G\right)$ for all successor ordinals α, and
3. $\zeta_\alpha G = \bigcup_{\beta<\alpha} \zeta_\beta G$ whenever α is a limit ordinal.

Each term of the transfinite upper central series has cardinality at most the cardinality of G. Consequently, there exists a least ordinal μ such that

$$\zeta_\mu G = \zeta_{\mu+1}G = \zeta_{\mu+2}G = \cdots.$$

The subgroup $\zeta_\mu G$ is called the *hypercenter* of G. If G coincides with some term of its transfinite upper central series, then it is called a *hypercentral group* (or *ZA-group*).

Clearly, every nilpotent group is hypercentral. On the other hand, not every hypercentral group is nilpotent. For example, let $gp(x)$ be a cyclic group of order 2. The semi-direct product

$$D_{2^\infty} = gp(x) \ltimes \mathbb{Z}_{2^\infty},$$

referred to as the *locally dihedral 2-group*, is hypercentral [30] but not nilpotent.

The next lemma deals with quotients of a P-torsion-free group by its (transfinite) upper central subgroups. Part of its proof is contained in the proof of Corollary 2.20. See [23] for complete details.

Lemma 5.9 (D. H. McLain) *Let G be any group. If G is P-torsion-free, then $G/\zeta_\alpha G$ is P-torsion-free for all ordinals $\alpha \geq 1$.*

Lemma 5.10 *Suppose that G is a P-torsion-free locally nilpotent group. If $H \leq G$ and $\overline{H} = G$, then $\zeta_\beta H = \zeta_\beta\left(\overline{H}\right) \cap H$ for any ordinal β.*

Proof The proof is done by transfinite induction. The result is clear for $\beta = 0$. Assume that the lemma holds for all ordinals $\alpha < \beta$.

- Suppose that β is a successor ordinal. By the induction hypothesis, we have $\zeta_{\beta-1}H = \zeta_{\beta-1}G \cap H$. Thus,

$$\left[\zeta_\beta G \cap H, \, H\right] \leq \left[\zeta_\beta G, \, G\right] \cap H \leq \zeta_{\beta-1}G \cap H = \zeta_{\beta-1}H,$$

and hence, $\zeta_\beta G \cap H \leq \zeta_\beta H$. On the other hand, Corollary 5.3 gives

$$\left[\zeta_\beta H, \, G\right] = \left[\zeta_\beta H, \, \overline{H}\right] \leq \overline{\left[\zeta_\beta H, \, \overline{H}\right]} \leq \overline{\left[\zeta_\beta H, \, H\right]} \leq \overline{\zeta_{\beta-1}H} \leq \overline{\zeta_{\beta-1}G} = \zeta_{\beta-1}G,$$

where the last equality is a consequence of Lemmas 5.1 and 5.9. Therefore, $\zeta_\beta H \leq \zeta_\beta G \cap H$. We conclude that $\zeta_\beta H = \zeta_\beta G \cap H$.
- If β is a limit ordinal, then

$$\zeta_\beta H = \bigcup_{\alpha<\beta} \zeta_\alpha H = \bigcup_{\alpha<\beta}(\zeta_\alpha G \cap H) = \zeta_\beta G \cap H.$$

This completes the proof.   □

Other results on isolators can be found in [10] and [18].

5.2   Extraction of Roots

The study of groups with unique roots dates back to the 1940s, when B. H. Neumann [24] showed that every group can be embedded in a group in which roots exist, but are not necessarily unique. Shortly afterwards, Mal'cev [22] proved that any torsion-free nilpotent group can be embedded in a nilpotent group of the same class in which roots not only exist, but are uniquely defined. Other major contributors toward the development of the theory of extraction of roots include Kontorovič [16] and Černikov (see [7] and [8]).

A major paper on root extraction is due to G. Baumslag [2], where he constructs the "freest possible" group in which roots exist and are uniquely defined. This is the so-called *free $\mathscr{D}$-group*. More recently, S. Majewicz and M. Zyman studied root extraction in nilpotent R-powered groups where R is a binomial domain (see [20] and [21]).

The purpose of this section is to acquaint the reader with some of the theory of root extraction in groups. In particular, we focus on the extraction of roots in nilpotent groups.

5.2.1 $\mathscr{U}_P$-Groups

Definition 5.10 A group G is called a $\mathscr{U}_P$-*group* if every element of G has at most one nth root for every P-number n. If P is the set of all primes, then G is termed a $\mathscr{U}$-*group*.

If $P = \{p\}$, then we simply write $\mathscr{U}_p$ instead of $\mathscr{U}_{\{p\}}$. There are several equivalent formulations of Definition 5.10 as follows:

1. A group is a $\mathscr{U}_P$-group if every element of the group has at most one pth root for every prime $p \in P$.
2. If G is a $\mathscr{U}_P$-group and g, $h \in G$ satisfy $g^n = h^n$ for any P-number n, then $g = h$.
3. A group G is a $\mathscr{U}_P$-group if for each P-number n and $g \in G$, the equation $g = x^n$ has at most one solution.

If $g \in G$ has a unique nth root, then we write it as $g^{1/n}$.

Proposition 5.1

 (i) A subgroup of a $\mathscr{U}_P$-group is a $\mathscr{U}_P$-group.
 (ii) Every P-torsion-free abelian group is a $\mathscr{U}_P$-group.
 (iii) Every $\mathscr{U}_P$-group is P-torsion-free. Equivalently, P-torsion groups cannot be $\mathscr{U}_P$-groups.

Proof

 (i) Let G be a $\mathscr{U}_P$-group, and let $H \leq G$. If $g \in H$ and n is a P-number, then either g has no nth root in G (nor in H) or $g^{1/n}$ exists in G. If $g^{1/n} \in H$, then g has a unique nth root in H; otherwise it has none. Thus, H is a $\mathscr{U}_P$-group.
 (ii) Suppose that g, $h \in G$ and $g^n = h^n$ for some P-number n. Then $\left(gh^{-1}\right)^n = 1$ because G is abelian. Since G is P-torsion-free, $gh^{-1} = 1$, and thus $g = h$.
 (iii) This part is the trivial half of Theorem 2.7. □

Remark 5.5 The converse of Proposition 5.1 (iii) is not true. For example, the group

$$G = \left\langle a, b \mid a^2 = b^2 \right\rangle$$

is torsion-free, but is not a $\mathscr{U}$-group.

Lemma 5.11 *Suppose that G and H are $\mathscr{U}_P$-groups and $\psi \in Hom(G, H)$. If g has an nth root in G, then $\psi(g)$ has an nth root in H.*

Proof Observe that

$$\psi(g) = \left(\psi\left(g^{1/n}\right)\right)^n.$$

Hence, $\psi\left(g^{1/n}\right)$ is the unique nth root of $\psi(g)$. □

In particular, if G is a $\mathscr{U}_P$-group and $g \in G$ has an nth root for some P-number n, then every conjugate of g also has an nth root.

Let G be any group and suppose that $g,\ h \in G$. If $[g,\ h] = 1$, then $[g^m,\ h^n] = 1$ for any $m,\ n \in \mathbb{Z}$. The converse holds for $\mathscr{U}_P$-groups when m and n are P-numbers.

Lemma 5.12 (P. G. Kontorovič) *Suppose that G is a $\mathscr{U}_P$-group and $g,\ h \in G$. If m and n are P-numbers and $[g^m,\ h^n] = 1$, then $[g,\ h] = 1$.*

Proof If $[g^m,\ h^n] = 1$, then

$$h^{-n}gh^n = h^{-n}(g^m)^{1/m}h^n = (h^{-n}g^m h^n)^{1/m} = (g^m)^{1/m} = g.$$

Hence, $[g,\ h^n] = 1$. Thus,

$$g^{-1}hg = g^{-1}(h^n)^{1/n}g = \left(g^{-1}h^n g\right)^{1/n} = (h^n)^{1/n} = h.$$

And so, $[g,\ h] = 1$. □

Corollary 5.5 *Let G be a $\mathscr{U}_P$-group and suppose that g_1 and g_2 are elements in G which have nth roots for some P-number n. If g_1 and g_2 commute, then $g_1^{1/n}g_2^{1/n}$ is the nth root of $g_1 g_2$.*

Proof Since g_1 and g_2 commute, their nth roots $g_1^{1/n}$ and $g_2^{1/n}$ also commute by Lemma 5.12. Thus,

$$\left(g_1^{1/n}g_2^{1/n}\right)^n = \left(g_1^{1/n}\right)^n \left(g_2^{1/n}\right)^n = g_1 g_2.$$

Therefore, the nth root of $g_1 g_2$ is $g_1^{1/n}g_2^{1/n}$. □

Lemma 5.13 *A direct product of a family of $\mathscr{U}_P$-groups is a $\mathscr{U}_P$-group.*

Proof Let I be a nonempty index set, and suppose that $(G_i)_{i \in I}$ is a family of $\mathscr{U}_P$-groups. Let G be the direct product of the G_i, and pick an element $g = (g_i)_{i \in I}$ in G. Let n be a P-number.

- If g_i has no nth root for some $i \in I$, then g has no nth root.
- If g_i has a unique nth root for every $i \in I$, then so does g:

$$g^{1/n} = \left(g_i^{1/n}\right)_{i \in I}.$$

Therefore, G is a $\mathscr{U}_P$-group. □

Remark 5.6 The (standard) wreath product of two $\mathscr{U}_P$-groups is again a $\mathscr{U}_P$-group. This was proven by G. Baumslag (see Theorem 18.1 of [2]). He also proved in [3] (Corollary 4.7) that the unrestricted wreath product of a nontrivial $\mathscr{U}$-group G by a $\mathscr{U}$-group H is a $\mathscr{U}$-group if and only if H is a torsion group.

5.2.2 $\mathscr{U}_P$-Groups and Quotients

$\mathscr{U}_P$-groups and P-isolated subgroups are naturally related to each other.

Lemma 5.14 *Let G be any group and $N \trianglelefteq G$. If G/N is a $\mathscr{U}_P$-group, then N is P-isolated in G.*

Proof The result follows from Proposition 5.1 (iii) and Lemma 5.1. □

Corollary 5.6 *If G is a nilpotent group and $N \trianglelefteq G$, then N is P-isolated in G if and only if G/N is a $\mathscr{U}_P$-group.*

Proof By Theorem 2.7, a nilpotent group is a $\mathscr{U}_P$-group if and only if it is P-torsion-free. Apply Lemmas 5.1 and 5.14. □

The quotient of a $\mathscr{U}_P$-group by its center is also a $\mathscr{U}_P$-group. The next lemma which is very useful in its own right, will be needed to establish this.

Lemma 5.15 *If G is a $\mathscr{U}_P$-group and S is a nonempty subset of G, then $C_G(S)$ is P-isolated in G. In particular, $Z(G)$ is P-isolated in G.*

Proof Suppose that $g^n \in C_G(S)$ for some P-number n. If $s \in S$, then $s^{-1}g^n s = g^n$ and thus $\left(s^{-1}gs\right)^n = g^n$. Since G is a $\mathscr{U}_P$-group, $s^{-1}gs = g$. And so, $g \in C_G(S)$. □

Corollary 5.7 *If G is a $\mathscr{U}_P$-group, then $G/Z(G)$ is also a $\mathscr{U}_P$-group. Thus, $Inn(G)$ is also a $\mathscr{U}_P$-group.*

Proof Suppose that $(gZ(G))^n = (hZ(G))^n$ for some P-number n and elements $gZ(G)$ and $hZ(G)$ in $G/Z(G)$. There exists $z \in Z(G)$ such that $g^n = h^n z$. Thus, g^n and h^n commute. By Lemma 5.12, g and h also commute. Therefore, $g^n = h^n z$ can be expressed as $\left(gh^{-1}\right)^n = z$. Since $Z(G)$ is P-isolated in G by Lemma 5.15, gh^{-1} is contained in $Z(G)$. Thus, $gZ(G) = hZ(G)$. The last statement follows from Corollary 1.1. □

Remark 5.7 In general, the quotient of a $\mathscr{U}_P$-group need not be a $\mathscr{U}_P$-group. For example, the additive group $\mathbb{Z}$ is torsion-free abelian and thus a $\mathscr{U}$-group by Proposition 5.1 (ii). However, the quotient $\mathbb{Z}/n\mathbb{Z}$, being isomorphic to $\mathbb{Z}_n$, is a finite cyclic group for any $n \in \mathbb{N}$. Such groups are not $\mathscr{U}$-groups by Proposition 5.1 (iii).

The converse of Corollary 5.7 is true for P-torsion-free groups.

Corollary 5.8 *If G is a P-torsion-free group and $G/Z(G)$ is a $\mathscr{U}_P$-group, then G is a $\mathscr{U}_P$-group.*

Proof Suppose that $g^n = h^n$ for some g, $h \in G$ and P-number n. In the factor group $G/Z(G)$, we have $(gZ(G))^n = (hZ(G))^n$. Thus, $gZ(G) = hZ(G)$ because $G/Z(G)$ is assumed to be a $\mathscr{U}_P$-group. Hence, there exists $z \in Z(G)$ such that $g = hz$. This implies that $g^n = h^n z^n$. Since $g^n = h^n$, we have $z^n = 1$, and consequently, $z = 1$ because G is P-torsion-free. Therefore, $g = h$. □

5.2.3 P-Torsion-Free Locally Nilpotent Groups

By Theorem 2.7, a nilpotent group is P-torsion-free if and only if it is a $\mathscr{U}_P$-group. This is also the case for locally nilpotent groups.

Theorem 5.7 (S. N. Černikov, A. I. Mal'cev) *A locally nilpotent group is a $\mathscr{U}_P$-group if and only if it is P-torsion-free.*

Proof Suppose that G is a P-torsion-free locally nilpotent group and $g^n = h^n$ for some g, $h \in G$ and P-number n. The subgroup $K = gp(g, h)$ of G is finitely generated (hence, nilpotent) and P-torsion-free. By Theorem 2.7, K is a $\mathscr{U}_P$-group. Hence, $g = h$ and G is a $\mathscr{U}_P$-group. The converse is just Proposition 5.1 (iii). □

Corollary 5.9 *Let G be a locally nilpotent group. A normal subgroup N of G is P-isolated in G if and only if G/N is a $\mathscr{U}_P$-group.*

Proof The result follows from Theorem 5.7, together with Lemmas 5.1 and 5.14. □

Corollary 5.10 *If G is a P-torsion-free (locally) nilpotent group, then $G/Z(G)$ is also P-torsion-free (locally) nilpotent.*

Proof This is a consequence of Theorems 2.7 and 5.7, together with Corollary 5.7 and Proposition 5.1 (iii). □

Corollary 5.11 *Let G be a P-torsion-free locally nilpotent group. If S is a nonempty subset of G, then $C_G(S)$ is P-isolated in G. In particular, $Z(G)$ is P-isolated in G.*

Proof The result is immediate from Theorem 5.7 and Lemma 5.15. □

5.2.4 Upper Central Subgroups of $\mathscr{U}_P$-Groups

Theorem 5.8 *Let G be a $\mathscr{U}_P$-group. For each ordinal $\alpha \geq 0$,*

 (i) *$\zeta_{\alpha+1}G$ is P-isolated in G;*
 (ii) *$G/\zeta_{\alpha+1}G$ is a $\mathscr{U}_P$-group (hence, P-torsion-free);*
 (iii) *$\zeta_{\alpha+1}G/\zeta_\alpha G$ is an abelian $\mathscr{U}_P$-group (hence, P-torsion-free).*

Proof The proof is done by transfinite induction on α. If $\alpha = 0$, then Lemma 5.15, Corollary 5.7, and Proposition 5.1 (i) and (iii) establish all three statements. Assume that the theorem holds for all $\alpha < \beta$.

- If $\beta - 1$ exists, then $G/\zeta_{\beta-1}G$ is a $\mathcal{U}_P$-group by (ii) above. By Lemma 5.15,

$$Z(G/\zeta_{\beta-1}G) = \zeta_\beta G/\zeta_{\beta-1}G$$

is P-isolated in $G/\zeta_{\beta-1}G$. Thus, (i) follows from Theorem 5.1. By Corollary 5.7, $G/\zeta_\beta G$ is a $\mathcal{U}_P$-group since

$$G/\zeta_\beta G \cong \frac{G/\zeta_{\beta-1}G}{Z(G/\zeta_{\beta-1})}$$

by the Third Isomorphism Theorem. Moreover, $G/\zeta_\beta G$ is P-torsion-free by Proposition 5.1 (iii). This establishes (ii). To prove (iii), note that $\zeta_{\alpha+1}G/\zeta_\alpha G$ is a $\mathcal{U}_P$-group since it is a subgroup of a $\mathcal{U}_P$-group and thus Proposition 5.1 (i) applies. We again invoke Proposition 5.1 (iii).

- If β is a limit ordinal, then

$$\zeta_\beta G = \bigcup_{\alpha < \beta} \zeta_\alpha G,$$

where $\zeta_\alpha G$ is P-isolated in G. Thus, $\zeta_\beta G$ is P-isolated in G by Lemma 5.4 (ii). Suppose that $\left(g\zeta_\beta G\right)^n = \left(h\zeta_\beta G\right)^n$ for some P-number n and g, $h \in G$. Then $g^n = h^n z$ for some $z \in \zeta_\beta G$. Thus, $z \in \zeta_\alpha G$ for some $\alpha < \beta$. Therefore, we have $\left(g\zeta_\alpha G\right)^n = \left(h\zeta_\alpha G\right)^n$. Since $G/\zeta_\alpha G$ is a $\mathcal{U}_P$-group, we obtain $g\zeta_\alpha G = h\zeta_\alpha G$, and consequently, $g\zeta_\beta G = h\zeta_\beta G$. This gives (ii), and (iii) follows from Proposition 5.1. □

Corollary 5.12 *Let G be a $\mathcal{U}_P$-group.*

1. *If G is nilpotent of class c, then for each $i = 0, 1, \ldots, c - 1$,*

   (i) $\zeta_{i+1}G$ *is P-isolated in G;*
   (ii) $G/\zeta_{i+1}G$ *is a nilpotent $\mathcal{U}_P$-group (equivalently, P-torsion-free);*
   (iii) $\zeta_{i+1}G/\zeta_i G$ *is an abelian $\mathcal{U}_P$-group (equivalently, P-torsion-free).*

2. *If G is locally nilpotent, then for each ordinal $\alpha \geq 0$,*

   (i) $\zeta_{\alpha+1}G$ *is P-isolated in G;*
   (ii) $G/\zeta_{\alpha+1}G$ *is a locally nilpotent $\mathcal{U}_P$-group (equivalently, P-torsion-free);*
   (iii) $\zeta_{\alpha+1}G/\zeta_\alpha G$ *is an abelian $\mathcal{U}_P$-group (equivalently, P-torsion-free).*

Proof The result follows at once from Theorems 2.7, 5.7, and 5.8. □

5.2.5   Extensions of $\mathcal{U}_P$-Groups

Lemma 5.16 *A central extension of a $\mathcal{U}_P$-group by a $\mathcal{U}_P$-group is a $\mathcal{U}_P$-group.*

Proof Suppose that G is a central extension of a $\mathcal{U}_P$-group Q by a $\mathcal{U}_P$-group N. Let g and h be elements of G such that $g^n = h^n$ for some P-number n. Passing to the quotient G/N, we obtain $(gN)^n = (hN)^n$. Since G/N is isomorphic to the $\mathcal{U}_P$-group Q, G/N is also a $\mathcal{U}_P$-group. This means that $gN = hN$ and thus $g = ha$ for some $a \in N$. Hence,

$$g^n = (ha)^n = h^n a^n$$

because $N \le Z(G)$. Therefore $h^n = h^n a^n$ and thus $a^n = 1$. Since N is a $\mathcal{U}_P$-group, $a = 1$. And so, $g = h$.   □

Lemma 5.17 *A locally nilpotent group which is an extension of a $\mathcal{U}_P$-group by a $\mathcal{U}_P$-group is itself a $\mathcal{U}_P$-group.*

Proof The result follows from Lemma 2.13 and Theorem 5.7.   □

In [2], G. Baumslag showed that an extension of a $\mathcal{U}_P$-group by a $\mathcal{U}_P$-group is not necessarily a $\mathcal{U}_P$-group. For example, let

$$B = \left\langle b_1, b_2, \ldots \mid b_{i+1}^2 = b_i, \ [b_i, b_j] = 1, \ i = 1, 2, \ldots, j = 1, 2, \ldots \right\rangle.$$

Clearly, B is a $\mathcal{U}_2$-group. Let A be the unrestricted direct product of $|B|$ copies of B indexed by the elements of B, where $|B|$ represents the cardinality of B. For each $b \in B$, we denote the bth copy of B by B_b. For every $a \in A$, we let $a_b \in B_b$ denote the bth component of a. By Lemma 5.13, A is also a $\mathcal{U}_2$-group. Define G to be the unrestricted wreath product of B by B :

$$G = B \wr B = \left\langle A, B \ \middle| \ (b')^{-1} a_b b' = a_{bb'} \ (a, b, b' \in B) \right\rangle$$

Thus, G is an extension of the $\mathcal{U}_2$-group B by the $\mathcal{U}_2$-group A.

We assert that G is not a $\mathcal{U}_2$-group. Choose an element $b \ne 1$ in B, and consider the element a^* of A whose component in B_{b^n} is the isomorphic image of $a^{(-1)^n}$ in B_{b^n}, where $1 \ne a \in B$, and whose components in all other $B_{b'}$ ($b' \in B$) are equal to the identity. Noting that $b^{-1} a^* b a^* = (a^*)^b a^*$, we have that the b^nth component of $(a^*)^b$ is $a_{b^n}^{(-1)^{n+1}}$, while the b^nth component of a^* is $a_{b^n}^{(-1)^n}$. Clearly,

$$a_{b^n}^{(-1)^{n+1}} a_{b^n}^{(-1)^n} = 1 \in B_{b^n}.$$

Thus, if we put $g = ba^*$, then

$$g^2 = ba^* ba^* = b^2 \left(b^{-1} a^* b a^* \right) = b^2.$$

However, $g = ba^* \ne b$ since $a^* \ne 1$. This proves the assertion.

5.2.6 P-Radicable and Semi-P-Radicable Groups

Definition 5.11 A group G is called a *P-radicable group* (or an $\mathcal{E}_P$-*group*) if every element of G has at least one nth root in G for every P-number n. Equivalently, G is P-radicable if any of the following holds:

1. Every element of G has at least one pth root in G for every prime $p \in P$;
2. For each P-number n and $g \in G$, the equation $g = x^n$ has at least one solution.

If P is the set of all primes, then G is called a *radicable group* (also referred to as an $\mathcal{E}$-*group* or a *complete group*). If $P = \{p\}$, then G is *p-radicable*.

Lemma 5.18 *A homomorphic image of a P-radicable group is P-radicable.*

Proof Let G be a P-radicable group, and let $\varphi \in Hom(G, H)$ for some group H. If n is a P-number and $g \in G$, then $\varphi(g)$ has an nth root in $\varphi(G)$. Indeed, there exists $h \in G$ such that $h^n = g$ because G is P-radicable. Thus,

$$\varphi(g) = \varphi(h^n) = (\varphi(h))^n,$$

and consequently, $\varphi(h)$ is an nth root of $\varphi(g)$ in $\varphi(G)$. $\qquad\square$

Remark 5.8 In general, a proper subgroup of a P-radicable group need not be P-radicable. For instance, the additive group $\mathbb{Q}$ is divisible, but the subgroup $\mathbb{Z}$ of $\mathbb{Q}$ is not. See Lemma 5.24.

Lemma 5.19 *A direct product of a family of P-radicable groups is P-radicable.*

Proof Let I be a nonempty index set. Let $(G_i)_{i \in I}$ be a family of P-radicable groups and suppose that G is a direct product of the G_i. Choose an element $g = (g_i)_{i \in I}$ in G. If n is any P-number, then there exists $h_i \in G_i$ such that $h_i^n = g_i$ for each $i \in I$. Thus,

$$g = (h_i^n)_{i \in I} = (h_i)_{i \in I}^n.$$

Hence, $(h_i)_{i \in I}$ is an nth root of g. $\qquad\square$

Remark 5.9 The wreath product of two P-radicable groups is not always P-radicable. However, one can obtain a P-radicable wreath product by imposing certain conditions on the given groups. For instance, if G and H are p-radicable groups for some prime p and G is nontrivial, then the unrestricted wreath product of G by H is p-radicable if and only if H does not contain elements of order p. This and related results can be found in [3].

Lemma 5.20 *If G is a P-radicable group and $\tau(G) \leq G$, then $\tau(G)$ is P-radicable.*

Proof Let $1 \neq g \in \tau(G)$ such that $g^n = 1$ for some $n \in \mathbb{N}$. If m is any P-number, then there is an element $h \in G$ such that $g = h^m$. Thus, $h^{mn} = 1$, so $h \in \tau(G)$. $\qquad\square$

A finite group cannot be radicable unless it is trivial.

Lemma 5.21 *Every nontrivial radicable group is infinite.*

Proof Assume that there exists a nontrivial finite radicable group G of order n. If $1 \neq g \in G$, then there exists $h \in G$ such that $g = h^n$. Since $|G| = n$, $h^n = 1$. This implies that $g = 1$, a contradiction.   □

In contrast to Lemma 5.21, a finite p-torsion-free group is always p-radicable for any prime p.

Lemma 5.22 *Every finite p-torsion-free group is p-radicable.*

Proof Let G be a finite p-torsion-free group with exponent n. Clearly, $n \neq p$ because G is p-torsion-free. Furthermore, p does not divide n. To see this, suppose that $n = pm$ for some $m \in \mathbb{N}$. If $1 \neq g \in G$, then

$$1 = g^n = g^{pm} = (g^m)^p .$$

Since G is p-torsion-free, it must be the case that $g^m = 1$. This contradicts the assumption that G has exponent n.

Now, since n and p are relatively prime, there are integers r and s such that $rn + sp = 1$. If $g \in G$, then

$$g = g^{rn+sp} = (g^r)^n (g^s)^p = (g^s)^p.$$

Therefore g has a pth root and thus G is p-radicable.   □

5.2.7   Extensions of P-Radicable Groups

Just as for $\mathcal{U}_P$-groups, an extension of a P-radicable group by a P-radicable group is not necessarily P-radicable. Counterexamples have been constructed using wreath products and are given by G. Baumslag in [3].

Lemma 5.23 *A central extension of a P-radicable group by a P-radicable group is P-radicable.*

Proof Let G be a central extension of a P-radicable group Q by another P-radicable group N. Let $g \in G$, and let n be a P-number. Since G/N is isomorphic to Q, G/N is also P-radicable. Thus, $gN \in G/N$ has an nth root in G/N, say hN. This means that $gN = (hN)^n = h^n N$. Hence, $g = h^n a$ for some $a \in N$. Now, a has an nth root in N, say a_0, since N is P-radicable. Therefore,

$$g = h^n a_0^n = (ha_0)^n$$

since $N \leq Z(G)$. Thus, g has an nth root in G.   □

The question of whether or not *P*-radicability is preserved under extensions for nilpotent groups will be answered in Corollary 5.16.

5.2.8 Divisible Groups

Abelian radicable groups are usually referred to as *divisible* and *P*-radicable abelian groups are called *P-divisible*. If G is an additively written *P*-divisible group, then

$$G = \{ng \mid g \in G\} = nG$$

for every *P*-number n. Equivalently, $G = pG$ for all primes $p \in P$.

Our intention here is to mention only those results on divisible groups which will be needed for our discussion on nilpotent *P*-radicable groups.

A divisible group which plays a major role in the study of abelian groups is the additive group of the rational numbers $\mathbb{Q}$. Clearly, $\mathbb{Q}$ is torsion-free.

Lemma 5.24 $\mathbb{Q}$ *is divisible.*

Proof Let p/q be an element of $\mathbb{Q}$. If $n \in \mathbb{N}$, then $p/qn \in \mathbb{Q}$ and $n(p/qn) = p/q$. Thus, p/qn is an nth root of p/q. □

Another important divisible group is the *p*-quasicyclic group $\mathbb{Z}_{p^\infty}$. Recall from Section 2.2 that for any prime p, the *p*-quasicyclic group (also referred to as the Prüfer *p*-group) is the group with additive presentation

$$\mathbb{Z}_{p^\infty} = \langle x_1, x_2, \ldots \mid px_1 = 0, \ px_{n+1} = x_n \ \text{for} \ n = 1, 2, \ldots \rangle. \tag{5.2}$$

It is clear from the presentation that $\mathbb{Z}_{p^\infty}$ is a *p*-group.

Lemma 5.25 $\mathbb{Z}_{p^\infty}$ *is divisible.*

Proof It suffices to exhibit a qth root for each generator x_i in the presentation (5.2), where q is any prime. If $q = p$, then we read off from (5.2) that x_{i+1} is the pth root of x_i for $i = 1, 2, \ldots$.

Suppose that $q \neq p$. A direct calculation shows that x_i has order p^i. Since p^i and q are relatively prime, there exist integers a and b such that $ap^i + bq = 1$. Thus,

$$x_i = \left(ap^i + bq\right)x_i = ap^i x_i + bqx_i = bqx_i = q\left(bx_i\right).$$

Hence, the qth root of x_i is bx_i. □

The next result is a fundamental structure theorem for divisible groups. A proof can be found in [15].

Theorem 5.9 *Let G be a divisible group.*

 (i) *If G is torsion-free, then it is a direct sum of isomorphic copies of $\mathbb{Q}$.*
 (ii) *If G is a torsion group, then it is a direct sum of p-quasicyclic groups for various primes p.*
 (iii) *If G is mixed, then it is a direct sum of isomorphic copies of $\mathbb{Q}$ and p-quasicyclic groups for various primes p.*

A finitely generated nontrivial abelian group can never be divisible. This is the point behind the next result.

Theorem 5.10 *No nontrivial divisible group is finitely generated.*

Proof Suppose that G is a nontrivial divisible group. Assume on the contrary, that G is finitely generated. By the Fundamental Theorem of Finitely Generated Abelian Groups, G is isomorphic to the direct product of a finite number of copies of $\mathbb{Z}$ and a finite abelian group. Clearly, no factors of $\mathbb{Z}$ can appear in G; otherwise, $\mathbb{Z}$ would be divisible, which is absurd. Hence, G is isomorphic to a finite abelian group, say H. Since G is divisible, Lemma 5.21 implies that H is trivial. We conclude that G must be trivial, a contradiction. $\square$

A proper subgroup of a divisible group is not necessarily divisible (see Remark 5.8). In particular, we have:

Lemma 5.26 $\mathbb{Q}$ *has no proper divisible subgroups.*

Proof Let $0 \neq H$ be a divisible subgroup of $\mathbb{Q}$. We claim that $H = \mathbb{Q}$. To begin with, we show that $\mathbb{Z}$ is a subgroup of H. Suppose that a/b is a nonzero element of H and assume without loss of generality, that $a > 0$. Then $a = b(a/b)$ is also contained in H. Since H is divisible, for any $n > 0$, there exists an element $h \in H$ such that $a = nh$. In particular, there exists an element $h \in H$ such that $a = ah$. This means that $1 \in H$ and consequently, $\mathbb{Z} \leq H$.

Next, suppose that $r/s \in \mathbb{Q}$. Clearly, $r \in H$ because $r \in \mathbb{Z}$. Since H is divisible, there exists an element $k \in H$ such that $r = sk$, where we may assume without loss of generality, that $k > 0$. It is evident that $k = r/s$. Thus, $r/s \in H$ and consequently, $H = \mathbb{Q}$. $\square$

5.2.9   Nilpotent P-Radicable Groups

In [7] and [8], S. N. Černikov made various contributions toward the development of the theory of hypercentral P-radicable groups. Our discussion will be restricted to nilpotent groups. The related results for hypercentral groups can be found in the papers of S. N. Černikov or the work of G. Baumslag [2].

Theorem 5.11 (S. N. Černikov) *If G is a P-radicable nilpotent group, then $Z(G)$ is P-isolated in G.*

Proof The proof is done by induction on the class c of G. The result is trivial when $c = 1$. Assume that it is true for $c > 1$. Let $1 \neq g \in G$, and suppose $g^n \in Z(G)$ for some P-number n. We claim that $g \in Z(G)$. By Lemmas 2.12 and 5.18, $G/Z(G)$ is a P-radicable nilpotent group of class $c - 1$. By induction, $(gZ(G))^n \in Z(G/Z(G))$ implies that $gZ(G) \in Z(G/Z(G))$ and thus $g \in \zeta_2 G$. Let $h \in G$ and suppose that $h_0^n = h$ for some $h_0 \in G$. Then

$$[g, h] = [g, h_0^n] = [g^n, h_0] = 1.$$

Therefore, $g \in Z(G)$.   □

Corollary 5.13 *If G is a P-radicable nilpotent group, then $G/Z(G)$ is a nilpotent $\mathscr{U}_P$-group (equivalently, P-torsion-free).*

Proof The result follows from Theorems 5.7 and 5.11, together with Lemma 5.1. □

Corollary 5.14 *If G is a P-radicable nilpotent group, then $\tau_P(G) \leq Z(G)$.*

Proof If $g \in \tau_P(G)$, then $g^n = 1$ for some P-number n. Hence, $g^n \in Z(G)$. Apply Theorem 5.11.   □

In the case that P is the set of all primes, we have:

Corollary 5.15 *If G is a radicable nilpotent group, then $\tau(G) \leq Z(G)$. Thus, every torsion radicable nilpotent group is abelian.*

Just as for certain $\mathscr{U}_P$-groups, the upper central subgroups of a P-radicable nilpotent group inherit various root properties from the group.

Theorem 5.12 *Suppose that G is a P-radicable nilpotent group of class c. For each $i = 0, 1, \ldots, c - 1$,*

(i) $G/\zeta_i G$ is P-radicable and nilpotent;
(ii) $\zeta_i G$ is P-radicable, nilpotent, and P-isolated in G;
(iii) $\zeta_{i+1} G/\zeta_i G$ is P-divisible.

Proof Lemma 5.18 immediately gives (i). The proofs of (ii) and (iii) are by induction on i.

- If $i = 0$, then (ii) is trivial. We prove (iii). By Theorem 5.11, $Z(G)$ is P-isolated in G. Let $g \in Z(G)$ and choose any P-number n. Since G is P-radicable, there exists $h \in G$ such that $h^n = g$. Thus, $h^n = Z(G)$ and consequently, $h \in Z(G)$. Thus, $Z(G)$ is P-radicable.
- Suppose that the theorem is true for $0 \leq i < j$. By Corollary 5.13, $G/Z(G)$ is a $\mathscr{U}_P$-group. Since

$$G/\zeta_j G \cong \frac{G/\zeta_{j-1} G}{Z(G/\zeta_{j-1} G)},$$

it follows by induction and Theorem 5.8 that $G/\zeta_j G$ is also $\mathcal{U}_P$-group. Therefore, $G/\zeta_j G$ is P-torsion-free by Proposition 5.1 (iii) and thus $\zeta_j G$ is P-isolated in G by Lemma 5.1. We need to show that $\zeta_j G$ is P-radicable. Suppose that $g \in \zeta_j G$ and n is a P-number. Since G is P-radicable, there exists $h \in G$ such that $g = h^n$. Furthermore, $h^n \in \zeta_j G$ and $\zeta_j G$ is P-isolated in G. And so, $h \in \zeta_j G$. Thus, $\zeta_j G$ is P-radicable and (ii) is proven. Now, $G/\zeta_{j-1} G$ is P-radicable by Lemma 5.18. By (ii) above, $Z(G/\zeta_{j-1} G) = \zeta_j G/\zeta_{j-1} G$ is P-radicable. This proves (iii).   □

Remark 5.10 Even though the center of a P-radicable nilpotent group is P-radicable by Theorem 5.12 (ii), a nilpotent group whose center is P-radicable is not necessarily P-radicable. For example, the semi-direct product

$$G = \left(\mathbb{Z}_{p^\infty} \oplus \mathbb{Z}_{p^\infty}\right) \rtimes_\varphi \mathbb{Z},$$

which is obtained by setting $A = \mathbb{Z}_{p^\infty}$ in Example 2.20, is a nilpotent group whose center is isomorphic to $\mathbb{Z}_{p^\infty}$, a divisible group. However, it is easy to see that G is not radicable.

Theorem 5.13 *If G is a P-radicable nilpotent group, then the terms of the lower central series of G are also P-radicable.*

Proof Suppose that G is of class c. We show first that the central subgroup $\gamma_c G$ is P-radicable. By definition,

$$\gamma_c G = gp\left([g, h] \mid g \in G, h \in \gamma_{c-1} G\right). \tag{5.3}$$

Let $[g, h]$ be one of the generators of $\gamma_c G$ in (5.3). If n is a P-number, then g has an nth root, say $\bar{g}$. By Lemma 1.13,

$$[g, h] = [\bar{g}^n, h] = [\bar{g}, h]^n.$$

Thus, $[g, h]$ has an nth root in $\gamma_c G$. Since $\gamma_c G$ is abelian, each of its elements has an nth root; that is, $\gamma_c G$ is P-radicable.

The rest of the proof is done by induction on the class c of G. The result is trivial if $c = 1$. Suppose that it is true for all P-radicable nilpotent groups of class less than $c > 1$. By Lemma 5.18, $G/\gamma_c G$ is a P-radicable group of class $c - 1$. By induction and Corollary 2.1, $\gamma_i(G/\gamma_c G) = \gamma_i G/\gamma_c G$ is P-radicable for $1 \leq i \leq c - 1$. Thus, $\gamma_i G$ is a central extension of a P-radicable group by another. By Lemma 5.23, $\gamma_i G$ is P-radicable.   □

In order to determine whether or not a given nilpotent group is P-radicable, one can inspect its abelianization. To prove this, we need a simple lemma.

Lemma 5.27 *If A is a P-divisible $\mathbb{Z}$-module, then so is $\bigotimes_{\mathbb{Z}}^k A$ for any $k \in \mathbb{N}$.*

Proof Let $a_1 \otimes \cdots \otimes a_k \in \bigotimes_{\mathbb{Z}}^k A$ and suppose that n is a P-number. Since A is P-divisible, there exists $b \in A$ such that $na_1 = b$. Then

$$n(a_1 \otimes \cdots \otimes a_k) = (na_1) \otimes \cdots \otimes a_k$$
$$= b \otimes \cdots \otimes a_k.$$

Thus, $a_1 \otimes \cdots \otimes a_k$ has an nth root in $\bigotimes_{\mathbb{Z}}^k A$.   □

Theorem 5.14 *A nilpotent group G is P-radicable if and only if $Ab(G)$ is a P-divisible group.*

Proof Suppose that G is of class c and $Ab(G)$ is P-divisible. We claim that $\gamma_c G$ is P-divisible. By Lemma 5.27, $\bigotimes_{\mathbb{Z}}^i Ab(G)$ is P-divisible for $i = 1, 2, \ldots$. By Corollary 2.10 and Lemma 5.18, each quotient $\gamma_i G/\gamma_{i+1} G$ is P-divisible. In particular, $\gamma_c G/\gamma_{c+1} G = \gamma_c G$ is P-divisible as claimed.

The rest of the proof is done by induction on c. If $c = 1$, then the result is evident. Suppose that it is true for all P-radicable nilpotent groups of class less than $c > 1$. By Lemma 2.8, $G/\gamma_c G$ has nilpotency class $c - 1$. Thus, $G/\gamma_c G$ is P-radicable by induction. Since $\gamma_c G$ is P-radicable and $\gamma_c G \leq Z(G)$, Lemma 5.23 proves the claim. The converse is immediate by Lemma 5.18.   □

Another type of group which arises in the study of extraction of roots is the so-called *semi-P-radicable group*.

Definition 5.12 A group G is called *semi-P-radicable* if $G = G^n$ for every P-number n. If P is the set of all primes, then G is called *semi-radicable* or *Černikov complete*.

Thus, G is semi-P-radicable if every element of G can be expressed in the form

$$g_1^n g_2^n \cdots g_k^n$$

for some elements $g_1, g_2, \ldots, g_k$ of G and any P-number n. Some elementary properties of semi-P-radicable groups are collected in the next lemma.

Lemma 5.28 *(i) Every P-radicable group is semi-P-radicable.*
(ii) Every semi-P-radicable abelian group is P-divisible.
(iii) Every homomorphic image of a semi-P-radicable group is semi-P-radicable.
(iv) Semi-P-radicability is preserved under extensions.

Proof

(i) The result is obvious.
(ii) Suppose that G is a semi-P-radicable abelian group, and let n be a P-number. If $g \in G$, then there exist elements $g_1, \ldots, g_k$ in G such that

$$g = g_1^n g_2^n \cdots g_k^n = (g_1 g_2 \cdots g_k)^n.$$

Therefore, g has an nth root and thus G is P-divisible.
(iii) Let $\varphi \in Hom(G, H)$, where G is semi-P-radicable and H is any group. Since $G = G^n$ for any P-number n, we have $\varphi(G) = \varphi(G^n) = (\varphi(G))^n$. Thus, $\varphi(G)$ is semi-P-radicable.

(iv) Let G be a group with $N \trianglelefteq G$ and suppose that N and G/N are semi-P-radicable. If $g \in G$, then there exist elements $g_1, \ldots, g_k \in G$ such that

$$gN = (g_1 N)^n (g_2 N)^n \cdots (g_k N)^n$$

for some P-number n. Hence, $g = g_1^n g_2^n \cdots g_k^n h$ for some $h \in N$. Since N is semi-P-radicable, there exist elements $h_1, \ldots, h_l \in G$ such that $h = h_1^n h_2^n \cdots h_l^n$. The result follows. $\qquad \square$

Remark 5.11 Not all semi-P-radicable groups are P-radicable. We present an example given in [30]. Suppose that $p \in P$ and $W = A \wr T$, where both A an T are isomorphic copies of $\mathbb{Z}_{p^\infty}$. Let

$$A = \left\langle a_1, a_2, \ldots \mid a_1^p = 1, a_2^p = a_1, a_3^p = a_2, \ldots \right\rangle$$

and

$$T = \left\langle t_1, t_2, \ldots \mid t_1^p = 1, t_2^p = t_1, t_3^p = t_2, \ldots \right\rangle$$

be presentations for A and T respectively. Then $W = B \rtimes T$, where $B = \prod_{t \in T} A^t$ and T acts on B by conjugation. Realize that this action induces a $\mathbb{Z}T$-module structure on the abelian group B. Since both B and T are clearly semi-p-radicable, W must also be semi-p-radicable. We prove that W is not p-radicable.

We argue that $t_1 a_1$ does not have a pth root in W. Suppose on the contrary that there exist elements $t \in T$ and $b \in B$ such that

$$(tb)^p = t_1 a_1.$$

A standard computation gives

$$t^p b^{t^{p-1} + t^{p-2} + \cdots + t + 1} = t_1 a_1.$$

Hence, $t^p = t_1$ and

$$b^{t^{p-1} + t^{p-2} + \cdots + t + 1} = a_1 \tag{5.4}$$

in B. For each $k = 0, \ldots, p-1$, let t^{kp} act on both sides of (5.4). This gives the following sequence of equations:

$$b^{t^{p-1} + t^{p-2} + \cdots + t + 1} = a_1$$

$$b^{t^{2p-1} + t^{2p-2} + \cdots + t^{p+1} + t^p} = a_1^{t^p}$$

$$b^{t^{3p-1} + t^{3p-2} + \cdots + t^{2p+1} + t^{2p}} = a_1^{t^{2p}}$$

$$\vdots$$

$$b^{t^{p^2-1} + t^{p^2-2} + \cdots + t^{(p-1)p+1} + t^{(p-1)p}} = a_1^{t^{(p-1)p}}.$$

Multiplying these equations together gives

$$b^{t^{p^2}-1+t^{p^2}-2+\cdots+t+1} = a_1^{t^{p^2}-p+t^{p^2}-2p+\cdots+t^{2p}+t^p+1} = d, \tag{5.5}$$

an element in B. By letting $t-1$ act across (5.5), we obtain

$$b^{t^{p^2}-1} = a_1^{t^{p^2}-p+1-t^{p^2}-p+t^{p^2}-2p+1-t^{p^2}-2p+\cdots+t^{2p+1}-t^{2p}+t^{p+1}-t^p+t-1} = d^{t-1}. \tag{5.6}$$

Since t has order p^2, $d^{t-1} = 1$. However, the exponent of a_1 in (5.6) is a nonzero element of the group ring $\mathbb{Z}T$. This gives a contradiction.

Theorem 5.15 *A nilpotent group G is semi-P-radicable if and only if it does not contain a proper normal subgroup of finite index equal to a P-number.*

Proof Suppose that G is *any* semi-P-radicable group. Assume on the contrary that G contains a proper normal subgroup N such that $[G : N] = n$, where n is a P-number. Then $(G/N)^n = N$ and thus G/N is not semi-P-radicable. On the other hand, G/N must be semi-P-radicable since it is a homomorphic image of G by Lemma 5.28. This gives a contradiction.

Conversely, suppose that G contains no proper subgroup of finite index equal to a P-number. Assume that G is not semi-P-radicable, so that $G^n \neq G$ for some P-number n. Since $G^n \trianglelefteq G$, G/G^n is nilpotent of order at most n. There are two cases to consider:

1. If G/G^n is abelian, then it is a direct sum of cyclic groups of prime order. Thus, there exists a normal subgroup H/G^n of G/G^n such that the quotient of G/G^n by H/G^n is cyclic of prime order. Since this quotient is isomorphic to G/H by the Third Isomorphism Theorem, H must be a proper subgroup of G with finite index equal to a P-number. This contradicts the assumption that G contains no proper subgroup of finite index equal to a P-number.
2. If G/G^n is not abelian, then $\zeta_2(G/G^n) \neq Z(G/G^n)$. By Theorem 2.30, there exists $\varphi \in Hom(G/G^n, Z(G/G^n))$ such that

$$\varphi(G/G^n) = K/G^n \neq G^n$$

for some subgroup K of G. Since K/G^n is abelian, it has a subgroup N/G^n of finite index equal to a P-number by the previous case. Thus, $\varphi^{-1}(N/G^n)$ has finite index equal to a P-number in G/G^n. Therefore, G has a subgroup of finite index equal to a P-number, a contradiction. □

A related result for the case when P is the set of all primes is:

Theorem 5.16 *A nilpotent group is semi-radicable if and only if it does not contain a proper subgroup of finite index.*

Notice that the normality condition is no longer required. The proof is similar to the one for Theorem 5.15, but uses Theorem 5.26.

Theorem 5.17 *Let G be a nilpotent group. The following are equivalent:*

(i) G has no proper normal subgroups of finite index equal to a P-number;
(ii) G is semi-P-radicable;
(iii) G is P-radicable.

Proof In light of Theorem 5.15 and the fact that every P-radicable group is semi-P-radicable, we only need to verify that (ii) implies (iii). If G is a semi-P-radicable nilpotent group, then so is $Ab(G)$. Since $Ab(G)$ is abelian, it is P-divisible. The result follows from Theorem 5.14.   □

We offer another proof of (ii) $\Rightarrow$ (iii) which can be found in [2]. The idea is essentially the same as the one given for Theorem 5.13. Let G be a semi-P-radicable nilpotent group of class c. We show first that $\gamma_c G$ is P-divisible. By definition,

$$\gamma_c G = gp\left([g, g'] \mid g \in G, \ g' \in \gamma_{c-1}G\right). \tag{5.7}$$

Choose a generator $h = [g, g']$ of $\gamma_c G$ in (5.7) and let n be a P-number. Since G is semi-P-radicable, we can write

$$g = g_1^n g_2^n \cdots g_k^n,$$

where $g_1, \ldots, g_k \in G$. Since $[g_i^n, g'] \in \gamma_c G \le Z(G)$ for each $i = 1, 2, \ldots, k$, Lemmas 1.4 (v) and 1.13 give

$$h = [g_1^n g_2^n \cdots g_k^n, g']$$
$$= ([g_1, g'][g_2, g'] \cdots [g_k, g'])^n.$$

It follows that h can be expressed as an nth power of an element in $\gamma_c G$. This procedure applies for all P-numbers n. Since $\gamma_c G$ is abelian, $\gamma_c G$ is P-divisible as claimed.

The rest of the proof is done by induction on c. The result is clearly true when $c = 1$. Suppose that $c > 1$ and assume that every semi-P-radicable nilpotent group of class less than c is P-radicable. By Lemma 5.28 (iii), $G/\gamma_c G$ is generated by its nth powers for any P-number n. Since $G/\gamma_c G$ is of class $c-1$, $G/\gamma_c G$ is P-radicable by induction. Therefore, G is P-radicable by Lemma 5.23.

Corollary 5.16 *A nilpotent group which is an extension of a P-radicable group by a P-radicable group is itself P-radicable.*

Proof The result is immediate from Lemma 5.28 (iv) and Theorem 5.17.   □

Remark 5.12 It follows from Lemma 5.18, Corollary 5.16, and Lemma 5.27 that P-radicability is property $\mathscr{P}$ for nilpotent groups (see Definition 2.15). This is related to Theorems 5.14 and 2.19.

5.2.10   The Structure of a Radicable Nilpotent Group

We restate Corollary 2.20 for the transfinite upper central series. A proof can be found in Theorem 2.25 of [29].

Theorem 5.18 (D. H. McLain) *Let G be any group. If $Z(G)$ is P-torsion-free, then $\zeta_{\alpha+1}G/\zeta_\alpha G$ is P-torsion-free for every ordinal α.*

The next theorem describes the structure for radicable nilpotent groups.

Theorem 5.19 (S. N. Černikov) *Let G be a radicable nilpotent group. There exists a well-ordered family of subgroups $\{A_\alpha \mid 0 \le \alpha \le \lambda\}$ of G for some ordinal λ such that:*

(1) A_0 is a central subgroup of G, as well as the direct product of p-quasicyclic groups for various primes p;
(2) A_α is isomorphic to $\mathbb{Q}$ whenever $0 < \alpha < \lambda$;
(3) if $B_0 = 1$ and $B_\beta = gp(A_\alpha \mid 0 \le \alpha < \beta)$, then B_β is normal in G, $B_\beta \cap A_\beta = 1$ for every $0 < \beta < \lambda$, and $B_\lambda = G$.

Proof We follow [17] and [30]. Set $A_0 = B_1 = \tau(G)$. By Lemma 5.20 and Corollary 5.15, A_0 is a divisible subgroup of $Z(G)$. It is also the direct product of p-quasicyclic groups for various primes p by Theorem 5.9 (ii).

Clearly, G/B_1 is torsion-free and nilpotent by Corollaries 2.5 and 2.15. By Theorems 5.12 (iii) and 5.18, the upper central factors of G/B_1 are torsion-free and divisible. Furthermore, each such factor is a direct sum of isomorphic copies of $\mathbb{Q}$ by Theorem 5.9 (i). Hence, one can refine the upper central series of G/B_1 and obtain an ascending transfinite central series

$$1 = B_0 < B_1 < \cdots < B_\lambda = G,$$

where $B_{\alpha+1}/B_\alpha \cong \mathbb{Q}$ for $\alpha > 0$. Thus, $B_{\alpha+1}/B_\alpha$ is torsion-free and divisible for $\alpha > 0$. By Corollary 5.16, each B_α is radicable.

We construct A_α. Let $g \in B_{\alpha+1} \setminus B_\alpha$ where $\alpha > 0$. Since $B_{\alpha+1}$ is radicable, it contains a set of elements $S = \{g_n \mid n \ge 1\}$ such that

$$g = g_1, \ g_1 = g_2^2, \ g_2 = g_3^3, \ \ldots, \ g_{m-1} = g_m^m, \ \ldots.$$

Put $A_\alpha = gp(S) \le B_{\alpha+1}$. Note that A_α is divisible. We claim that $B_\alpha \cap A_\alpha = 1$. Assume that there exists an element $1 \ne a \in B_\alpha \cap A_\alpha$. Since $a \in A_\alpha$, $a = g_k^r$ for some $r \in \mathbb{N}$ and $k \ge 1$. Hence, $g_k^r \in B_\alpha$ and thus $(g_k B_\alpha)^r = B_\alpha$ in $B_{\alpha+1}/B_\alpha$. Therefore, $g_k B_\alpha = B_\alpha$ because $B_{\alpha+1}/B_\alpha$ is torsion-free. This means that $g_k \in B_\alpha$. However, a computation shows that $g = g_k^{k!} \in B_\alpha$. This contradicts the assumption that $g \in B_{\alpha+1} \setminus B_\alpha$. Consequently, $B_\alpha \cap A_\alpha = 1$ as claimed. Thus, A_α is isomorphic to $gp(g_nB_\alpha \mid n \geq 1) = A_\alpha/B_\alpha$, which is a subgroup of $B_{\alpha+1}/B_\alpha \cong \mathbb{Q}$. Since A_α is also divisible, it must be isomorphic to $\mathbb{Q}$ itself by Lemma 5.26. Therefore,

$$B_{\alpha+1}/B_\alpha \cong \mathbb{Q} \cong A_\alpha.$$

And so, $B_{\alpha+1} = A_\alpha B_\alpha$ and $B_\alpha = gp(A_\gamma \mid 0 \leq \gamma < \alpha)$.   □

5.2.11   The Maximal P-Radicable Subgroup

Let G be a nilpotent group. If $\{H_i \mid i \in I\}$ is the collection of all semi-P-radicable subgroups of G, then $gp(H_i \mid i \in I)$ is also a semi-P-radicable subgroup of G. This, together with Theorem 5.17, implies that every nilpotent group has a unique maximal P-radicable subgroup for any nonempty set of primes P. This unique maximal P-radicable subgroup of G is denoted by $\varrho_P(G)$ (or $\varrho(G)$ when P is the set of all primes). If $P = \{p\}$, then we write $\varrho_p(G)$. Evidently, $\varrho_P(G)$ always exists since the trivial group is P-radicable.

A thorough study of the maximal P-radicable subgroup of a nilpotent group is given by R. B. Warfield, Jr. in [33]. We give only a brief survey of this work.

Definition 5.13 A group G is called *P-reduced* if it contains no nontrivial semi-P-radicable subgroups.

By Theorem 5.17, a P-reduced nilpotent group cannot have nontrivial P-radicable subgroups. Furthermore, if G is a nilpotent group, then $G/\varrho_P(G)$ is P-reduced by Corollary 5.16.

One can determine if a nilpotent group is P-reduced by examining its center.

Theorem 5.20 *Let G be a nilpotent group. Then G is P-reduced if and only if $Z(G)$ is P-reduced.*

Proof If G is P-reduced, then $Z(G)$ is clearly P-reduced. The converse is proven by contradiction. Suppose that $Z(G)$ is P-reduced and assume that G is not P-reduced. By Theorem 5.17, there exists a nontrivial P-radicable subgroup H of G. Hence, there exists an integer $n \geq 0$ such that H is a subgroup of $\zeta_{n+1}G$ but not ζ_nG. Thus, $\zeta_{n+1}G/\zeta_nG$ contains the nontrivial P-radicable subgroup $H\zeta_nG/\zeta_nG$. Choose an element $h\zeta_nG \in H\zeta_nG/\zeta_nG$ such that $h \in H$ but $h \notin \zeta_nG$. By Theorem 2.30, there exists a homomorphism

$$\varphi : \zeta_{n+1}G/\zeta_nG \to Z(G)$$

such that $\varphi(h\zeta_nG) \neq 1$. Consequently, *im* $\varphi \neq 1$. Furthermore, $\varphi\left(H\zeta_nG/\zeta_nG\right)$ is P-radicable by Lemma 5.18. This contradicts the hypothesis that $Z(G)$ is P-reduced. Therefore, G is P-reduced.   □

The maximal P-radicable subgroup of a P-torsion nilpotent group is always central.

Lemma 5.29 *If G is a P-torsion nilpotent group, then $\varrho_P(G) < Z(G)$.*

Proof The proof is done by induction on the class c of G. If $c = 1$, then the result is obvious. Suppose that the lemma holds for $c > 1$ and let $g \in G$. Since G is a P-torsion group, there exists a P-number n such that $g^n = 1$. Hence, $g^n \in Z(G)$ and thus $G/Z(G)$ is a P-torsion group. It is also nilpotent of class $c - 1$ by Lemma 2.12. By induction, $\varrho_P(G/Z(G)) < Z(G/Z(G))$ or equivalently, $[\varrho_P(G), G] \leq Z(G)$. Choose an element $h \in \varrho_P(G)$. There exists an element $h_0 \in \varrho_P(G)$ such that $h = h_0^n$. By Lemma 1.13,

$$[h, g] = \left[h_0^n, g \right] = [h_0, g^n] = 1.$$

Therefore, $h \in Z(G)$ and consequently, $\varrho_P(G) < Z(G)$. □

Theorem 5.21 *Every finitely generated P'-torsion-free abelian group has a trivial maximal P-radicable subgroup.*

Proof Let G be a finitely generated P'-torsion-free abelian group. By the Fundamental Theorem of Finitely Generated Abelian Groups, there exists a set of distinct primes $\{p_1, \ldots, p_k\}$ and positive integers $n_1, \ldots, n_k$ such that G is isomorphic to the direct sum of a finite number of copies of $\mathbb{Z}$ and a finite abelian group of order $p_1^{n_1} \cdots p_k^{n_k}$. The primes $p_1, \ldots, p_k$ must lie in P since G is assumed to be P'-torsion-free.

Suppose that $\varrho_P(G) \neq 1$ and let $1 \neq g_0 \in \varrho_P(G)$. Put $m = p_1 \cdots p_k$. Since m is a P-number, there exist elements $g_1, g_2, \ldots$ in $\varrho_P(G)$ such that

$$g_0 = mg_1, \; g_1 = mg_2, \ldots, \; g_l = mg_{l+1}, \ldots \qquad (5.8)$$

Consider the subgroup

$$H = gp(g_0, g_1, g_2, \ldots, g_l, \ldots)$$

of G. Since G is finitely generated, H is also finitely generated. Thus, $H = gp(g_i)$ for some $i \geq 0$. Now, $g_{i+1} \in H$ implies that $g_{i+1} = tg_i$ for some integer t. Hence, $g_i = (tm)g_i$ by (5.8) and consequently, $(tm - 1)g_i = 0$. This means that $tm - 1$ is a P-number and must be divisible by at least one of the primes $p_1, \ldots, p_k$. This is impossible since m is divisible by every such prime. □

Theorem 5.22 *If G is a finitely generated nilpotent group, then $\varrho_P(G) = \tau_{P'}(G)$.*

Proof Let $g \in \tau_{P'}(G)$. There exists a P'-number m such that $g^m = 1$. If n is a P-number, then there exist integers r and s satisfying $rm + sn = 1$. Hence,

$$g = g^{rm+sn} = (g^m)^r (g^s)^n = (g^s)^n.$$

Therefore, $g \in \varrho_P(G)$ and thus $\tau_{P'}(G) \leq \varrho_P(G)$. This being the case, we may assume that $\tau_{P'}(G) = 1$.

We claim that $\varrho_P(G) = 1$. In light of Theorem 5.20, it suffices to show that $\varrho_P(Z(G)) = 1$. Well, $Z(G)$ is P'-torsion-free by assumption and finitely generated by Theorem 2.18. Thus, $\varrho_P(Z(G)) = 1$ by Theorem 5.21.   □

Corollary 5.17 *If G is a finitely generated nilpotent group, then $\varrho_P(G)$ is a finite P'-torsion group.*

Proof The result is a consequence of Theorem 5.22 and Corollary 2.16.   □

We investigate a certain subgroup of a nilpotent group which is closely related to the maximal P-radicable subgroup. To begin with, let G be any (not necessarily nilpotent) group. Pick a prime p and consider the descending series

$$G \geq G^p \geq G^{p^2} \geq \cdots \geq G^{p^k} \geq \cdots \geq \bigcap_{n=1}^{\infty} G^{p^n} = G^{p^{\infty}}. \tag{5.9}$$

Since the conjugate of a p^kth power is again a p^kth power, each G^{p^k} is a normal subgroup of G. In additive notation, we write the last equality of (5.9) as

$$\bigcap_{n=1}^{\infty} p^n G = p^{\infty} G.$$

The definition of semi-p-radicability may be given in terms of $G^{p^{\infty}}$.

Lemma 5.30 *A group G is semi-p-radicable if and only if $G = G^{p^{\infty}}$.*

Two relationships between $G^{p^{\infty}}$ and $\varrho_p(G)$ are apparent.

1. $\varrho_p(G)$ is always a subgroup of $G^{p^{\infty}}$. For suppose that $g \in \varrho_p(G)$, then $g = g_1^p$ for some $g_1 \in \varrho_p(G)$. Moreover, $g_1 = g_2^p$ for some $g_2 \in \varrho_p(G)$. Thus, $g = g_2^{p^2}$. Continuing in this way, we see that for any $n > 0$, there exists $g_n \in \varrho_p(G)$ such that $g = g_n^{p^n}$. This means that $g \in G^{p^{\infty}}$.
2. If $G^{p^{\infty}}$ is a p-radicable subgroup of G, then it must equal $\varrho_p(G)$. This is due to the fact that $\varrho_p(G)$ is maximal with respect to p-radicability.

In general, $G^{p^{\infty}}$ need not be a p-radicable subgroup of G. Consider, for example, the abelian p-group G with presentation (in additive notation)

$$G = \langle x_1, x_2, \ldots \mid px_1 = 0, \ p^n x_{n+1} = x_1 \ \text{for} \ n = 1, 2, \ldots \rangle.$$

The subgroup $C = gp(x_1)$ is a cyclic group of order p and thus not p-radicable. We claim that $p^{\infty} G = C$. It is easy to see that any nontrivial subgroup of G must contain C. Consequently, $C \leq p^{\infty} G$. Now, the factor group G/C is a direct product of cyclic groups. Hence,

$$\bigcap_{n=1}^{\infty} p^n (G/C) = C.$$

It follows that $p^{\infty} G = C$ as claimed.

Theorem 5.23 *If G is a p-torsion-free nilpotent group, then $G^{p^\infty} = \varrho_p(G)$.*

Proof Suppose that G has nilpotency class c and let $g \in G^{p^\infty}$. For every integer $n \geq 0$, we have $g \in G^{p^n}$. Choose n sufficiently large so that $n > f(p, c)$, where $f(p, c)$ is the integer guaranteed by Lemma 4.4. There exists $h \in G^{p^n}$ such that

$$g = h^{p^{n-f(p,\, c)}}.$$

We need to prove that $h \in G^{p^\infty}$, establishing that h is a pth root of g in G^{p^∞}.

Set $m = n - f(p, c)$ and let $l \in \mathbb{N}$. Since $g \in G^{p^\infty}$ and $n + l > f(p, c)$, there exists an element $k \in G$ such that $g = k^{p^{m+l}}$. Then

$$k^{p^{m+l}} = \left(k^{p^l} \right)^{p^m} = h^{p^m}.$$

Since G is p-torsion-free, $k^{p^l} = h$ by Theorem 2.7. Hence, h has a p^lth root for *every* $l \in \mathbb{N}$. And so, $h \in G^{p^\infty}$. □

Corollary 5.18 *If G is a torsion-free nilpotent group, then $G^{p^\infty} = \varrho_p(G)$ for any prime p.*

Proof The result is an immediate consequence of Theorem 5.23. □

The next corollary follows at once from Theorems 5.22 and 5.23.

Corollary 5.19 *Let p be any prime. If G is a finitely generated torsion-free nilpotent group, then $G^{p^\infty} = 1$.*

If G is any finitely generated nilpotent group (not necessarily torsion-free), then G^{p^∞} is a finite p'-group. This can be deduced from the next theorem.

Theorem 5.24 *Let G be a nilpotent group of class c and let p be any prime. If $\tau_p(G)^{p^n} = 1$ for some $n \in \mathbb{N}$, then $G^{p^\infty} = \varrho_p(G)$.*

Proof The case when G is p-torsion-free is taken care of by Theorem 5.23. Suppose that G has p-torsion elements and let n be as in the hypothesis. We claim that $G^{p^{n+f(p,\, c)}}$ is p-torsion-free, where $f(p, c)$ is the integer guaranteed in Lemma 4.4. Let $m = n + f(p, c)$ and suppose that $g \in \tau_p(G) \cap G^{p^m}$, then $g \in \tau_p(G)$ implies that $g^{p^t} = 1$ for some integer $t \geq 0$ and $g \in G^{p^m}$ implies that

$$g = h^{p^{m-f(p,\, c)}} = h^{p^n}$$

for some $h \in G$ by Lemma 4.4. Hence,

$$g^{p^t} = \left(h^{p^n} \right)^{p^t} = h^{p^{n+t}} = 1$$

and thus $h \in \tau_p(G)$. By the hypothesis, $h^{p^n} = 1$ and consequently, $g = 1$. This proves the claim. By the first case, we have $\left(G^{p^m} \right)^{p^\infty} = \varrho_p \left(G^{p^m} \right)$. However,

$$\left(G^{p^m}\right)^{p^\infty} = \bigcap_{n=1}^{\infty} \left(G^{p^m}\right)^{p^n} = \bigcap_{n=1}^{\infty} G^{p^{m+n}} = G^{p^\infty}.$$

Furthermore,

$$\varrho_p\left(G^{p^m}\right) = \varrho_p(G).$$

To see this, suppose that $g \in \varrho_p(G)$. Since g has a p^nth root for every $n \in \mathbb{N}$, there exists $h \in G$ such that

$$g = h^{p^{m+l}} = \left(h^{p^m}\right)^{p^l}$$

for all $l \in \mathbb{N}$. Clearly, $h^{p^m} \in G^{p^m}$. Thus, g has p^lth roots in G^{p^m}, that is, $g \in \varrho_p\left(G^{p^m}\right)$. The reverse inclusion is obvious. This completes the proof.   □

Corollary 5.20 *If G is a finitely generated nilpotent group and p is any prime, then G^{p^∞} is a finite p'-group.*

Proof By Corollary 2.16, $\tau_p(G)$ is a finite p-group. Thus, $G^{p^\infty} = \varrho_p(G)$ by Theorem 5.24. The result follows from Corollary 5.17.   □

5.2.12   Residual Properties

If G is a nilpotent group, then information about G^{p^∞} for a given prime p and $\varrho_P(G)$ for a nonempty set of primes P can be exploited to answer questions about the residual properties of G. These properties, in turn, can be used to prove certain embedding theorems of nilpotent groups into radicable nilpotent groups. Residual properties also play a significant role in the context of M. Dehn's decision problems, which we discuss in Section 7.1.

Definition 5.14 (P. Hall) Let $\mathscr{Q}$ be a property of groups. A group G is said to be *residually* $\mathscr{Q}$ if for every $1 \neq g \in G$, there exists a normal subgroup N_g of G such that $g \notin N_g$ and G/N_g has property $\mathscr{Q}$.

Our main interest is when $\mathscr{Q}$ is the property of being finite, or the property of being a finite P-group for a nonempty set of primes P.

There are several equivalent versions of Definition 5.14. The first one illustrates that a residually $\mathscr{Q}$ group has many images with property $\mathscr{Q}$ and thus can be recovered from groups that have this property.

Lemma 5.31 *A group G is residually $\mathscr{Q}$ if and only if for every $1 \neq g \in G$, there exists a group H that has property $\mathscr{Q}$ and an epimorphism $\varphi : G \to H$ such that $\varphi(g) \neq 1$.*

Proof Suppose that G is residually $\mathcal{Q}$ and $1 \neq g \in G$. There exists a normal subgroup N_g of G with $g \notin N_g$ such that G/N_g has property $\mathcal{Q}$. Consider the natural epimorphism $\pi : G \to G/N_g$. Clearly, $\pi(g) \neq 1$ in G/N_g because $\pi(g) = gN_g \neq N_g$. Hence, $H = G/N_g$ satisfies the required criteria.

Conversely, let $1 \neq g \in G$ and suppose that H is a group with property $\mathcal{Q}$. Further, suppose that $\varphi : G \to H$ is an epimorphism such that $\varphi(g) \neq 1$. Since $g \notin \ker \varphi$ and

$$G/\ker \varphi \cong \varphi(G) = H,$$

the subgroup $N_g = \ker \varphi$ satisfies the conditions of Definition 5.14. $\square$

Definition 5.15 Let $\{G_i \mid i \in I\}$ be a family of groups for some nonempty index set I. Denote the unrestricted direct product of the G_i by $\overline{\prod}_{i \in I} G_i$ and let π_i be the projection map of $\overline{\prod}_{i \in I} G_i$ onto G_i, that is,

$$\pi_i(g_1, \ldots, g_i, \ldots) = g_i.$$

If H is a subgroup of $\overline{\prod}_{i \in I} G_i$, then the restriction map

$$\pi_i|_H : H \to G_i$$

is called the *projection* of H to G_i. The subgroup H is termed the *subcartesian product* of the G_i if $\pi_i(H) = G_i$ for all $i \in I$.

The next lemma gives other equivalent definitions of residually $\mathcal{Q}$. This can be found in [9] and [26].

Lemma 5.32 *Let $\mathcal{Q}$ be a property of groups and let G be any group. The following are equivalent:*

(1) G is residually $\mathcal{Q}$;
(2) Let Λ be a nonempty index set. There exists a family $\{N_\lambda \mid \lambda \in \Lambda\}$ of normal subgroups of G such that G/N_λ has property $\mathcal{Q}$ for all $\lambda \in \Lambda$ and $\cap_{\lambda \in \Lambda} N_\lambda = 1$;
(3) G is a subcartesian product of groups having property $\mathcal{Q}$.

Proof $(1) \Rightarrow (2)$: For each $1 \neq g \in G$, choose a normal subgroup N_g of G such that $g \notin N_g$ and G/N_g has property $\mathcal{Q}$. Such a subgroup exists by Definition 5.14 since G is residually $\mathcal{Q}$. It is easy to see that the family of these N_g satisfies (2).
$(2) \Rightarrow (1)$: The result is obvious.
$(2) \Rightarrow (3)$: Let $\{N_\lambda \mid \lambda \in \Lambda\}$ be a family of normal subgroups such that G/N_λ has property $\mathcal{Q}$ for all $\lambda \in \Lambda$ and $\cap_{\lambda \in \Lambda} N_\lambda = 1$. Define a mapping

$$\alpha : G \to \overline{\prod_{\lambda \in \Lambda}}(G/N_\lambda) \quad \text{by} \quad \alpha(g) = \varphi_g, \quad \text{where} \quad \varphi_g(\lambda) = gN_\lambda.$$

Clearly, $\varphi_{gh}(\lambda) = \varphi_g(\lambda)\varphi_h(\lambda)$ for any g, $h \in G$. Thus, α is a homomorphism. In fact, α is a monomorphism. For suppose that $\alpha(g) = \alpha(h)$ for some g, $h \in G$, then $\varphi_g = \varphi_g$ implies that $gN_\lambda = hN_\lambda$ for all $\lambda \in \Lambda$. Thus, $gh^{-1} \in N_\lambda$ for all $\lambda \in \Lambda$ and consequently,

$$gh^{-1} \in \bigcap_{\lambda \in \Lambda} N_\lambda = 1.$$

Hence, $g = h$. Furthermore, the projection of $\alpha(G)$ onto a factor of $\overline{\prod}_{\lambda \in \Lambda}(G/N_\lambda)$ equals the whole factor. Thus, (3) holds.

(3) $\Rightarrow$ (2) : Suppose that G is a subgroup of $\overline{\prod}_{\lambda \in \Lambda}G_\lambda$, where each G_λ has property $\mathscr{Q}$ and each projection map satisfies $\pi_\lambda(G) = G_\lambda$. Set $N_\lambda = G \cap ker\, \pi_\lambda$. Clearly, $N_\lambda \trianglelefteq G_\lambda$ and $\cap_{\lambda \in \Lambda}N_\lambda = 1$ because $\cap_{\lambda \in \Lambda}ker\, \pi_\lambda = 1$. Furthermore, an application of the Second Isomorphism Theorem gives

$$G/N_\lambda = G/(G \cap ker\, \pi_\lambda) \cong G\, ker\, \pi_\lambda/ker\, \pi_\lambda \cong \pi_\lambda(G) = G_\lambda.$$

And so, G/N_λ has property $\mathscr{Q}$.   □

It is clear that every finite group is residually finite and every residually finite p-group is residually finite for any prime p.

Lemma 5.33 *An infinite cyclic group is a residually finite p-group for any prime p.*

Proof Let $G = gp(g)$ be an infinite cyclic group. For every $k \in \mathbb{N}$, define the subgroups $G_k = gp\left(g^{p^k}\right)$ of G. Then $\bigcap_{k \in \mathbb{N}} G_k = 1$ and G/G_k is a finite p-group of order p^k. Apply Lemma 5.32.   □

Since finite groups are residually finite, it follows from Lemma 5.33 that every cyclic group is residually finite.

Lemma 5.34 *Let $\mathscr{Q}$ be a property of groups. If G_1, ..., G_k are residually $\mathscr{Q}$ groups, then $G_1 \times \cdots \times G_k$ is residually $\mathscr{Q}$.*

Proof Let $g = (g_1, \ldots, g_k) \in G_1 \times \cdots \times G_k$ be a nonidentity element. There exists a natural number $l \in \{1, \ldots, k\}$ such that $g_l \neq 1$. The projection map

$$\psi_l : G_1 \times \cdots \times G_k \to G_l$$

maps g onto $g_l \neq 1$ in G_l. By hypothesis, G_l is residually $\mathscr{Q}$. Thus, there exists a homomorphism φ from G_l to some group G with property $\mathscr{Q}$ such that $\varphi(g_l) \neq 1$ in G. We conclude that $\varphi \circ \psi_l$ maps g to a nonidentity element of G.   □

By Lemmas 5.33 and 5.34, we have:

Theorem 5.25 *Every finitely generated abelian group is residually finite.*

The study of residual finiteness naturally leads to the study of subgroups of finite index in a group. We collect some results on such subgroups which will be needed later. See [31] and [32] for more background material.

Definition 5.16 Let G be a group with $H \le G$. The subgroup

$$\bigcap_{g \in G} gHg^{-1}$$

of H is called the *normal core* of H, denoted by *core(H)*.

The normal core of H can be realized as the kernel of the homomorphism induced by the natural action of G on the left coset space G/H. More precisely, suppose that $C = \{xH \mid x \in G\}$ is the set of left cosets of H in G and let S_C denote the symmetric group on C. One can show that *core(H)* is the kernel of the homomorphism

$$\varphi : G \to S_C \text{ defined by } \varphi(g) = \sigma_g, \text{ where } \sigma_g(xH) = gxH.$$

Furthermore, *core(H)* turns out to be the largest normal subgroup of G contained in H. If $[G : H] = n$, then we identify S_C with S_n, the symmetric group on the set $\{1, 2, \ldots, n\}$.

Theorem 5.26 *Let G be a group and $H \le G$. If H has finite index in G, then H contains a normal subgroup of finite index in G. In particular, if $[G : H] = n$, then $[G : core(H)] \le n!$.*

Proof We use the same notations as above. Suppose that $H < G$ and $[G : H] = n$ for some $n \in \mathbb{N}$. Since C has n elements, S_C has $n!$ elements. By the First Isomorphism Theorem,

$$G/core(H) \cong \varphi(G) \le S_C.$$

Thus, $[G : core(H)]$ divides $|S_C| = n!$.   □

Theorem 5.27 (Poincaré) *Let G be any group. If H and K are subgroups of finite index in G, then $H \cap K$ has finite index in G.*

Proof Suppose that $[G : H] = m$ and $[G : K] = n$ for some $m, n \in \mathbb{N}$. We first show that $x(H \cap K) = xH \cap xK$ for any left coset of $H \cap K$. Let $a \in xH \cap xK$. Since $a \in xH$, there exists $h \in H$ such that $a = xh$. Hence, $xh \in xK$ and thus $h \in K$. Therefore, $h \in H \cap K$ and consequently, $a \in x(H \cap K)$. Similarly, $x(H \cap K) \subseteq xH \cap xK$.

Next, let $\{x_1 H, \ldots, x_m H\}$ and $\{y_1 K, \ldots, y_n K\}$ be the left cosets of H and K respectively. We have shown that any left coset of $H \cap K$ is the intersection of a left coset $x_i H$ of H with a left coset $y_j K$ of K. Hence, $H \cap K$ has at most mn left cosets. And so, $[G : H \cap K] < \infty$.   □

Theorem 5.28 *If G is a finitely generated group and $n \in \mathbb{N}$, then G contains only a finite number of subgroups of index n.*

Proof Suppose that G is generated by $\{x_1, \ldots, x_m\}$. If H is a subgroup of index n in G, then by our discussion proceeding Definition 5.16, there exists $\varphi \in Hom(G, S_n)$ with kernel *core(H)*. Since φ is determined by the elements

$$\varphi(x_1), \ \ldots, \ \varphi(x_m)$$

and $|S_n| = n! < \infty$, there are only a finite number of homomorphisms from G into S_n. This means that there are only finitely many normal subgroups which can be the normal core of a subgroup of index n in G. Furthermore, any such normal subgroup can be the normal core of only a finite number of subgroups of index n in G. This follows from the fact that if $N = core(K)$ for some $K < G$ and $[G : K] = n$, then the factor group G/N is finite by Theorem 5.26 and $K/N \leq G/N$. Therefore, there are only finitely many subgroups of index n in G. $\square$

Theorem 5.29 *If G is a finitely generated group and $[G : H] < \infty$ for some $H \leq G$, then there exists a characteristic subgroup I of G such that $I \leq H$ and $[G : I] < \infty$.*

Proof Suppose that $[G : H] = n$ and define I to be the intersection of all subgroups of index n in G. Clearly, I is a subgroup of H. By Theorem 5.28, there are only finitely many subgroups of index n in G. Thus, I is a finite intersection of subgroups of finite index. It follows from Theorem 5.27 that $[G : I] < \infty$. It remains to show that I is characteristic in G. Let $\varphi \in Aut(G)$. Since I is a subgroup of H, $\varphi(I)$ is a subgroup of $\varphi(H)$. Now, $[G : \varphi(H)] = n$ because every automorphism of G preserves the index of a subgroup. Therefore, I contains $\varphi(H)$. And so, $\varphi(I) < I$. $\square$

We return to our discussion of residual properties. A major contribution due to K. A. Hirsch is that finitely generated nilpotent groups are residually finite. In fact, he showed that polycyclic groups are residually finite [14]. We now set out to prove this. The main ingredient is the next theorem due to P. Hall.

Theorem 5.30 *A cyclic extension of a finitely generated residually finite group is residually finite.*

Proof Let G be a group and suppose that H is a finitely generated normal subgroup of G. Further, suppose that H is residually finite and

$$G = gp(a, \ H)$$

for some $a \in G$. Thus, G/H is a cyclic group with generator aH. We claim that G is residually finite. Let $1 \neq g \in G$.

Case (i): Suppose that $g \notin H$, so that $gH \neq H$ in G/H. Since G/H is cyclic, it is residually finite (see Lemma 5.33). Hence, there exists a normal subgroup N_g/H of G/H such that $gH \notin N_g/H$ and $(G/H) / (N_g/H)$ is finite. Thus, $g \notin N_g$ and by the Third Isomorphism Theorem, G/N_g is finite.

Case (ii): Suppose that $g \in H$. Since H is residually finite, there exists a normal subgroup K of H such that $g \notin K$ and $[H : K] < \infty$. By Theorem 5.29, there is a characteristic subgroup C of H such that $C \leq K$, $[H : C] < \infty$ and $g \notin C$. By Lemma 1.8, we have that $C \vartriangleleft G$.

- Suppose that G/H is finite. Since $G/H \cong (G/C)/(H/C)$ by the Third Isomorphism Theorem and H/C is finite, G/C is also finite. Furthermore, $g \notin C$. Thus, we simply take $N_g = C$.
- If G/H is infinite cyclic, then $gp(aH) = gp(a)H$ with $H \cap gp(a) = 1$. Put

$$\overline{H} = H/C, \ \overline{G} = G/C, \ \overline{g} = gC, \ \text{and} \ \overline{a} = aC.$$

Let $\overline{D} = C_{\overline{G}}(\overline{H})$. Since $[H : C] < \infty$, $\overline{H}$ is a finite normal subgroup of $\overline{G}$. Thus, $N_{\overline{G}}(\overline{H}) = \overline{G}$ and $Aut(\overline{H})$ is finite. By Theorem 1.3, $\overline{D}$ must be of finite index in $\overline{G}$. Thus, there exists $m > 0$ such that $\overline{a}^m \in \overline{D}$.

We claim that $\overline{a}^m$ has infinite order. Assume, on the contrary, that $\overline{a}^m$ has finite order k for some $k \in \mathbb{N}$. Then $((aC)^m)^k = C$ and thus $a^{mk} \in C$. This means that $a^{mk} \in H$, contradicting the hypothesis that $gp(aH)$ is infinite cyclic. And so, $\overline{a}^m$ has infinite order as claimed.

Now, $\overline{g}$ is an element of the finite group H/C and thus $\overline{g} \notin gp(\overline{a}^m)$. Moreover, $gp(\overline{a}^m) \lhd \overline{G}$ because the elements of $\overline{H}$ commute with $\overline{a}^m$ and $\overline{a}$ clearly commutes with $\overline{a}^m$. Finally, $\overline{G}/gp(\overline{a}^m)$ is of finite index in $\overline{G}$ because $\overline{H}$ is finite.

We conclude that $\overline{G}/gp(\overline{a}^m)$ is finite and the image of g under the composition of natural maps

$$G \to \overline{G} \to \overline{G}/gp(\overline{a}^m)$$

is not the identity. By Lemma 5.31, G is residually finite. □

Corollary 5.21 (K. A. Hirsch) *Every polycyclic group is residually finite.*

Theorem 5.31 *Finitely generated nilpotent groups are residually finite.*

Proof By Theorem 4.4, every finitely generated nilpotent group is polycyclic. Apply Corollary 5.21. □

Corollary 5.22 *Let G be a finitely generated nilpotent group and choose a non-identity element $g \in G$. If every nontrivial normal subgroup contains g, then G is a finite p-group for some prime p.*

Proof We prove that G is finite by contradiction. Assume that G is infinite. According to Theorem 5.31, there exists a normal subgroup N of G such that $g \notin N$ and $[G : N] < \infty$. Since G is infinite, N is nontrivial. Thus, $g \in N$ by hypothesis, a contradiction. Therefore, G must be finite. By Theorem 2.13 (vi), G is a direct product of its Sylow subgroups. Since g is contained in every nontrivial normal subgroup, G has only one Sylow subgroup. □

Theorem 5.32 (K. W. Gruenberg) *Every finitely generated torsion-free nilpotent group is residually a finite p-group for every prime p.*

The original proof of this theorem can be found in [9]. We give a different proof based on our earlier discussion of the subgroup G^{p^∞} of a group G. This approach is taken by R. B. Warfield, Jr. in [33].

Proof Fix a prime p and let G be a finitely generated torsion-free nilpotent group. By Corollary 5.19,

$$G^{p^\infty} = \bigcap_{i=1}^{\infty} G^{p^i} = 1.$$

For each $i \in \mathbb{N}$, the quotient G/G^{p^i} is a finitely generated nilpotent p-group and thus a finite p-group according to Theorem 2.25. The result follows from Lemma 5.32.□

Theorem 5.33 *Let G be a finitely generated nilpotent group. Then G is a residually finite P-torsion group if and only if it is P'-torsion-free.*

Proof Suppose that G is a residually finite P-torsion group and assume, on the contrary, that there is an element $g \neq 1$ in G such that $g^n = 1$ for some P'-number n. There exists a normal subgroup N of G such that $g \notin N$ and G/N is a finite P-torsion group. Now, $g^n = 1$ implies $(gN)^n = N$ in G/N. Since G/N is P-torsion, this can happen only if $gN = N$. This contradicts the assumption that $g \notin N$. Thus, G must be P'-torsion-free.

Conversely, suppose that G is P'-torsion-free. Every torsion element of G must have order a P-number and thus $\tau(G) = \tau_P(G)$. By Corollary 2.16, $\tau(G)$ is a finite P-torsion group. Hence,

$$\tau(G) = \tau_{p_1}(G) \times \cdots \times \tau_{p_t}(G)$$

for some finite set of distinct primes $Q = \{p_1, \ldots, p_t\} \subseteq P$ by Theorem 2.13 (vi). Clearly, G is Q'-torsion-free. Let $m = p_1^{r_1} \cdots p_t^{r_t}$ be the order of $\tau(G)$ for some $r_1, \ldots, r_t \in \mathbb{N}$. Since $\tau_{p_j}(G)$ has order a positive power of p_j for each $p_j \in Q$, $\varrho_{p_j}(G) = G^{p_j^\infty}$ by Theorem 5.24. Furthermore, $\varrho_Q(G) = 1$ by Theorem 5.22 because G is Q'-torsion-free. Since $\varrho_Q(G) = \bigcap_{p \in Q} \varrho_p(G)$, we have

$$\bigcap_{i=1}^{\infty} G^{m^i} = \bigcap_{i=1}^{\infty} G^{\left(p_1^{r_1} \cdots p_t^{r_t}\right)^i} = \bigcap_{i=1}^{\infty} G^{p_1^{r_1 i} \cdots p_t^{r_t i}}$$

$$= \bigcap_{i=1}^{\infty} \left(\bigcap_{p_j \in Q} G^{p_j^{r_j i}} \right) = \bigcap_{p_j \in Q} \left(\bigcap_{i=1}^{\infty} G^{p_j^{r_j i}} \right)$$

$$= \bigcap_{p \in Q} G^{p^\infty} = \bigcap_{p \in Q} \varrho_p(G) = \varrho_Q(G) = 1.$$

Since each quotient G/G^{m^i} is a finite P-torsion group, the result follows at once from Lemma 5.32 (2).   □

In [6], G. Baumslag showed that Theorem 5.32 can be generalized. We begin with a lemma.

Lemma 5.35 *Let G be a finitely generated torsion-free nilpotent group. Suppose that H is an isolated subgroup of G and let $Y \leq Z(G)$. If $I(HY, G) = G$, then $H \trianglelefteq G$.*

Proof The proof is done by induction on the Hirsch length r of G. If $r = 1$, then G is abelian and the result is trivial.

Suppose that $r > 1$ and assume that the lemma is true for all finitely generated nilpotent groups of Hirsch length at most $r - 1$. We claim that $Z(H) \leq Z(G)$. By Theorem 5.7, G is a $\mathscr{U}$-group. Hence, all centralizers are isolated in G according to Lemma 5.15. In particular, $Z(H)$ is isolated in G because $Z(H) = H \cap C_H(G)$ and H is isolated in G by hypothesis.

Choose any element $g \in G$. By hypothesis, $g \in I(HY, G)$. According to Theorem 5.2, there exists $m \in \mathbb{N}$ such that $g^m \in HY$. This implies that g^m centralizes $Z(H)$. For suppose that $h \in Z(H) = H \cap C_H(G)$. Since $g^m \in HY \leq HZ(G)$, there exist elements $h_1 \in H$ and $z \in Z(G)$ such that $g^m = h_1 z$. Clearly, h commutes with h_1 because $h \in C_H(G)$. Since z is central in G, h commutes with g^m. And so, g^m centralizes $Z(H)$ as asserted. Hence, g centralizes $Z(H)$ because $Z(H)$ is isolated in G. Since g is an arbitrary element of G, we have that $Z(H) \leq Z(G)$ as claimed.

Now, $G/Z(H)$ is torsion-free by Lemma 5.1. Since G has Hirsch length r, $G/Z(H)$ has Hirsch length less than r by Theorem 4.7. Thus, $H/Z(H) \trianglelefteq G/Z(H)$ by induction. And so, $H \trianglelefteq G$. $\square$

Theorem 5.34 *Let G be a finitely generated torsion-free nilpotent group and let H be an isolated subgroup of G. For any given prime p,*

$$\bigcap_{i=1}^{\infty} G^{p^i} H = H.$$

Note that Theorem 5.32 follows from this if we put $H = 1$.

Proof We proceed by induction on the class c of G. If $c = 1$, then G is a free abelian group of finite rank. Since H is isolated in G, the quotient G/H is torsion-free by Lemma 5.1. Hence, G/H is free abelian and thus $G = H \times K$ for some $K \leq G$. We claim that

$$G^{p^i} H = H \times K^{p^i} \qquad (i = 1, 2, \ldots).$$

First, notice that $HK^{p^i} = H \times K^{p^i}$ because $H \cap K^{p^i} = 1$. Clearly, $HK^{p^i} \leq G^{p^i} H$. We assert that $G^{p^i} H \leq K^{p^i} H$. If $g^{p^i} h$ is a generator of $G^{p^i} H$, then

$$g^{p^i} h = (h_1 k_1)^{p^i} h = h_1^{p^i} h k_1^{p^i}$$

for some $h_1 \in H$ and $k_1 \in K$. Thus, $g^{p^i} h \in K^{p^i} H$ and the assertion is proved. Since K is free abelian of finite rank, $\bigcap_{i=1}^{\infty} K^{p^i} = 1$ by Corollary 5.19. This proves the theorem when $c = 1$.

Suppose that $c > 1$ and assume that the result is true for all finitely generated torsion-free nilpotent groups of class less than c. Put

$$Z = Z(G), \quad I = I(HZ, G), \quad \text{and} \quad L = \bigcap_{i=1}^{\infty} G^{p^i} H.$$

We claim that $L = H$. By Corollary 5.10, G/Z is torsion-free nilpotent. Furthermore, I/Z is isolated in G/Z because I is isolated in G. By induction,

$$\bigcap_{i=1}^{\infty} (G/Z)^{p^i} I/Z = I/Z.$$

It follows that $H \leq L \leq I$. Now, I is a finitely generated torsion-free nilpotent group and H is isolated in I. By Lemma 5.35, H is normal in I. Moreover, Lemma 5.1 implies that I/H is torsion-free because H is isolated in I.

Let $\ell H \in L/H$ and $i \in \mathbb{N}$. There exist elements $g_1, \ldots, g_t \in G$ and $h \in H$ such that

$$\ell = g_1^{p^i} \cdots g_t^{p^i} h.$$

If i is sufficiently large, then Lemma 4.4 guarantees that $g_1^{p^i} \cdots g_t^{p^i}$ can be written as a p^jth power, say g^{p^j}, where j tends to infinity with i. Thus, $g^{p^j} = \ell h^{-1} \in I$. Since I is isolated in G, $g \in I$. It follows that ℓH has a p^nth root in I/H for every n. If it were the case that $\ell H \neq H$, then there would exist a properly increasing infinite series of subgroups in the finitely generated torsion-free nilpotent group I/H, contradicting Theorem 2.18. Consequently, ℓH must equal H and thus $L = H$ as desired. $\square$

We give an alternative proof of Corollary 5.4 which invokes Theorem 5.34. This appears in [6].

Theorem 5.35 (V. M. Gluškov) *Let G be finitely generated torsion-free nilpotent group. If H is an isolated subgroup of G, then $N_G(H)$ is isolated in G.*

Proof Assume on the contrary, that $N_G(H)$ is not isolated in G. There exists a prime p and an element $g \in G$ such that $g^p \in N_G(H)$ but $g \notin N_G(H)$. Consequently, there is an element h in H such that $g^{-1}hg \notin H$. By Theorem 5.34,

$$g^{-1}hg \notin \bigcap_{t=1}^{\infty} G^{q^t} H$$

for any prime q. Thus, there exists $t \in \mathbb{N}$ such that

$$g^{-1}hg \notin G^{q^t} H. \tag{5.10}$$

In particular, choose $q \neq p$. There exist $m, n \in \mathbb{Z}$ such that $1 = mp + nq^t$. Thus,

$$gG^{q^t} = g^{mp+nq^t}G^{q^t} = (g^p)^m(g^n)^{q^t}G^{q^t} = (g^p)^mG^{q^t}.$$

Hence, there exists $k \in G^{q^t}$ such that $gk^{-1} = (g^p)^m$. Now, $g^p \in N_G(H)$ implies that $(g^p)^m \in N_G(H)$, so that gk^{-1} belongs to $N_G(H)$. And so,

$$\left(gk^{-1}\right)^{-1}hgk^{-1} = k\left(g^{-1}hg\right)k^{-1} \in H.$$

This means that $g^{-1}hg \in k^{-1}Hk \leq G^{q^t}H$, contradicting (5.10). $\qquad\square$

We end our discussion of residual properties with a theorem of G. Higman [11].

Theorem 5.36 *Let P be any infinite set of primes. If G is a finitely generated nilpotent group, then $\bigcap_{p \in P} G^p$ is finite. If G is also torsion-free, then $\bigcap_{p \in P} G^p = 1$.*

Proof The proof is done by induction on the class c of G. If $c = 1$, then G is abelian and the result follows from the Fundamental Theorem of Finitely Generated Abelian Groups.

Suppose that $c > 1$ and set $K = \bigcap_{p \in P} G^p$. Let $\pi : G \rightarrow G/\gamma_c G$ be the natural homomorphism. By Theorem 2.4 and Lemma 2.8,

$$\pi(K) = K\gamma_c G/\gamma_c G = \bigcap_{p \in P}\left(G^p\gamma_c G/\gamma_c G\right)$$

is nilpotent of class at most $c - 1$ and thus finite by the induction hypothesis. Since $K\gamma_c G/\gamma_c G \cong K/(K \cap \gamma_c G)$ by the Second Isomorphism Theorem, $K/(K \cap \gamma_c G)$ is finite. It suffices to prove that $K \cap \gamma_c G$ is finite.

Assume on the contrary, that g is an element of $K \cap \gamma_c G$ of infinite order. Define the set

$$Q = \{p \in P \mid p > c\}.$$

Clearly, Q is infinite since P is infinite by hypothesis. Furthermore, $g \in K \leq G^p$ for each $p \in Q$. By Lemma 3.2, g has a pth root in G, say h_p, for each $p \in Q$. Now, $g \in \gamma_c G$ and $\gamma_c G$ is a central subgroup of G. Thus, $gp(g)$ is a normal subgroup of G. Put $N = gp(g)$. The element $h_p N$ has order p in G/N. Thus, G/N contains infinitely many elements, each having order a prime in Q. This is impossible since G/N is finitely generated nilpotent. Therefore, g has finite order. Thus, $K \cap \gamma_c G$ contains only elements of finite order and the result follows. $\qquad\square$

For more on residual properties of nilpotent groups, see [34].

5.2.13 $\mathscr{D}_P$-*Groups*

Recall from Definition 4.9 that a group G is called a $\mathscr{D}$-*group* or a $\mathbb{Q}$-*powered group* if every element has a unique nth root for each natural number $n > 1$. If P is a set of primes, then G is called a $\mathscr{D}_P$-*group* if every element has a unique nth root for every P-*number* n.

A $\mathscr{D}$-group ($\mathscr{D}_P$-group) is simply a radicable $\mathscr{U}$-group (P-radicable $\mathscr{U}_P$-group respectively). We write "$\mathscr{D}_p$-group" instead of "$\mathscr{D}_{\{p\}}$-group" whenever $P = \{p\}$. In a $\mathscr{D}$-group ($\mathscr{D}_P$-group), the equation $g = x^n$ has exactly one solution for every $g \in G$ and every natural number $n > 1$ (P-number n respectively).

Our goal here is to give some essential results on $\mathscr{D}_P$-groups. In particular, we look at extensions of $\mathscr{D}_P$-groups. We begin with a theorem on finite P-torsion groups.

Theorem 5.37 *Every finite P-torsion group is a $\mathscr{D}_{P'}$-group.*

Proof Let G be a finite P-torsion group. Since every P-torsion group is P'-torsion-free, G is a $\mathscr{U}_{P'}$-group by Theorem 2.7.

We claim that G is P'-radicable. Let g be an element of G of order n and suppose that m is a P'-number. Since n is a P-number, m and n are relatively prime. Thus, there exists integers a and b such that $am + bn = 1$. Then

$$g = g^{am+bn} = (g^a)^m (g^n)^b = (g^a)^m.$$

And so, g has an mth root, namely g^a. Since m is a P'-number, G is P'-radicable as claimed. $\square$

The next theorem is immediate from Proposition 5.1 (i) and Theorems 5.12 (ii) and 5.13.

Theorem 5.38 *The upper and lower central subgroups of a nilpotent $\mathscr{D}_P$-group are $\mathscr{D}_P$-groups.*

Theorem 5.39 *If G is a P-radicable nilpotent group of class c, then $G/\zeta_{i+1}G$ and $\zeta_{i+1}G/\zeta_i G$ are nilpotent $\mathscr{D}_P$-groups for $i = 0, 1, \ldots, c - 1$.*

Proof By Theorem 5.12 (i), $G/\zeta_{i+1}G$ is P-radicable and nilpotent. If $i = 0$, then $G/Z(G)$ is a $\mathscr{U}_P$-group by Corollary 5.13. Since

$$G/\zeta_{i+1}G \cong \frac{G/Z(G)}{\zeta_i(G/Z(G))}$$

by the Third Isomorphism Theorem, it follows from Theorem 5.8 (ii) that $G/\zeta_{i+1}G$ is a nilpotent $\mathscr{D}_P$-group. Hence, $\zeta_{i+1}G/\zeta_i G$ is an abelian $\mathscr{U}_P$-group by Proposition 5.1 (i). Therefore, $\zeta_{i+1}G$ is P-radicable and nilpotent by Theorem 5.12 (ii). This implies that $\zeta_{i+1}G/\zeta_i G$ is P-radicable. And so, $\zeta_{i+1}G/\zeta_i G$ is a $\mathscr{D}_P$-group. $\square$

Theorem 5.40 *Suppose that G is a nilpotent $\mathscr{D}_P$-group and $N \trianglelefteq G$. Then N is a $\mathscr{D}_P$-group if and only if G/N is a $\mathscr{D}_P$-group.*

Proof Let N be a $\mathscr{D}_P$-group. By Lemma 5.38, N is P-isolated in G. Therefore, G/N is a $\mathscr{U}_P$-group by Corollary 5.6. It is also P-radicable by Lemma 5.18.

Conversely, suppose that G/N is a $\mathscr{D}_P$-group. By Corollary 5.6, N is P-isolated in G. The result follows from Lemma 5.38. $\square$

Lemma 5.36 *A direct product of a family of $\mathscr{D}_P$-groups is a $\mathscr{D}_P$-group.*

Proof Apply Lemmas 5.13 and 5.19. □

5.2.14 Extensions of $\mathscr{D}_P$-Groups

The example after Lemma 5.17 demonstrates that an extension of a $\mathscr{D}_P$-group by a $\mathscr{D}_P$-group is not always a $\mathscr{D}_P$-group. However, we have:

Lemma 5.37 *A central extension of a $\mathscr{D}_P$-group by a $\mathscr{D}_P$-group is a $\mathscr{D}_P$-group.*

Proof The result follows from Lemmas 5.16 and 5.23. □

We wish to prove that if a nilpotent group is an extension of a $\mathscr{D}_P$-group by a $\mathscr{D}_P$-group, then it is also a $\mathscr{D}_P$-group. In order to establish this fact, two preparatory lemmas are needed.

Lemma 5.38 *Let G be a $\mathscr{D}_P$-group. A subgroup H of G is P-isolated in G if and only if it is a $\mathscr{D}_P$-group.*

Proof Suppose that H is P-isolated in G. In light of Proposition 5.1 (i), we only need to show that H is P-radicable. Let $h \in H$, and let n be a P-number. Since G is P-radicable, there exists $g \in G$ such that $h = g^n$. Since H is P-isolated in G and $g^n \in H$, we have $g \in H$. Hence, g is an nth root of h in H.

Conversely, suppose that H is a $\mathscr{D}_P$-subgroup of G and $g^n \in H$ for some $g \in G$ and P-number n. Since H is P-radicable, g^n has an nth root in H, say h. This means that $g^n = h^n$ in H. However, H is also a $\mathscr{U}_P$-group, so $g = h$. Consequently, $g \in H$ and H is P-isolated in G. □

Lemma 5.39 *Let G be a nilpotent $\mathscr{U}_P$-group and let N be a normal $\mathscr{D}_P$-subgroup of G. If $g \in G$, $n \in N$, and m is a P-number, then the element $\tilde{g} = g^m n$ has an mth root in G.*

Proof Suppose that G is of class c. Set $N_i = N \cap \zeta_i G$, where $0 \leq i \leq c$. We claim that $N_1 = N \cap Z(G)$ is a $\mathscr{D}_P$-subgroup of G. Observe that Theorem 5.8 (i) guarantees that $Z(G)$ is P-isolated in G since G is a $\mathscr{U}_P$-group. Furthermore, N_1 is P-isolated in G and therefore in N, according to Lemma 5.2 (ii). By Lemma 5.38, N_1 is a $\mathscr{D}_P$-subgroup of G as claimed.

Now, there exists a least integer t satisfying $N_t = N$ since $N_c = N$. We proceed by induction on t. If $t = 1$, then $N = N \cap Z(G)$ and thus $N \leq Z(G)$. Since N is a $\mathscr{D}_P$-group, there exists $n_0 \in N$ such that $n = n_0^m$. Hence,

$$\tilde{g} = g^m n = g^m n_0^m = (g n_0)^m.$$

And so, $\tilde{g}$ has an mth root in G. This gives the basis step of a proof by induction.

By applying the induction hypothesis to G/N_1, we conclude that $\tilde{g}N_1$ has an mth root in G/N_1. Thus, $\tilde{g} = h^m n_1$ for some $h \in G$ and $n_1 \in N_1$. Since N_1 is a $\mathscr{D}_P$-group, there exists $n_2 \in N_1$ such that $n_2^m = n_1$. Therefore,

$$\tilde{g} = h^m n_1 = h^m n_2^m = (hn_2)^m$$

because $n_2 \in Z(G)$. Hence, $\tilde{g}$ has an mth root in G.   □

Theorem 5.41 *If G is a nilpotent group which is an extension of a $\mathscr{D}_P$-group by a $\mathscr{D}_P$-group, then G is also a $\mathscr{D}_P$-group.*

This is also true for locally nilpotent groups (see Theorem 23.2 in [2]).

Proof Suppose that $N \trianglelefteq G$. By Lemma 5.17, G is a $\mathscr{U}_P$-group. We claim that G is P-radicable. Let m be a P-number and $\tilde{g} \in G$. Since G/N is a $\mathscr{D}_P$-group, $\tilde{g}N$ has a unique mth root in G/N, say gN. Thus, there exists $n \in N$ such that $\tilde{g} = g^m n$. The result follows from Lemma 5.39.   □

The next theorem is a variation of Theorem 11.5 in G. Baumslag's work [2]. It is also proven by P. Ribenboim in [28] (Proposition 6.18).

Theorem 5.42 *Let G be a nilpotent group with $H \leq G$. Suppose that P_1 and P_2 are sets of primes and $P_1 \cap P_2 = \emptyset$. If H is a $\mathscr{U}_{P_1}$-group and $G = I_{P_2}(H, G)$, then G is also a $\mathscr{U}_{P_1}$-group. If in addition, H is P_1-radicable, then G is a $\mathscr{D}_{P_1}$-group.*

Proof We first prove that G is a $\mathscr{U}_{P_1}$-group. According to Theorem 2.7, it suffices to show that G is a P_1-torsion-free. Let $g \in G$ and suppose that n is a P_1-number such that $g^n = 1$. We assert that $g = 1$. By hypothesis, $G = I_{P_2}(H, G)$. Hence, there exists a P_2-number m such that $g^m \in H$. Thus,

$$\left(g^m\right)^n = \left(g^n\right)^m = 1 = 1^n.$$

Since H is $\mathscr{U}_{P_1}$-group, $g^m = 1$. Now, m and n are relatively prime since $P_1 \cap P_2 = \emptyset$. Therefore, there exist $a, \ b \in \mathbb{Z}$ such that $am + bn = 1$. And so,

$$g = g^{am+bn} = \left(g^m\right)^a \left(g^n\right)^b = 1,$$

proving the assertion.

Next, suppose that H is also P_1-radicable. Let $g \in G$ and suppose that n is a P_1-number. It is enough to show that g has an nth root. As before, there exists a P_2-number m such that $g^m \in H$. Since H is P_1-radicable, g^m has an nth root in H. Hence, $g^m = h^n$ for some $h \in H$. Now,

$$\left(g^{-1}hg\right)^n = g^{-1}h^n g = g^{-1}g^m g = g^m = h^n. \tag{5.11}$$

Since G is a $\mathscr{U}_{P_1}$-group, (5.11) becomes $g^{-1}hg = h$. Thus, g and h commute. Since m and n are relatively prime, there exist integers a and b such that $am + bn = 1$. Hence,

$$g = g^{am+bn} = \left(g^m\right)^a g^{bn} = \left(h^n\right)^a g^{bn} = h^{an} g^{bn} = \left(h^a g^b\right)^n.$$

And so, g has an nth root.   □

Remark 5.13 In light of Theorem 2.7, one could replace "$\mathscr{U}_{P_1}$-group" with "P_1-torsion-free group" in Theorem 5.42.

5.2.15 Some Embedding Theorems

There are several instances in which one can embed a nilpotent group into a radicable (locally) nilpotent group. One result in this direction was already discussed in Chapter 4 (see Theorem 4.15). Our intention is to present some other classical results.

Let G be a torsion-free nilpotent group and let $g \in G$. In [22], A. I. Mal'cev proved that for any prime p, G can be embedded in a torsion-free nilpotent group H of the same nilpotency class as G in such a way that g has a pth root in H. His proof relied on methods involving Lie algebras. Using residual properties, G. Baumslag proved this without the assumption of torsion-freeness (see [4]).

Theorem 5.43 *Let G be a finitely generated nilpotent group and let p be any prime. For any element $g \in G$, there exists a finitely generated nilpotent group H_g such that G embeds in H_g and g has a pth root in H_g.*

Proof Let $x \in G$ and $x \notin \tau_p(G)$. We assert that there is a normal subgroup N_x of G such that $x \notin N_x$ and $G_x = G/N_x$ is a finite p-torsion-free group. First, notice that $\tau_p(G)$ is a finite p-group by Corollary 2.16. Thus, the factor group $G/\tau_p(G)$ is a finitely generated p-torsion-free nilpotent group by Corollary 2.15. Furthermore, $G/\tau_p(G)$ is a residually finite p'-group by Theorem 5.33. Since $x \notin \tau_p(G)$, we have that $x\tau_p(G) \neq \tau_p G$. Hence, there exists a normal subgroup $N_x/\tau_p G$ of $G/\tau_p G$ satisfying the property that

$$\frac{G/\tau_p G}{N_x/\tau_p G} \cong G/N_x$$

is a finite p'-group and $x\tau_p G \notin N_x/\tau_p G$. This means that $x \notin N_x$ and G/N_x is finite and p-torsion-free. This proves the assertion.

We claim that G contains a p-torsion-free characteristic subgroup of finite index. Suppose that $\tau_p(G) = \{g_1, \ldots, g_k\}$. By Theorem 5.31, there exist normal subgroups $N_1, \ldots, N_k$ of finite index in G such that $g_1 \notin N_1, \ldots, g_k \notin N_k$. Then

$$M = \bigcap_{1 \leq i \leq k} N_i$$

is a normal subgroup of G of finite index by Theorem 5.27. Furthermore, M is p-torsion-free because each g_i is excluded from at least one of the $N_1, \ldots, N_k$, and thus $g_i \notin M$ for each $i = 1, \ldots, k$. The existence of a p-torsion-free characteristic subgroup of finite index in G follows from Theorem 5.29, proving the claim.

Let $N(p)$ be a p-torsion-free characteristic subgroup of finite index in G and set $G(p) = G/N(p)$. By Lemma 5.32, G can be embedded in the direct product of $G(p)$ and an unrestricted direct product:

$$D = G(p) \times \overline{\prod_{x \notin \tau_p(G)}} G_x.$$

Now, $G(p)$ can be embedded in a finite nilpotent group J so that every element of $G(p)$ has a pth root in J (see [1]). Furthermore, for each $x \notin \tau_p(G)$, every element of G_x has a pth root in G_x because G_x is finite p-torsion-free by Lemma 5.22. Let

$$D^* = J \times \overline{\prod_{x \notin \tau_p(G)}} G_x.$$

By Remark 2.9, D^* is nilpotent. Since D embeds in D^* and the factors in a direct product commute, every element of D must have a pth root in D^*.

Let g_0 be a pth root of g in D^*. Consider the finitely generated subgroup $H_g = gp(G, g_0)$ of D^*. Clearly, G embeds in H_g and H_g is nilpotent because D^* is. Moreover, g has a pth root in H_g.   □

Remark 5.14 If G is p-torsion-free, then H_g can be chosen p-torsion-free. To see this, let g_0 be a pth root of g in D^* and define the subgroup $K_g = gp(G, g_0)$ of D^*. If we put $H_g = K_g/\tau_p(K_g)$, then H_g is a p-torsion-free group in which G embeds. Furthermore, g has a pth root in H_g. Similarly, if G is torsion-free, then H_g can be chosen to be torsion-free.

Theorem 5.44 (G. Baumslag) *Every finitely generated nilpotent group can be embedded in a locally nilpotent radicable group.*

Proof Arrange the set of all primes in a sequence

$$p_1, p_2, p_3, \ldots, \tag{5.12}$$

where (5.12) satisfies the condition that if p is any prime in (5.12) and $m \in \mathbb{N}$, then there exists a natural number $n > m$ such that $p_n = p$. Thus, each prime in (5.12) occurs infinitely many times.

Let G be a finitely generated nilpotent group. We construct a countable ascending sequence of finitely generated nilpotent groups

$$G_1 \le G_2 \le G_3 \le \cdots \tag{5.13}$$

as follows: Set $G_1 = G$. Suppose that $G_1, G_2, \ldots, G_n$ in (5.13) have been defined and that the elements of G_i ($i = 1, \ldots, n$) have already been explicitly enumerated in a nonterminating sequence:

$$u_{i1}, \ u_{i2}, \ u_{i3}, \ \ldots \quad (i = 1, \ 2, \ \ldots, \ n).$$

We allow repetitions of group elements in an enumeration, so the various sequences can therefore be chosen infinite, as required. By Theorem 5.43, we adjoin one root at a time and take G_{n+1} to be any finitely generated nilpotent group containing G_n such that

$u_{11}, \ u_{12}, \ \ldots, \ u_{1 \ n}$ have p_nth roots in G_{n+1},

$u_{21}, \ u_{22}, \ \ldots, \ u_{2 \ n-1}$ have p_{n-1}th roots in $G_{n+1}, \ \ldots$, and

u_{n1} has a p_1th root in G_{n+1}.

Therefore, we can define all the groups G_i inductively in this way. Let

$$G^* = \bigcup_{n=1}^{\infty} G_n.$$

We claim that G^* is a locally nilpotent radicable group. It is obviously locally nilpotent because (5.13) is an ascending series of finitely generated nilpotent subgroups of G^*. We show that it is radicable. Let $g \in G^*$ and let p be any prime. There exists $m \in \mathbb{N}$ such that $g \in G_m$. Now, by the choice of the sequence of primes (5.12) and the groups $G_1, \ G_2, \ \ldots$, it follows that g has a pth root in G_l for a sufficiently large l. Thus, g has a pth root in G^*.   □

By using certain embedding techniques described by B. H. Neumann in [25], one can obtain:

Theorem 5.45 *Every locally nilpotent group can be embedded in a locally nilpotent radicable group.*

A free nilpotent group can always be embedded in a nilpotent $\mathscr{D}_P$-group. This is the content of the next theorem.

Theorem 5.46 *Every free nilpotent group can be embedded in a nilpotent $\mathscr{D}_P$-group of the same class.*

We give a proof due to Baumslag [5] which uses a method that is similar to the one given in Theorem 5.44. The original proof by A. I. Mal'cev uses Lie ring methods [22].

Proof Let G be a free nilpotent group of class c, freely generated by the set X. Set $G = G_1$ and $X = X_1$. Suppose that for each $n \in \mathbb{N}$, G_n is a free nilpotent group, freely generated by X_n, and the cardinality of each X_n is equal to the cardinality of X_1. Just as in the proof of Theorem 5.44, we define

$$p_1, \ p_2, \ p_3, \ \ldots \tag{5.14}$$

to be an infinite sequence of primes in P, chosen so that for any $p \in P$ and any positive integer n, there exists an integer $m \geq n$ such that $p_m = p$. Next, define

$$H_n = gp\,(x^{p_{n-1}} \mid x \in X_n) \le G_n \qquad (n \ge 2).$$

By an unpublished theorem of G. Baumslag, H_n is free nilpotent of class c, freely generated by

$$X_n^{p_{n-1}} = \{x^{p_{n-1}} \mid x \in X_n\}.$$

It is clear that the cardinalities of $X_n^{p_{n-1}}$, X_n, and X_{n-1} all coincide. Consequently, we may identify G_{n-1} with H_n for $n = 2,\ 3,\ \ldots$. Put

$$G^* = \bigcup_{n=1}^{\infty} G_n.$$

Evidently, G^* is nilpotent of class c. By the choice of the sequence (5.14), every element of G^* can be written as a product of pth powers for any $p \in P$. Therefore, G^* is semi-P-radicable and thus P-radicable by Theorem 5.17. It is also true that G^* is torsion-free since each G_n is torsion-free. Thus, G^* is a $\mathscr{U}_P$-group by Theorem 2.7. Therefore, G^* is a nilpotent $\mathscr{D}_P$-group of class c.   □

5.3 The P-Localization of Nilpotent Groups

In this section, we offer a brief introduction to the theory of P-localization of nilpotent groups. We refer the reader to [12, 13, 28], and [33] for fine accounts on the subject. This material is closely related to root extraction in nilpotent groups.

In the last two sections, we have insisted that P be a nonempty set of primes. For reasons that will become evident, we now declare that P can be any proper subset of the set of primes, including the empty set. In this situation, P' cannot be empty. In particular, P' is the set of all primes if P is the empty set.

5.3.1 P-Local Groups

Our discussion of the P-localization theory of nilpotent groups begins with the definition of a P-local group.

Definition 5.17 A group G is called P-*local* if the set map

$$\psi : G \to G \text{ defined by } \psi(g) = g^n$$

is a bijection for all P'-numbers n.

Note that ψ need not be a homomorphism of G, but merely a bijective map from G to itself. If P consists of a single prime p in Definition 5.17, then we refer to G as *p-local*.

The injectivity of ψ implies that G is a $\mathscr{U}_{P'}$-group and its surjectivity implies that G is P'-radicable. Thus, G is P-local if and only if it is a $\mathscr{D}_{P'}$-group. For convenience, some of the earlier results on extraction of roots will be reformulated in terms of P-local groups.

We begin by giving a description of P-local abelian groups. For a fixed set of primes P, consider the ring

$$\mathbb{Z}_P = \{m/n \in \mathbb{Q} \mid n \neq 0 \text{ is a } P'\text{-number}\}.$$

Thus, $\mathbb{Z}_P$ consists of those rational numbers whose denominators are relatively prime to the elements of P. If P is the empty set, then $\mathbb{Z}_P$ is the field of rational numbers $\mathbb{Q}$. We readily observe that a P-local abelian group is one that allows an action by the ring $\mathbb{Z}_P$. In terms of modules, this gives our first proposition.

Proposition 5.2 *An abelian group is P-local if and only if it is a $\mathbb{Z}_P$-module.*

By Proposition 5.2, the P-local subgroups of an abelian group A are simply the $\mathbb{Z}_P$-submodules of A. In general, a subgroup of a P-local group need not be P-local.

The next result connects isolator theory with P-local groups and their P-local subgroups. It is merely a translation of Lemma 5.38.

Lemma 5.40 *Let G be a P-local group and $H \leq G$. Then H is P-local if and only if H is P'-isolated in G.*

Lemma 5.41 *Every P-local group is P'-torsion-free.*

Proof Since P-local groups are $\mathscr{U}_{P'}$-groups, they are P'-torsion-free by Proposition 5.1 (iii). □

The next lemma is a trivial consequence of Lemma 5.14.

Lemma 5.42 *Let N be a normal subgroup of a group G. If G/N is P-local, then N is P'-isolated in G.*

Since our goal here is to offer an overview of the localization theory for nilpotent groups, we next focus on groups which are both P-local and nilpotent.

Theorem 5.47 *If G is a nilpotent group which is an extension of a P-local nilpotent group by a P-local nilpotent group, then G is P-local.*

Proof The result is just Theorem 5.41 in the context of P-local groups. □

Theorem 5.48 *Let G be a P-local nilpotent group and $N \trianglelefteq G$. Then N is P-local if and only if G/N is P-local.*

Proof See Theorem 5.40. □

In a nilpotent $\mathscr{D}_P$-group, the lower and upper central subgroups are themselves $\mathscr{D}_P$-groups according to Theorem 5.38. In the language of P-local groups, we have:

Theorem 5.49 *The lower and upper central subgroups of a P-local nilpotent group are P-local.*

The next two lemmas are of a general nature. No nilpotency is assumed.

Lemma 5.43 *A direct product of a family of P-local groups is P-local.*

Proof See Lemma 5.36. □

Lemma 5.44 *The intersection of a collection $\{G_i \mid i \in I\}$ of P-local subgroups of a P-local group G is P-local.*

Proof Let $K = \bigcap_{i \in I} G_i$. Since G is P-local, it is a $\mathscr{U}_{P'}$-group. By Proposition 5.1 (i), K also is a $\mathscr{U}_{P'}$-group.

We show that K is P'-radicable. Suppose that $g \in K$ and n is a P'-number. Since $g \in G_i$ for each $i \in I$ and G_i is P-local (thus P'-radicable), there exists an element $h_i \in G_i$ such that $g = h_i^n$ for each $i \in I$. Hence, $h_i^n = h_j^n$ for all $i, j \in I$. Since G is a $\mathscr{U}_{P'}$-group, $h_i = h_j$ for all $i, j \in I$. If we set $h = h_i$ for all $i \in I$, then $h \in K$ and $g = h^n$. Therefore, K is P'-radicable and thus P-local. □

Lemma 5.44 motivates the next definition.

Definition 5.18 Suppose that G is a P-local group, and let S be a nonempty subset of G. The *P-local subgroup of G generated by S* is the intersection of all P-local subgroups of G containing S.

The next result is a mere reformulation from isolator theory.

Theorem 5.50 *Suppose that G is a P-local group and $H \leq G$. If M is the P-local subgroup of G generated by H, then $M = I_{P'}(H, G)$.*

Proof By Definition 5.2, $I_{P'}(H, G)$ is the intersection of all P'-isolated subgroups of G containing H, while M is the intersection of all P-local subgroups containing H. The result follows from Lemma 5.40. □

Corollary 5.23 *Suppose that G is a P-local nilpotent group and $H \leq G$. Let M be the P-local subgroup of G generated by H. For every $g \in M$, there is a P'-number n such that $g^n \in H$.*

Proof The result is a consequence of Theorems 5.2 and 5.50. □

Corollary 5.24 *Let G be a P-local nilpotent group. If $N < G$ and N generates G as a P-local group, then any nontrivial subgroup of G intersects N nontrivially.*

Proof Let H be a nontrivial subgroup of G and choose $1 \neq g \in H$. By Corollary 5.23, there exists a P'-number n such that $g^n \in N$ and thus $g^n \in N \cap H$. As G is P-local, it is also P'-torsion-free by Lemma 5.41. Thus, $g^n \neq 1$. □

Corollary 5.25 *Let G be a P-local nilpotent group. If H is a subgroup of G which generates G as a P-local group, then G and H have the same nilpotency class.*

Proof By Theorem 5.50, $G = I_{P'}(H, G)$. Apply Theorem 5.2. □

Suppose that $\varphi : G \to H$ is a homomorphism between *P*-local nilpotent groups and N is a subgroup of G that generates G as a *P*-local group. Then φ is completely determined by its action on N. We record this as a lemma.

Lemma 5.45 *Let* $\varphi,\ \psi\ :\ G \to H$ *be two homomorphisms of P-local nilpotent groups. Suppose that N is a subgroup of G that generates G as a P-local group. If φ and ψ coincide on N, then $\varphi = \psi$.*

Proof Let $g \in G$. By Corollary 5.23, there exists an *P'*-number n such that $g^n \in N$. Thus

$$(\varphi(g))^n = \varphi(g^n) = \psi(g^n) = (\psi(g))^n.$$

Since H is *P*-local, $\varphi(g) = \psi(g)$. □

5.3.2 Fundamental Theorem of P-Localization of Nilpotent Groups

The *P*-localization of a group in a given subcategory is defined next. We assume familiarity with the basic definitions from category theory. The reader may wish to consult Chapter IV of [19] for background material.

Definition 5.19 Let $\mathscr{S}$ be a subcategory of the category of groups. A homomorphism

$$e : G \to G_P$$

in $\mathscr{S}$ is called a *P-localizing map* (also referred to as a *P-localization map*) if G_P is *P*-local and for any *P*-local group K in $\mathscr{S}$ and every homomorphism $\psi : G \to K$, there exists a unique homomorphism $\varphi : G_p \to K$ such that $\psi = \varphi \circ e$. The group G_P is termed the *P-localization* of G.

Lemma 5.46 *Let $\mathscr{S}$ be a subcategory of the category of groups and suppose that $\mathscr{S}$ admits a P-localizing map. Given a homomorphism $\psi : G \to K$ in $\mathscr{S}$, there exists a unique homomorphism $\psi_P : G_P \to K_P$ making the diagram*

$$
\begin{array}{ccc}
G & \xrightarrow{\ \psi\ } & K \\
\downarrow{\scriptstyle e} & & \downarrow{\scriptstyle e} \\
G_P & \xrightarrow{\ \psi_P\ } & K_P
\end{array}
$$

commute.

Proof The composite map

$$e \circ \psi : G \to K_P$$

is clearly a homomorphism from G to a P-local group K_P. Applying Definition 5.19 to $e \circ \psi$ and G gives the result.   □

The diagram in Lemma 5.46 gives a functor L from $\mathscr{S}$ to itself. The pair (L, e) is called a *localization theory* in $\mathscr{S}$.

In [12], P. Hilton used group cohomology theory to prove the existence of a localization theory for nilpotent groups (see also [13]). A different proof which utilizes some of the results that we have provided thus far in this section was carried out by R. B. Warfield, Jr. in [33]. We follow R. B. Warfield's work.

Theorem 5.51 (Fundamental Theorem of P-Localization) *Let G be a nilpotent group of class c. There exists a P-local nilpotent group G_P of class at most c and a P-localization map $e : G \to G_P$.*

Proof Let I be a nonempty index set. Consider, up to isomorphism, all pairs of the form

$$S_i = (H_i, \sigma_i) \quad (i \in I),$$

where H_i is a P-local nilpotent group, $\sigma_i : G \to H_i$ is a homomorphism, and $\sigma_i(G)$ generates H_i as a P-local group. Clearly, at least one such pair exists: take H_1 to be the trivial group and $\sigma_1 : G \to H_1$ to be the trivial homomorphism.

Let H be the direct product of the H_i, and define the map

$$\sigma : G \to H \text{ by } g \mapsto (\sigma_i(g))_{i \in I}.$$

For each $i \in I$, the class of $\sigma_i(G)$ is at most c. Moreover, since $\sigma_i(G)$ generates H_i as a P-local group, their classes are equal by Corollary 5.25. Thus, H is of class at most c by Remark 2.9. Furthermore, H is P-local since each H_i is P-local. This is immediate from the definitions.

Let G_P be the P-local subgroup of H which is generated P-locally by $\sigma(G)$. We rename the map σ as e and regard it as a homomorphism from G to G_P. We claim that e is a P-localization map. Suppose that K is any P-local group and $\psi : G \to K$ is a homomorphism. If W is the P-local subgroup of K which is generated by $\psi(G)$, then there exists $i_0 \in I$ such that, up to isomorphism,

$$S_{i_0} = \left(H_{i_0}, \sigma_{i_0}\right) = (W, \psi).$$

The homomorphism

$$\mu : G_P \to H_{i_0} \text{ defined by } (h_i)_{i \in I} \mapsto h_{i_0}$$

satisfies $\mu \circ e = \sigma_{i_0}$. We show that μ is unique with respect to this property. Let $\chi : G_P \to K$ be a homomorphism such that $\chi \circ e = \sigma_{i_0}$. For every $g \in G$,

$$\mu(e(g)) = \chi(e(g)),$$

and thus μ and χ coincide on $e(G)$. However, $e(G)$ generates G_P as a P-local group. By Lemma 5.45, $\mu = \chi$ on G_P. Thus, e is a P-localization map. $\square$

Corollary 5.26 *If G is a nilpotent group with P-localization map $e : G \to G_P$, then $e(G)$ generates G_P as a P-local group.*

Proof The result follows immediately from the proof of Theorem 5.51. $\square$

In order to give some intuition behind Theorem 5.51, we construct the P-localization of an abelian group. Consider the subring $\mathbb{Z}_P$ of the ring $\mathbb{Q}$ discussed at the beginning of this section. It is a P-local abelian group and the obvious embedding of $\mathbb{Z}$ into $\mathbb{Z}_P$ given by $n \mapsto n/1$ is a P-localization map. Let A be any abelian group and define $A_P = A \otimes \mathbb{Z}_P$. By Proposition 5.2, A_P is a P-local abelian group. The homomorphism $a \mapsto a \otimes 1$ gives a P-localization map $e : A \to A_P$, yielding a localization theory in the category of abelian groups.

The P-localization of a subgroup of a P-local nilpotent group can be described in terms of isolators.

Theorem 5.52 *Let G be a P-local nilpotent group. If $N \leq G$, then N_P is isomorphic to the P'-isolator of N in G.*

Proof Since G is P-local, $G_P = G$. Let $i : N \to G$ be the natural inclusion of N in G and contemplate the P-localization diagram guaranteed by Lemma 5.46:

$$
\begin{array}{ccc}
N & \xrightarrow{\ i\ } & G \\
\downarrow{\scriptstyle e} & & \downarrow{\scriptstyle e} \\
N_P & \xrightarrow{\ i_P\ } & G_P = G
\end{array}
$$

It is clear from the diagram that $e : N \to N_P$ is a monomorphism. We claim that i_P is also a monomorphism. To see this, let $x \in ker\ i_P$. By Corollary 5.26, the image of e generates N_P as a P-local group. Thus, Corollary 5.23 guarantees the existence of a P'-number n such that $x^n \in im\ e$. And so, there exists a unique $y \in N$ such that $e(y) = x^n$. Now, $x \in ker\ i_P$ implies that $x^n \in ker\ i_P$ and thus $(i_P \circ e)(y) = 1$. Hence, $y = 1$ and i_P is a monomorphism as claimed. Thus, the P-local subgroup of G generated by N is isomorphic to N_P. The result follows from Theorem 5.50. $\square$

5.3.3   P-Morphisms

Our next definition contains useful generalizations of the more standard notions involving group homomorphisms. These are convenient for the study of P-localization, as illustrated by P. Hilton and others in [12] and [13].

Definition 5.20 Let G and H be groups. A homomorphism $\varphi : G \to H$ is called

  (i) *P-injective* if $ker\ \varphi = \{g \in G \mid g$ is P'-torsion$\}$;
 (ii) *P-surjective* if for all $h \in H$, there exists a P'-number n such that $h^n \in \varphi(G)$;
(iii) a *P-isomorphism* if φ is both P-injective and P-surjective.

For a single prime p, the usual notions are p-injective, p-surjective, and p-isomorphism.

In [12] and [13], various properties of P-morphisms are established. For example, if a homomorphism between P-*local* groups is given, then the standard definition of an injective (surjective) homomorphism is equivalent to the notion of a P-injective (P-surjective) homomorphism. This is demonstrated in the next lemma.

Lemma 5.47 *Let* $\varphi : G \to H$ *be a homomorphism of P-local groups.*

 (i) If φ is P-injective, then φ is injective.
(ii) If φ is P-surjective, then φ is surjective.

Proof

  (i) If φ is P-injective and $g \in ker\ \varphi$, then there exists a P'-number n such that $g^n = 1$. Since G is P-local, $g = 1$. Therefore, φ is injective.
 (ii) Suppose that φ is P-surjective. If $h \in H$, then there exists $g \in G$ and a P'-number n such that $\varphi(g) = h^n$. Since G is P-local, there is a unique $x \in G$ such that $x^n = g$. Thus,

$$\varphi(g) = \varphi(x^n) = (\varphi(x))^n = h^n.$$

As H is P-local, $\varphi(x) = h$. Thus, φ is surjective.   □

An important result for nilpotent groups is that the P-localizing map is always a P-isomorphism.

Theorem 5.53 *Let P be a nonempty set of primes. For any finitely generated nilpotent group G, the P-localizing map $e : G \to G_P$ is a P-isomorphism.*

Proof The proof is based on [33]. By Corollaries 5.23 and 5.26, e is P-surjective. We only need to show that e is P-injective. Let $1 \neq g$ be a P'-torsion element of G. There exists a P'-number n such that $g^n = 1$, so that

$$(e(g))^n = e(g^n) = 1.$$

Since G_P is P-local, $e(g) = 1$. Therefore, $g \in ker\ e$ and thus all P'-torsion elements of G lie in $ker\ e$.

We assert that every element in $ker\ e$ is P'-torsion; that is, the induced map

$$\hat{e} : G/\tau_{P'}(G) \to G_P$$

is an embedding. Since $G/\tau_{P'}(G)$ is P'-torsion-free, this is equivalent to showing that $e : G \to G_P$ is an embedding, where G is assumed to be P'-torsion-free. By Theorem 5.33 and Lemma 5.32, G is a subgroup of a direct product of finite P-torsion groups. The proof of Lemma 5.32 shows that each factor of the direct product is nilpotent of class at most the class of G. Furthermore, G is P-local by Theorem 5.37. Since such a direct product is itself P-local according to Lemma 5.43, G must embed in its P-localization. $\square$

Remark 5.15 Theorem 5.53 holds for arbitrary nilpotent groups and not only the finitely generated ones. The argument in this case uses direct limits and systems. See Theorem 8.9 of [33]. We can also allow P to be the empty set. See the comment before Corollary 5.28.

The next two corollaries can be found in P. Ribenboim's work [28]. They also appear in [33].

Corollary 5.27 *Let G be a nilpotent group and let P_1 and P_2 be any two sets of primes. Then $\left(G_{P_1}\right)_{P_2}$ is a $(P_1 \cap P_2)$-local group.*

Proof Fix a prime $p \in (P_1 \cap P_2)'$. We claim that every element of $\left(G_{P_1}\right)_{P_2}$ has a unique pth root in $\left(G_{P_1}\right)_{P_2}$. Observe that

$$(P_1 \cap P_2)' = P_1' \cup P_2' = P_2' \cup (P_2 \cap P_1').$$

If $p \in P_2'$, then the claim is certainly true since $\left(G_{P_1}\right)_{P_2}$ is P_2-local. Suppose, therefore, that $p \in P_2 \cap P_1'$ and let

$$e : G_{P_1} \to \left(G_{P_1}\right)_{P_2}$$

be the P_2-localization map. We establish that $\left(G_{P_1}\right)_{P_2}$ is a $\mathcal{D}_p$-group by using Theorem 5.42, where we take the disjoint sets of primes to be $\{p\}$ and P_2' and the groups to be $e(G_{P_1})$ and $\left(G_{P_1}\right)_{P_2}$ respectively. By Theorem 5.50 and Corollary 5.26,

$$\left(G_{P_1}\right)_{P_2} = I_{P_2'}\left(e(G_{P_1}),\ \left(G_{P_1}\right)_{P_2}\right).$$

Since G_{P_1} is P_1-local and $p \in P_1'$, G_{P_1} is p-radicable. Hence, $e(G_{P_1})$ is also p-radicable by Lemma 5.18.

We assert that $e(G_{P_1})$ is a $\mathcal{U}_p$-group. In light of Corollary 5.6, it is enough to show that $ker\ e$ is p-isolated. Suppose that $g^p \in ker\ e$ for some $g \in G_{P_1}$. By Theorem 5.53, every element of $ker\ e$ is P_2'-torsion. Thus, there exists $q \in P_2'$ such that $\left(g^p\right)^q = 1$ and consequently, $\left(g^q\right)^p = 1$. Since $g^q \in G_{P_1}$ and $p \in P_1'$, we have that $g^q = 1$

because G_{P_1} is P_1'-torsion-free. Therefore, $(e(g))^q = 1$ and thus $g \in \ker e$ because $q \in P_2'$ and $(G_{P_1})_{P_2}$ is P_2'-torsion-free. And so, $\ker e$ is p-isolated as asserted. The result follows from Theorem 5.42.   □

Corollary 5.28 *Let P_1 and P_2 be any two sets of primes and let G be a nilpotent group. Then $(G_{P_1})_{P_2} \cong G_{P_1 \cap P_2}$.*

Proof Clearly, every $(P_1 \cap P_2)$-local group is P_1-local and P_2-local. In particular, $G_{P_1 \cap P_2}$ is P_1-local and P_2-local. Let

$$e_1 : G \to G_{P_1}, \quad e_2 : G_{P_1} \to (G_{P_1})_{P_2}, \quad \text{and} \quad \psi : G \to G_{P_1 \cap P_2}$$

be P_1-, P_2- and $(P_1 \cap P_2)$-localization maps respectively. By Theorem 5.51, there exists a unique homomorphism

$$\varphi : G_{P_1} \to G_{P_1 \cap P_2}$$

such that $\varphi \circ e_1 = \psi$. The same theorem guarantees the existence of a unique homomorphism

$$\beta : (G_{P_1})_{P_2} \to G_{P_1 \cap P_2}$$

such that $\beta \circ e_2 = \varphi$. Hence, $\psi = \beta \circ e_2 \circ e_1$. Now, $(G_{P_1})_{P_2}$ is a $(P_1 \cap P_2)$-local group by Corollary 5.27. By Theorem 5.51, there exists a unique homomorphism

$$\kappa : G_{P_1 \cap P_2} \to (G_{P_1})_{P_2}$$

such that $\kappa \circ \psi = e_2 \circ e_1$. Thus, $\psi = \beta \circ \kappa \circ \psi$ and by uniqueness, $\beta \circ \kappa$ equals the identity map on $G_{P_1 \cap P_2}$. We need to show that $\kappa \circ \beta$ equals the identity map on $(G_{P_1})_{P_2}$. To do this, we first observe that

$$\kappa \circ \beta \circ e_2 \circ e_1 = \kappa \circ \psi = e_2 \circ e_1.$$

The homomorphisms $\kappa \circ \beta \circ e_2$ and e_2 coincide on $e_1(G)$. Since $(G_{P_1})_{P_2}$ is P_1-local, it follows from Lemma 5.45 and Corollary 5.26 that $\kappa \circ \beta \circ e_2 = e_2$ on G_{P_1}. Moreover, $(G_{P_1})_{P_2}$ is P_2-local and $\kappa \circ \beta$ coincides with the identity map on $e_2(G_{P_1})$. Once again, Lemma 5.45 and Corollary 5.26 imply that $\kappa \circ \beta$ equals the identity map on $(G_{P_1})_{P_2}$ as desired.   □

As we mentioned in Remark 5.15, it follows from Corollary 5.28 that the set of primes P in Theorem 5.53 can also be taken to be empty.

Corollary 5.29 *If G is a nilpotent group with P-localization G_P and P-localization map $e : G \to G_P$, then $\tau(G_P) = e(\tau(G))$. Thus, all torsion from the localization G_P comes from G.*

Proof It is clear that $\tau(G_P) \geq e(\tau(G))$. We claim that $\tau(G_P) \leq e(\tau(G))$. Let $g \in \tau(G_P)$. There exists a *P*-number m such that $g^m = 1$. By Corollary 5.26, $e(G)$ generates G_P as a *P*-local group. Thus, there exists a P'-number n such that $g^n \in e(G)$ by Corollary 5.23. Since m and n are coprime, there exist integers a and b such that $am + bn = 1$. Hence,

$$g = g^{am+bn} = (g^m)^a (g^n)^b = (g^n)^b.$$

And so, $g \in e(G)$. Suppose that $g = e(h)$ for some $h \in G$. We assert that $h \in \tau(G)$. Observe that

$$e(h^m) = (e(h))^m = g^m = 1$$

and thus h^m is in the kernel of e. By Theorem 5.53, there is a P'-number k such that $(h^m)^k = 1$. Hence, $h^{mk} = 1$ and consequently, $h \in \tau(G)$ as asserted. $\qquad\square$

The next result is immediate from Theorem 5.53 and Corollary 5.29.

Corollary 5.30 *If G is a torsion-free nilpotent group, then its P-localization G_P is torsion-free and its P-localization map $e : G \rightarrow G_P$ is an embedding.*

We end this section by considering the question of whether a nilpotent group is determined (up to isomorphism) by its p-localizations, where p varies over the set of all primes. In other words, given that G and H are nilpotent groups and $G_p \cong H_p$ for all prime numbers p, does it follow that G is isomorphic to H?

The examples given below, which are due to V. N. Remeslennikov (see [27] and p. 8 in [6]), show that this is not true in general, not even for the finitely generated case. As noted in [6], the original motivation for these examples was to show that it is not possible to classify finitely generated torsion-free nilpotent groups by their finite homomorphic images.

We give presentations of V. N. Remeslennikov's examples within the category of class 4 nilpotent groups. Let

$$S = \left\langle x, y \mid [y, x, y, y]^3 [y, x, x, y][y, x, x, x]^2 = 1 \right\rangle$$

and

$$T = \left\langle x, y \mid [y, x, y, y]^6 [y, x, x, y][y, x, x, x] = 1 \right\rangle.$$

It is possible to show that $S_p \cong T_p$ for every p, but that S and T are not isomorphic (see Corollary 3.2.3 in [35]). Another set of examples exhibiting this behavior were constructed by G. Mislin and can be found on p. 69 of [33].

References

1. G. Baumslag, Wreath products and p-groups. Proc. Camb. Philos. Soc. **55**, 224–231 (1959). MR0105437
2. G. Baumslag, Some aspects of groups with unique roots. Acta Math. **104**, 217–303 (1960). MR0122859
3. G. Baumslag, Roots and wreath products. Proc. Camb. Philos. Soc. **56**, 109–117 (1960). MR0113932
4. G. Baumslag, A generalisation of a theorem of Mal'cev. Arch. Math. (Basel) **12**, 405–408 (1961). MR0142636
5. G. Baumslag, Some remarks on nilpotent groups with roots. Proc. Am. Math. Soc. **12**, 262–267 (1961). MR0123609
6. G. Baumslag, *Lecture Notes on Nilpotent Groups*. Regional Conference Series in Mathematics, No. 2 (American Mathematical Society, Providence, RI, 1971). MR0283082
7. S.N. Černikov, Complete groups possessing ascending central series. Rec. Math. [Mat. Sbornik] N.S. **18**(60), 397–422 (1946). MR0018646
8. S.N. Černikov, On the theory of complete groups. Mat. Sbornik N.S. **22**(64), 319–348 (1948). MR0024906
9. K.W. Gruenberg, Residual properties of infinite soluble groups. Proc. Lond. Math. Soc. (3) **7**, 29–62 (1957). MR0087652
10. P. Hall, *The Edmonton Notes on Nilpotent Groups*. Queen Mary College Mathematics Notes. Mathematics Department (Queen Mary College, London, 1969). MR0283083
11. G. Higman, A remark on finitely generated nilpotent groups. Proc. Am. Math. Soc. **6**, 284–285 (1955). MR0069176
12. P. Hilton, Localization and cohomology of nilpotent groups. Math. Z. **132**, 263–286 (1973). MR0322074
13. P. Hilton, G. Mislin, J. Roitberg, *Localization of Nilpotent Groups and Spaces* (North-Holland Publishing Co., Amsterdam, 1975). MR0478146
14. K.A. Hirsch, On infinite soluble groups (II). Proc. Lond. Math. Soc. **S2–44**(5), 336. MR1576691
15. I. Kaplansky, *Infinite Abelian Groups* (University of Michigan Press, Ann Arbor, 1954). MR0065561
16. P. Kontorovič, On the theory of noncommutative groups without torsion. Dokl. Akad. Nauk SSSR (N.S.) **59**, 213–216 (1948). MR0023826
17. A.G. Kurosh, *The Theory of Groups* (Chelsea Publishing Co., New York, 1960). MR0109842. Translated from the Russian and edited by K. A. Hirsch. 2nd English ed. 2 volumes
18. J.C. Lennox, D.J.S. Robinson, *The Theory of Infinite Soluble Groups*. Oxford Mathematical Monographs (The Clarendon Press/Oxford University Press, Oxford, 2004). MR2093872
19. S. Mac Lane, G. Birkhoff, *Algebra*, 3rd edn. (Chelsea Publishing Co., New York, 1988). MR0941522
20. S. Majewicz, M. Zyman, On the extraction of roots in exponential A-groups. Groups Geom. Dyn. **4**(4), 835–846 (2010). MR2727667
21. S. Majewicz, M. Zyman, On the extraction of roots in exponential A-groups II. Commun. Algebra **40**(1), 64–86 (2012). MR2876289
22. A.I. Mal'cev, Nilpotent torsion-free groups. Izv. Akad. Nauk. SSSR. Ser. Mat. **13**, 201–212 (1949). MR0028843
23. D. McLain, A class of locally nilpotent groups, Ph.D. Dissertation, Cambridge University, 1956
24. B.H. Neumann, Adjunction of elements to groups. J. Lond. Math. Soc. **18**, 4–11 (1943). MR0008808
25. B.H. Neumann, An embedding theorem for algebraic systems. Proc. Lond. Math. Soc. (3) **4**, 138–153 (1954). MR0073548
26. H. Neumann, *Varieties of Groups* (Springer, New York, 1967). MR0215899

27. V.N. Remeslennikov, Conjugacy of subgroups in nilpotent groups. Algebra i Logika Sem. **6**(2), 61–76 (1967). MR0218459
28. P. Ribenboim, Equations in groups, with special emphasis on localization and torsion. II. Portugal. Math. **44**(4), 417–445 (1987). MR0952790
29. D.J.S. Robinson, *Finiteness Conditions and Generalized Soluble Groups. Part 1* (Springer, New York, 1972). MR0332989
30. D.J.S. Robinson, *Finiteness Conditions and Generalized Soluble Groups. Part 2*, (Springer, New York, 1972). MR0332990
31. J.S. Rose, *A Course on Group Theory* (Dover Publications, New York, 1994). MR1298629. Reprint of the 1978 original [Dover, New York; MR0498810 (58 #16847)]
32. W.R. Scott, *Group Theory* (Prentice-Hall, Inc., Englewood Cliffs, NJ, 1964). MR0167513
33. R.B. Warfield, Jr., *Nilpotent Groups*. Lecture Notes in Mathematics, vol. 513 (Springer, Berlin, 1976). MR0409661
34. B.A.F. Wehrfritz, A note on residual properties of nilpotent groups. J. Lond. Math. Soc. (2) **5**, 1–7 (1972). MR0302767
35. M. Zyman, *IA-Automorphisms and Localization of Nilpotent Groups* (ProQuest LLC, Ann Arbor, MI, 2007). MR2711065

Chapter 6
"The Group Ring of a Class of Infinite Nilpotent Groups" by S. A. Jennings

This chapter is based on a seminal paper entitled "The Group Ring of a Class of Infinite Nilpotent Groups" by S. A. Jennings [7]. In Section 6.1, we consider the group ring of a finitely generated torsion-free nilpotent group over a field of characteristic zero. We prove that its augmentation ideal is residually nilpotent. We introduce the dimension subgroups of a group in Section 6.2. These subgroups are defined in terms of the augmentation ideal of the corresponding group ring. We prove that the nth dimension subgroup coincides with the isolator of the nth lower central subgroup. This is a major result involving a succession of clever reductions where nilpotent groups play a prominent role. Section 6.3 deals with nilpotent Lie algebras. We show that there exists a nilpotent Lie algebra over a field of characteristic zero which is associated with a finitely generated torsion-free nilpotent group. As it turns out, the underlying vector space of this Lie algebra has dimension equal to the Hirsch length of the given group.

6.1 The Group Ring of a Torsion-Free Nilpotent Group

We begin with some definitions regarding group rings.

6.1.1 Group Rings and the Augmentation Ideal

Let G be a group and let R be a commutative ring with unity. The *group ring RG* of G over R is defined to be the set of all finite formal sums

$$\sum_{g \in G} r_g g \quad (r_g \in R),$$

together with the addition and multiplication rules

$$\left(\sum_{g\in G} r_g g\right) + \left(\sum_{g\in G} s_g g\right) = \sum_{g\in G} (r_g + s_g)g$$

and

$$\left(\sum_{g\in G} r_g g\right)\left(\sum_{h\in G} s_h h\right) = \sum_{g,\,h\in G} (r_g s_h)gh.$$

It is clear that RG is indeed a ring under the prescribed operations. We point out that the group ring RG can also be regarded as a free R-module with a basis consisting of the elements of G since R acts naturally on RG.

We will abuse notation and denote the unity element in R, the unity element in RG, and the group identity in G by 1. If $g \in G$ and $r \in R$, then it is conventional to write the elements $1g \in RG$ and $r1 \in RG$ simply as g and r respectively.

In this chapter, R will be a commutative ring with unity unless otherwise stated.

Definition 6.1 Let G be a group. The kernel of the ring homomorphism

$$\pi : RG \to R \text{ given by } \pi\left(\sum_{g\in G} r_g g\right) = \sum_{g\in G} r_g$$

is called the *augmentation ideal* of RG.

The augmentation ideal of RG is denoted by $A_R(G)$, or just $A(G)$ when R is understood from the context. Thus,

$$A(G) = \left\{\sum_{g\in G} r_g g \in RG \,\middle|\, \sum_{g\in G} r_g = 0\right\}.$$

Since π is a ring epimorphism, the quotient ring $RG/A(G)$ is isomorphic to R. Furthermore, $A(G)$ is an R-submodule of RG because it is an ideal of RG.

It is clear that $g - 1 \in A(G)$ for any $1 \neq g \in G$. The next lemma shows that such elements form an R-basis for $A_R(G)$.

Lemma 6.1 *Let G be any group. The set*

$$S = \{g - 1 \mid g \in G,\ g \neq 1\}$$

is an R-basis for $A_R(G)$. Thus, $A_R(G)$ is a free R-module with basis S.

Proof Let $x = \sum_{g\in G} r_g g \in A_R(G)$. Then $\sum_{g\in G} r_g = 0$. Thus,

$$x = x - \left(\sum_{g\in G} r_g\right)1 = \sum_{g\in G} r_g g - \left(\sum_{g\in G} r_g\right)1 = \sum_{g\in G} r_g(g-1).$$

Hence, S spans $A_R(G)$.

We need to show that S is linearly independent. Suppose that

$$\sum_{1\neq g\in G} r_g(g-1) = 0.$$

Then

$$\sum_{g\in G} r_g g = \left(\sum_{g\in G} r_g\right)1. \tag{6.1}$$

Since G is a basis for the R-module RG and every g appearing in (6.1) is distinct from the identity, $r_g = 0$ for all of these g. The result follows. $\qquad\square$

As in Definition 2.6, let $A_R^n(G)$ (or $A^n(G)$ when R is understood) denote the ideal of RG generated by all products of n elements of the augmentation ideal of RG.

Corollary 6.1 *Let G be any group. The set*

$$S = \{(g_1 - 1)(g_2 - 1)\cdots(g_n - 1) \mid 1 \neq g_i \in G\}$$

spans $A^n(G)$.

Proof This is an immediate consequence of Lemma 6.1. $\qquad\square$

Lemma 6.2 *Let G be a group and $H \leq G$. For all $n \in \mathbb{N}$, $A^n(H) \subseteq A^n(G)$.*

Proof The result follows from the fact that $A(H) \subseteq A(G)$. $\qquad\square$

Let G be any group with $N \trianglelefteq G$. The natural homomorphism $\pi : G \to G/N$ induces a ring homomorphism

$$\varrho_N : RG \to R(G/N) \text{ defined by } \varrho_N\left(\sum_{g\in G} a_g g\right) = \sum_{g\in G} a_g(gN).$$

We record two useful facts about the map ϱ_N in the next lemma.

Lemma 6.3 *The kernel of ϱ_N is the two-sided ideal of RG given by*

$$RG \cdot A(N) = A(N) \cdot RG,$$

and $\varrho_N(A^n(G)) = A^n(G/N)$ for any $n \in \mathbb{N}$.

Proof We prove that $\ker \varrho_N = RG \cdot A(N)$. Let $x = \sum_i a_i g_i$ be an element of RG, where $a_i \in R$ and $g_i \in G$. For each g_k in this expression for x, we let $\bar{g}_k \in G$ be a representative of the left coset $g_k N$. We can write

$$x = \sum_{k,\, l} a'_{kl} \bar{g}_k n_l \quad \left(a'_{kl} \in R,\ \bar{g}_k \in G,\ n_l \in N \right).$$

Now,

$$\varrho_N(x) = \sum_{k,\, l} a'_{kl} \bar{g}_k N = \sum_{k} \left(\sum_{l} a'_{kl} \right) \bar{g}_k N \in R[G/N].$$

We readily observe that $x \in ker\, \varrho_N$ if and only if $\sum_l a'_{kl} = 0$ for every k.

Suppose now that $x \in ker\, \varrho_N$. For each k, we have $\sum_l a'_{kl} \bar{g}_k = 0$. Thus,

$$\sum_l a'_{kl} \bar{g}_k n_l = \sum_l a'_{kl} \bar{g}_k n_l - \sum_l a'_{kl} \bar{g}_k$$

$$= \sum_l a'_{kl} \bar{g}_k (n_l - 1)$$

for each k. Hence,

$$x = \sum_{k,\, l} a'_{kl} \bar{g}_k n_l = \sum_{k} \left(\sum_{l} a'_{kl} \bar{g}_k (n_l - 1) \right).$$

It follows that $x \in RG \cdot A(N)$.

Next, we show that $RG \cdot A(N) \subseteq ker\, \varrho_N$. Since ϱ_N maps every element of $A(N)$ to zero, $A(N) \subseteq \ker \varrho_N$. If $x \in RG \cdot A(N)$, then

$$x = \sum_i \varrho_i \sigma_i \quad (\varrho_i \in RG,\ \sigma_i \in A(N)).$$

Therefore, $x \in ker\, \varrho_N$ and thus $ker\, \varrho_N = RG \cdot A(N)$. By taking right coset representatives, one can prove in the same way that $ker\, \varrho_N = A(N) \cdot RG$. Finally, Corollary 6.1 gives $\varrho_N (A^n(G)) = A^n(G/N)$ for any $n \in \mathbb{N}$. $\qquad\qquad \square$

6.1.2 Residually Nilpotent Augmentation Ideals

Our goal is to topologize the group ring FG, where G is a finitely generated torsion-free nilpotent group and F is a field of characteristic zero. This will be achieved once we have proven that the powers of the augmentation ideal satisfy certain criteria.

Definition 6.2 Let G be any group. The augmentation ideal of RG is called *residually nilpotent* if

$$\bigcap_{n=1}^{\infty} A^n(G) = 0.$$

The main theorem of this section is:

Theorem 6.1 (S. A. Jennings) *Let G be a finitely generated torsion-free nilpotent group and let F be a field of characteristic zero. For every $n \in \mathbb{N}$, $A^{n+1}(G)$ is a proper ideal of $A^n(G)$. Furthermore, $A(G)$ is residually nilpotent.*

The original proof which we offer here relies on the fact that G has a Mal'cev basis and is adopted from [2]. We begin by establishing some preliminary lemmas. Suppose that G has Hirsch length r and let $\{g_1, \ldots, g_r\}$ be a Mal'cev basis for G. Every nontrivial element of G can be written in the form

$$\left(g_{i_1}^{\pm 1}\right)^{t_1} \left(g_{i_2}^{\pm 1}\right)^{t_2} \cdots \left(g_{i_s}^{\pm 1}\right)^{t_s},$$

where $1 \le i_1 < i_2 < \cdots < i_s \le r$ and $t_i \in \mathbb{N}$. Hence, each element of

$$S = \{g - 1 \mid g \in G, \ g \ne 1\}$$

is expressible in the form

$$\left(g_{i_1}^{\pm 1}\right)^{t_1} \left(g_{i_2}^{\pm 1}\right)^{t_2} \cdots \left(g_{i_s}^{\pm 1}\right)^{t_s} - 1 \quad \text{for } 1 \le i_1 < i_2 < \cdots < i_s \le r. \qquad (6.2)$$

Using the identity

$$xy - 1 = (x - 1)(y - 1) + (x - 1) + (y - 1), \qquad (6.3)$$

together with the Binomial Theorem, we can express every element of S as a sum of products of expressions of the form $g_{i_j}^{\pm 1} - 1$, where $g_{i_j} \ne 1$, in the right order; that is, each $g - 1 \in S$ can be written as a sum of elements of the form

$$\prod_{k=1}^{s} \left(g_{i_k}^{\pm 1} - 1\right)^{\alpha_k} \quad \text{for } 1 \le i_1 < i_2 < \cdots < i_s \le r \text{ and } \alpha_k \in \mathbb{N}, \qquad (6.4)$$

and none of the g_{i_k} equals the identity. By Lemma 6.1, the elements given in (6.4) span $A(G)$. Moreover, these elements are linearly independent over F. To see this, notice that any two different elements of the form (6.4) have different leading terms

$$g_{i_1}^{\pm \alpha_1} g_{i_2}^{\pm \alpha_2} \cdots g_{i_s}^{\pm \alpha_s}.$$

It follows that a nontrivial F-linear combination of elements of the form (6.4) can't equal zero. We summarize our discussion as a lemma:

Lemma 6.4 *The formally distinct products of the form*

$$\left(g_{i_1}^{\pm 1} - 1\right)^{\alpha_1} \left(g_{i_2}^{\pm 1} - 1\right)^{\alpha_2} \cdots \left(g_{i_s}^{\pm 1} - 1\right)^{\alpha_s},$$

where $1 \le i_1 < i_2 < \cdots < i_s \le r$ *and* $\alpha_j > 0$, *form a basis for the vector space* $A(G)$ *over* F.

Next, consider any product of the form

$$p = \left(g_{l_1}^{\pm 1} - 1\right)\left(g_{l_2}^{\pm 1} - 1\right) \cdots \left(g_{l_s}^{\pm 1} - 1\right), \tag{6.5}$$

where $l_j \in \{1, \ldots, r\}$. Define the *weight* $w(p)$ of p in (6.5) by

$$w(p) = 2^{l_1} + 2^{l_2} + \cdots + 2^{l_s}.$$

A *straight product* is a product of the form

$$\prod_{i=1}^{r} \left(g_i^{-1} - 1\right)^{\alpha_i} \left(g_i - 1\right)^{\beta_i}, \tag{6.6}$$

where α_i, β_i are nonnegative integers and $1 \le i \le r$. Observe that

$$w\left(\prod_{i=1}^{r} \left(g_i^{-1} - 1\right)^{\alpha_i} \left(g_i - 1\right)^{\beta_i}\right) = \sum_{i=1}^{r} \left(\alpha_i + \beta_i\right)2^i.$$

Lemma 6.5 *Each product of the form*

$$\left(g_{l_1}^{\pm 1} - 1\right)\left(g_{l_2}^{\pm 1} - 1\right)\cdots\left(g_{l_s}^{\pm 1} - 1\right),$$

where $l_j \in \{1, \ldots, r\}$, *can be expressed as a* $\mathbb{Z}$-*linear combination of straight products of no less weight.*

Proof We use a collection process to write the given expression as a $\mathbb{Z}$-linear combination of straight products. Observe first that if $p < q$, then

$$\left[g_q^{\pm 1}, g_p^{\pm 1}\right] = \prod_{u=q+1}^{r} \left(g_u^{\pm 1}\right)^{\delta_u}$$

for suitable positive integers $\delta_{q+1}, \ldots, \delta_r$ (see Remark 4.4). Hence, by (6.3), we have that

$$\left[g_q^{\pm 1}, g_p^{\pm 1}\right] - 1$$

is a $\mathbb{Z}$-linear combination of straight products involving only $g_{q+1}, \ldots, g_r$ and each straight product arising in this sum has weight at least 2^{q+1}. Furthermore,

$$w\left(\left(g_q^{\pm 1} - 1\right)\left(g_p^{\pm 1} - 1\right)\right) = 2^q + 2^p < 2^q + 2^q = 2^{q+1}.$$

It follows from the identity

$$(b-1)(a-1) = (a-1)(b-1) + ([b, a] - 1) + (a-1)([b, a] - 1)$$
$$+ (b-1)([b, a] - 1) + (a-1)(b-1)([b, a] - 1)$$

that

$$\left(g_q^{\pm 1} - 1\right)\left(g_p^{\pm 1} - 1\right) = \left(g_p^{\pm 1} - 1\right)\left(g_q^{\pm 1} - 1\right) + \cdots ,$$

where "$\cdots$" is a $\mathbb{Z}$-linear combination of straight products, each of whose weight exceeds $2^p + 2^q$.   □

We prove Theorem 6.1 by induction on the Hirsch length r of G. If $r = 1$, then G is infinite cyclic. Assume that $G = gp(g)$. In this case, there exists a ring isomorphism between the group ring FG and the ring of Laurent polynomials $F[t, t^{-1}]$ induced by the mapping $g \mapsto t$. It follows that FG is a PID and thus a UFD. Since FG is a commutative PID, there exists $a \in A(G)$ such that

$$A^m(G) = \langle a^m \rangle = \{ra^m \mid r \in FG\}$$

for every $m \in \mathbb{N}$. Suppose that there exists $0 \neq x \in \bigcap_{m=1}^{\infty} A^m(G)$. Then x can be written uniquely as a product of at least m primes in FG for every $m \in \mathbb{N}$. Since this is impossible, we must have

$$\bigcap_{m=1}^{\infty} A^m(G) = 0.$$

This completes the basis of induction.

Suppose that $r > 1$ and assume that the theorem is true for all finitely generated torsion-free nilpotent groups of Hirsch length less than r. Assume that $0 \neq x \in \bigcap_{q=1}^{\infty} A^q(G)$. We may assume without loss of generality that if $x = \sum r_j h_j$, then each group element h_j, when expressed in the canonical form $g_1^{\alpha_1} \cdots g_r^{\alpha_r}$ for some integers α_i, involves only positive powers of g_1. By Lemma 6.4, we can express x as a unique F-linear combination of the form

$$x = \sum_{\mu \in F}\left(\mu \prod \left(g_{i_j}^{\pm 1} - 1\right)^{\varepsilon_j}\right), \tag{6.7}$$

where $\varepsilon_j \in \mathbb{N}$ and the terms $\left(g_{i_j}^{\pm 1} - 1\right)$ in each product occur in the order induced by the Mal'cev basis. By the proof of Lemma 6.4, only powers of $(g_1 - 1)$, and not $\left(g_1^{-1} - 1\right)$, appear in (6.7). By collecting the "coefficients" of the powers of $(g_1 - 1)$, we obtain

$$x = x_0 + (g_1 - 1)^{m_1} x_1 + \cdots + (g_1 - 1)^{m_d} x_d,$$

where $0 < m_1 < \cdots < m_d$ and each x_i is an F-linear combination of products of the form

$$\left(g_{i_1}^{\pm 1} - 1\right)^{\eta_1} \cdots \left(g_{i_s}^{\pm 1} - 1\right)^{\eta_s}, \quad \text{where } 2 \leq i_1 < i_2 < \cdots < i_s \leq r \text{ and } \eta_j \geq 0.$$

Let $G_1 = gp(g_2, \ldots, g_r)$. We see that $x_i = f_i + \tilde{g}_i$ for some $f_i \in F$ and $\tilde{g}_i \in A(G_1)$, where $i = 0, 1, \ldots, d$. Note that f_0 must be zero. We claim that $f_i = 0$ for all $i = 0, 1, \ldots, d$. Set $H = gp(g_1)$ and define a homomorphism

$$\Phi : G \to H \text{ by } \Phi(g_1) = g_1 \text{ and } \Phi(g_i) = 1 \text{ if } i > 1.$$

Then Φ induces an F-linear map $\widehat{\Phi} : FG \longrightarrow FH$. Observe that

$$\widehat{\Phi}(x) = f_0 + (g_1 - 1)^{m_1} f_1 + \cdots + (g_1 - 1)^{m_d} f_d,$$

where $0 < m_1 < \cdots < m_d$. Thus, $\widehat{\Phi}(x) \in \bigcap_{q=1}^{\infty} A^q(H)$ because $x \in \bigcap_{q=1}^{\infty} A^q(G)$. Since H has Hirsch length equal to 1, we use the basis of induction to see that $A(H)$ is residually nilpotent. And so, $\widehat{\Phi}(x) = 0$. However, the elements

$$(g_1 - 1)^{m_1}, \ (g_1 - 1)^{m_2}, \ \ldots, \ (g_1 - 1)^{m_d}$$

are linearly independent over F for $0 < m_1 < \cdots < m_d$ by Lemma 6.4. Thus,

$$f_0 = \cdots = f_d = 0,$$

and consequently, $x_i \in A(G_1)$ for $0 \leq i \leq d$.

Now, G_1 is a finitely generated torsion-free nilpotent group of Hirsch length less than r by Theorem 4.7. By induction, $A(G_1)$ is residually nilpotent. Thus, if $x_0 \neq 0$, then there exists k_0 such that $x_0 \notin A^{k_0}(G_1)$. If $x_0 = 0$, then we may assume that $x_1 \neq 0$ and choose k_1 so that $x_1 \notin A^{k_1}(G_1)$. Set

$$k = \begin{cases} k_0 & \text{if } x_0 \neq 0 \\ k_1 + m_1 & \text{if } x_0 = 0, \end{cases}$$

and notice that $x \in A^{2^r k}(G)$ because $x \in A^n(G)$ for all n. By Lemma 6.4, it follows that x can be expressed as an F-linear combination of products of the form

$$\left(g_{i_1}^{\pm 1} - 1\right) \cdots \left(g_{i_u}^{\pm 1} - 1\right), \tag{6.8}$$

where each such product contains at least $2^r k$ factors. Observe that the products in (6.8) need not be straight. However, according to Lemma 6.5, each such product can be straightened without decreasing the weight. Consequently, we may write

$$x = h_1 p_1 + \cdots + h_t p_t,$$

where $h_j \in F$ and p_j is a straight product

$$\prod_{i=1}^{r} \left(g_i^{-1} - 1\right)^{\alpha_i} \left(g_i - 1\right)^{\beta_i} \tag{6.9}$$

of weight at least $2^r k$. Hence, in every straight product that occurs we have

$$\sum_{i=1}^{r} (\alpha_i + \beta_i) 2^i \geq 2^r k$$

and thus

$$\sum_{i=1}^{r} (\alpha_i + \beta_i) \geq k.$$

We conclude that each of the straight products in (6.9) has at least k factors.

Suppose that $p_1, \ldots, p_t$ are enumerated as such:

1. if $1 \leq j \leq v$, then p_j is of the form (6.9) satisfying $\alpha_1 = \beta_1 = 0$ and;
2. if $v < j \leq t$, then p_j is of the form (6.9) satisfying either $\alpha_1 \neq 0$ or $\beta_1 \neq 0$.

Setting the two expressions for x equal to each other leads to

$$x_0 + (g_1 - 1)^{m_1} x_1 + \cdots + (g_1 - 1)^{m_d} x_d = h_1 p_1 + \cdots + h_v p_v + h_{v+1} p_{v+1} +$$
$$\cdots + h_t p_t,$$

where $x_i \in A(G_1)$ for $0 \leq i \leq d$ and $p_l \in A(G_1)$ for $1 \leq l \leq v$. Since

$$\left(g_1 - 1\right)\left(g_1^{-1} - 1\right) = -\left(g_1 - 1\right) - \left(g_1^{-1} - 1\right),$$

it follows from Lemma 6.4 that

$$x_0 = h_1 p_1 + \cdots + h_v p_v \tag{6.10}$$

and

$$(g_1 - 1)^{m_1} x_1 + \cdots + (g_1 - 1)^{m_d} x_d = h_{v+1} p_{v+1} + \cdots + h_t p_t. \tag{6.11}$$

There are two cases to consider, depending on the value of x_0.

1. If $x_0 \neq 0$, then each straight product $p_1, \ldots, p_v$ in (6.10) has at least k factors of the form $\left(g_i^{\pm 1} - 1\right)$, where $2 \leq i \leq r$. We conclude that $x_0 \in A^k(G_1) = A^{k_0}(G_1)$, a contradiction.

2. If $x_0 = 0$, then each straight product $p_{v+1}, \ldots, p_t$ in (6.11) takes the form

$$p_j = \left(g_1^{-1} - 1\right)^{\lambda_j} \left(g_1 - 1\right)^{\mu_j} q_j$$
$$= (g_1 - 1)^{\lambda_j + \mu_j} (-g_1)^{-\lambda_j} q_j,$$

where $q_j \in A(G_1)$ and the second equality follows from the Binomial Theorem.

Assume that $\lambda_j + \mu_j < m_1$ for any $j = v + 1, \ldots, t$ and let $\kappa = min(\lambda_j + \mu_j)$ for $j = v + 1, \ldots, t$. Since FG is an integral domain, we can cancel the factor $(g_1 - 1)^\kappa$ across (6.11). The resulting equation contradicts Lemma 6.4. Thus, we must have $\lambda_j + \mu_j \geq m_1$ for $v + 1 \leq j \leq t$. By canceling $(g_1 - 1)^{m_1}$ across (6.11), we end up with an expression for x_1 as a sum of products of elements of the form $\left(g_i^{\pm 1} - 1\right)$ for $2 \leq i \leq r$, each product having at least $(k - m_1)$ factors. Consequently, $x_1 \in A^{k-m_1}(G_1) = A^{k_1}(G_1)$, a contradiction.

This completes the proof of Theorem 6.1.

Remark 6.1 Theorem 6.1 is true even when G is not finitely generated. This was proven by M. Lazard (see [2]).

6.1.3 E. Formanek's Generalization

In [3], E. Formanek generalized Theorem 6.1 and gave a shorter proof. He showed that the field F can be replaced by any associative, but not necessarily commutative, ring with unity and the result is still true. This observation was also made by B. Hartley in [5].

E. Formanek's proof invokes two lemmas, one due to Swan [17] and the other due to himself [3]. Both of them rely on defining a certain action of a group ring on another group ring. We describe this action, which will be implicit in the proofs of both lemmas.

In what follows, R is any associative (not necessarily commutative) ring with unity. Let G be a nilpotent group and let $1 \neq H \lhd G$ such that $G/H \cong \mathbb{Z}$. Fix a generator t of $\mathbb{Z}$. If x is an element of G which maps to t, then we may define an action of the group ring RG on the group ring RH. Let x act on RH by

$$x \cdot \sum_{i=1}^{n} r_i h_i = \sum_{i=1}^{n} r_i x^{-1} h_i x \quad (r_i \in R, \ h_i \in H),$$

and let H act on RH by left multiplication:

$$h \cdot \sum_{i=1}^{n} r_i h_i = \sum_{i=1}^{n} r_i h h_i \quad (r_i \in R, \ h_i \in H).$$

These induce an action of G on RH in the obvious way and thus an action of RG on RH. Of course, this induced action depends on the chosen x. Consequently, RH becomes a left RG-module.

Lemma 6.6 (R. Swan) *Suppose that $G = gp(H, x)$ is a nilpotent group of class c, where $H \lhd G$ and*

$$G/H \cong gp(x) \cong \mathbb{Z}.$$

If $m \in \mathbb{N}$, then $A^{m^c}(G) \cdot RH \subseteq A^m(H)$.

Proof The proof is adopted from [11]. We proceed by induction on c. If $c = 1$, then G is abelian. In this case, $h^x = h$ for all $h \in H$. Thus, x acts trivially on RH. Furthermore, G is a direct sum of H and $gp(x)$. It follows that $A^m(G) \cdot RH = A^m(H)$.

Suppose that $c > 1$ and assume that the lemma is true for all nilpotent groups of class less than c. Let $L = \gamma_c G$. By Lemma 1.6, $\gamma_2 G \trianglelefteq H$ and thus

$$L \trianglelefteq \gamma_2 G \trianglelefteq H.$$

In addition, $L \leq Z(G)$ since G is of class c. Now, both G/L and H/L have nilpotency class less than c. By induction,

$$A^{m^{c-1}}(G/L) \subseteq R(H/L) \subseteq A^m(H/L). \tag{6.12}$$

Since $RHA(L)$ is the kernel of the natural map $RH \to R(H/L)$ by Lemma 6.3, it follows from (6.12) that

$$A^{m^{c-1}}(G) \cdot RH \subseteq A^m(H) + RHA(L). \tag{6.13}$$

Now, x acts trivially on L because L is abelian. From this, we see that RH is a left RG- and right RL-bimodule, where the right action is right multiplication. This, together with (6.13) and the fact that $A(H)$ is an RG-submodule of RH give

$$A^{2m^{c-1}}(G) \cdot RH = A^{m^{c-1}}(G)A^{m^{c-1}}(G) \cdot RH$$

$$\subseteq A^{m^{c-1}}(G) \cdot [A^m(H) + RHA(L)]$$

$$\subseteq A^{m^{c-1}}(G) \cdot A^m(H) + A^{m^{c-1}}(G) \cdot (RHA(L))$$

$$\subseteq A^m(H) + \left(A^{m^{c-1}}(G) \cdot RH \right) A(L)$$

$$\subseteq A^m(H) + (A^m(H) + RHA(L))A(L)$$

$$= A^m(H) + RHA^2(L).$$

Continuing in this way, we find that

$$A^{rm^{c-1}}(G) \cdot RH \subseteq A^m(H) + RHA^r(L) \quad \text{for} \ \ 1 \le r \le m.$$

In particular,

$$A^{m^c}(G) \cdot RH \subseteq A^m(H) + RHA^m(L).$$

Since $L \le H$, $RHA^m(L) \subseteq A^m(H)$. Therefore, $A^{m^c}(G) \cdot RH \subseteq A^m(H)$.   □

Lemma 6.7 (E. Formanek) *Suppose that G is a torsion-free nilpotent group. Let $1 \ne H \lhd G$ such that $G/H \cong \mathbb{Z}$ and let $0 \ne \alpha \in RG$. There exists an element $x \in G$ which maps to a generator of $\mathbb{Z}$ such that $\alpha \cdot RH \ne 0$, where the module action $RG \times RH \to RH$ is defined relative to x.*

Proof Let $\alpha = \sum_{g \in G} r_g g$, where $r_g \in R$. We may assume without loss of generality that the coefficient r_1 of the identity element $1 \in G$ is not zero. Let

$$1, g_1, \ldots, g_n$$

be the distinct group elements arising in α. We claim that each g_i is contained in at most one cyclic group generated by an element of G which maps to a generator t of $\mathbb{Z}$. Suppose that a and b are distinct elements of G which both map to t. We assert that $gp(a) \cap gp(b) = 1$. Assume on the contrary, that $a^i = b^j$ for some $i, j \in \mathbb{Z}$. Then $t^i = t^j$ and thus $i = j$. Since G is torsion-free, Theorem 2.7 gives $a = b$, a contradiction. This proves the assertion.

Now, if x is an element of G which maps to t, then the element hx of G also maps to t for all $h \in H$. Since H is infinite, this means that there are infinitely many elements of G that map to t. By our previous discussion, it follows that the element x which maps to t can be chosen so that no g_i lies in $gp(x)$. Then

$$g_1 = h_1 x^{i_1}, \quad g_2 = h_2 x^{i_2}, \quad \ldots, \quad g_n = h_n x^{i_n}$$

for some nonidentity elements $h_1, h_2, \ldots, h_n \in H$ and $i_1, i_2, \ldots, i_n \in \mathbb{Z}$. Consider the action of α on $1 \in RH$:

$$\alpha \cdot 1 = \left(r_1 1 + \sum_{j=1}^{n} r_{g_j} g_j \right) \cdot 1$$

$$= r_1 (1 \cdot 1) + \sum_{j=1}^{n} r_{g_j} \left(g_j \cdot 1 \right)$$

$$= r_1 + \sum_{j=1}^{n} r_{g_j} \left(h_j x^{t_j} \cdot 1 \right)$$

$$= r_1 + \sum_{j=1}^{n} r_{g_j} h_j.$$

Since $r_1 \neq 0$ by assumption and $h_j \neq 1$ for $j = 1, 2, \ldots, n$, it must be the case that $\alpha \cdot 1 \neq 0$ as required.   □

We now prove the generalization of Theorem 6.1.

Theorem 6.2 (E. Formanek) *If G is a finitely generated torsion-free nilpotent group and R is any (not necessarily commutative) ring with unity, then $A(G)$ is residually nilpotent.*

Proof We proceed by induction on the Hirsch length r of G. If $r = 1$, then G is infinite cyclic. Suppose that G is generated by g and let $x \in \bigcap_{n=1}^{\infty} \in A^n(G)$. We claim that $x = 0$. Choose $m \in \mathbb{N}$. Since $x \in A^m(G)$, we can write x uniquely as

$$x = \alpha_1 \left(g^{\pm 1} - 1 \right)^{m_1} + \alpha_2 \left(g^{\pm 1} - 1 \right)^{m_2} + \cdots + \alpha_k \left(g^{\pm 1} - 1 \right)^{m_k}$$

for $\alpha_i \in R$ and $0 < m \leq m_1 < \cdots < m_k$ by Lemma 6.4. Let t be an indeterminate. There exists a ring isomorphism from RG to the ring of Laurent polynomials $R\left[t, t^{-1} \right]$ induced by the map $g \mapsto t$. Under this isomorphism, x is mapped to a Laurent polynomial of (negative or positive) degree of at least m. Since $x \in \bigcap_{n=1}^{\infty} \in A^n(G)$, it must be that $x = 0$. This completes the basis of induction.

Suppose that $r > 1$ and let H be a normal subgroup of G of Hirsch length $r - 1$ such that $G/H \cong \mathbb{Z}$. By the induction hypothesis, $\bigcap_{n=1}^{\infty} A^n(H) = 0$. Assume that $A(G)$ is not residually nilpotent and let $0 \neq \alpha \in \bigcap_{n=1}^{\infty} A^n(G)$. By Lemma 6.7, there exists $x \in G$ which maps to a generator t of $\mathbb{Z}$ such that $\alpha \cdot RH \neq 0$ under the RG action on RH relative to x. However,

$$\alpha \cdot RH \subseteq \bigcap_{n=1}^{\infty} A^n(H)$$

by Lemma 6.6. Thus, $\alpha \cdot RH = 0$, a contradiction.   □

By modifying the procedure above, E. Formanek extended Theorem 6.2 to certain nilpotent $\mathscr{D}$-groups. Define the $\mathbb{Q}$-*rank* of a nilpotent $\mathscr{D}$-group G of class c by

$$\sum_{i=1}^{c} [\gamma_i G / \gamma_{i+1} G : \mathbb{Q}],$$

where $\gamma_i G / \gamma_{i+1} G$ is viewed as a vector space over $\mathbb{Q}$. If F is a field of characteristic zero and G is a nilpotent $\mathscr{D}$-group of finite $\mathbb{Q}$-rank, then $A(G)$ turns out to be residually nilpotent. See [3] for details.

6.2   The Dimension Subgroups

In this section, we introduce the notion of a dimension subgroup. We prove that each dimension subgroup coincides with the isolator of a lower central subgroup when considered over a field of characteristic zero. The proof relies on the methods from the previous section and is laid out by P. Hall in [4]. A general study of dimension subgroups and the augmentation ideal of group rings over other rings can be found in [11] and [12].

6.2.1   Definition and Properties of Dimension Subgroups

We remind the reader that R is a commutative ring with unity unless otherwise stated. Let G be any group. Define the set $D_n(R, G)$ for each $n \in \mathbb{N}$ by

$$D_n(R, G) = G \cap \left(1 + A_R^n(G)\right)$$
$$= \{g \in G \mid g - 1 \in A_R^n(G)\}.$$

Note that $D_1(R, G) = G$.

Lemma 6.8 *For each* $n \in \mathbb{N}$, $D_n(R, G)$ *is a normal subgroup of* G.

Proof If g and h are elements of $D_n(R, G)$, then

$$gh^{-1} - 1 = ((g - 1) - (h - 1))h^{-1} \in A^n(G).$$

Therefore, $gh^{-1} \in D_n(R, G)$ and thus $D_n(R, G) \leq G$. To establish normality, observe that if $x \in G$, then

$$x^{-1}gx - 1 = x^{-1}(g - 1)x \in A^n(G).$$

And so, $x^{-1}gx \in D_n(R, G)$.   □

In fact, $D_n(R, G)$ is the kernel of the homomorphism induced on G by the natural homomorphism $RG \to RG/A_R^n(G)$. We call $D_n(R, G)$ the nth *dimension subgroup* of G over R. The descending series

$$G = D_1(R, G) \geq D_2(R, G) \geq \cdots$$

is called the *dimension series* of G over R. We will write $D_n(G)$ instead of $D_n(R, G)$ when R is understood.

Theorem 6.3 *If G is any group, then $\left[D_i(G), D_j(G)\right] \leq D_{i+j}(G)$ for any $i, j \in \mathbb{N}$. If G is nilpotent, then its dimension series is central.*

Proof If $g \in D_i(G)$ and $h \in D_j(G)$, then

$$[g, h] - 1 = g^{-1}h^{-1}((g-1)(h-1) - (h-1)(g-1)) \in A^{i+j}(G).$$

Thus, $\left[D_i(G), D_j(G)\right] \leq D_{i+j}(G)$. In particular, $\left[D_i(G), G\right] \leq D_{i+1}(G)$. If G happens to be nilpotent, then its dimension series is central by Lemma 2.1. □

Corollary 6.2 *If G is any group, then $\gamma_n G \leq D_n(G)$ for each $n \in \mathbb{N}$.*

Proof This is immediate from Theorem 6.3. □

Corollary 6.3 *Let G be a torsion-free nilpotent group of class c. If R is any ring with unity, then $D_k(R, G) = 1$ for some $k \geq c + 1$. Consequently, G embeds in $1 + A_R(G)/A_R^k(G)$ for some $k \geq c + 1$.*

Proof By Theorem 6.2 and Corollary 6.2, $D_k(R, G) = 1$ for some $k \geq c+1$. Define a group homomorphism

$$\psi : G \to 1 + A_R(G)/A_R^k(G) \text{ by } \psi(g) = 1 + \left((g - 1) + A_R^k(G)\right).$$

Clearly, the kernel of ψ equals $D_k(R, G)$. Thus, $\ker \psi = 1$ and consequently, ψ is an embedding. □

Remark 6.2 Let G be a finitely generated torsion-free nilpotent group of class c and let R be a binomial ring. For any $k \in \mathbb{N}$, define an R-action on $1 + A_R(G)/A_R^k(G)$ by

$$\bar{g}^\lambda = 1 + \lambda(\bar{g} - 1) + \binom{\lambda}{2}(\bar{g} - 1)^2 + \cdots + \binom{\lambda}{k-1}(\bar{g} - 1)^{k-1},$$

where

$$\bar{g} = 1 + \left((g - 1) + A_R^k(G)\right) \in 1 + A_R(G)/A_R^k(G)$$

for every $g \in G$ and $\lambda \in R$. Then $1 + A_R(G)/A_R^k(G)$, equipped with this R-action, becomes a nilpotent R-powered group according to Example 4.9. It follows from Corollary 6.3 that G embeds into this nilpotent R-powered group.

Lemma 6.9 *If G is any group and $H \leq G$, then for any $n \in \mathbb{N}$, $D_n(H) \leq D_n(G)$.*

Proof This a direct consequence of Lemma 6.2. □

Lemma 6.10 *Let G be any group and suppose that $N \trianglelefteq G$. If $N \leq D_n(G)$ for some $n \in \mathbb{N}$, then $D_n(G/N) = D_n(G)/N$.*

Proof Let $\varrho_N : RG \to R(G/N)$ be the ring homomorphism induced by the natural homomorphism $\pi : G \to G/N$. By Lemma 6.3,

$$\varrho_N\left(A^n(G)\right) = A^n(G/N).$$

We first show that $D_n(G/N) \leq D_n(G)/N$. If $gN \in D_n(G/N)$, then

$$gN - N \in A^n(G/N).$$

A pre-image of this element under ϱ_N is $g - 1$. Since $g - 1 \in A^n(G)$, $g \in D_n(G)$. Hence, $gN \in D_n(G)/N$. Therefore, $D_n(G/N) \leq D_n(G)/N$.

We establish the reverse inclusion. If $gN \in D_n(G)/N$, then $g \in D_n(G)$. Hence, $g - 1 \in A^n(G)$ and consequently,

$$\varrho_N(g - 1) = gN - N \in A^n(G/N).$$

Therefore, $gN \in D_n(G/N)$ and thus $D_n(G)/N \leq D_n(G/N)$. □

Lemma 6.11 *If G is a group and F is a field of characteristic zero, then $G/D_n(G)$ is torsion-free for each $n \in \mathbb{N}$. Equivalently, $D_n(G)$ is isolated in G.*

Proof The proof is done by induction on n. The case $n = 1$ is trivial. Assume that $G/D_n(G)$ is torsion-free for $n > 1$ and suppose that $g \in G$ such that $g^s \in D_{n+1}(G)$ for some $s \in \mathbb{N}$. We need to show that $g \in D_{n+1}(G)$. Well, $D_{n+1}(G) \leq D_n(G)$ implies that $g^s \in D_n(G)$, so $g \in D_n(G)$ by induction. Thus, $g - 1 \in A^n(G)$ and

$$(g - 1)^2 \in A^n(G)A^n(G) \subseteq A^{n+1}(G).$$

Now,

$$g^s = ((g - 1) + 1)^s = 1 + s(g - 1) + (g - 1)^2 \sum_{k=2}^{s} \binom{s}{k}(g - 1)^{k-2}.$$

Hence,

$$s(g - 1) = (g^s - 1) - (g - 1)^2 \sum_{k=2}^{s} \binom{s}{k}(g - 1)^{k-2} \tag{6.14}$$

and the right side of (6.14) is contained in $A^{n+1}(G)$. Since F has characteristic zero and $s \neq 0$, s must be invertible in F. Therefore, $g - 1 \in A^{n+1}(G)$.   □

We now come to the main theorem of this section. We follow the proof given by P. Hall in [4].

Theorem 6.4 (S. A. Jennings) *If G is any group and F is a field of characteristic zero, then for any $n \in \mathbb{N}$,*

$$D_n(F, G) = \{g \in G \mid g^m \in \gamma_n G \text{ for some integer } m > 0\}.$$

Thus, $D_n(F, G)$ is the isolator of $\gamma_n G$ in G.

Clearly, $G/\gamma_n G$ is a nilpotent group for any $n \geq 1$. Let $\tau(G/\gamma_n G) = \overline{\gamma}_n G/\gamma_n G$ be the torsion subgroup of $G/\gamma_n G$ and set $D_n(F, G) = D_n(G)$. It is easy to see that $\overline{\gamma}_n G$ is the isolator of $\gamma_n G$ in G. We need to establish that $D_n(G) = \overline{\gamma}_n G$.

Suppose that $g \in \overline{\gamma}_n G$. There exists $m \in \mathbb{N}$ such that $g^m \in \gamma_n G$. Thus, $g^m \in D_n(G)$ by Corollary 6.2. And so, $g \in D_n(G)$ by Lemma 6.11. Therefore, $\overline{\gamma}_n G \leq D_n(G)$. Establishing the reverse inclusion

$$\overline{\gamma}_n G \geq D_n(G) \tag{6.15}$$

is far more involved and requires substantial work. We will do this in stages by making various reductions.

Reduction (1): We may assume that G is finitely generated.

Suppose that (6.15) holds for any finitely generated group and let N be any group. We show that $\overline{\gamma}_n N \geq D_n(N)$.

If $g \in D_n(N)$, then $1 - g \in A^n(N)$. Thus, $1 - g$ is a finite F-linear combination of products of the form

$$(1 - g_{i_1}) \cdots (1 - g_{i_n})$$

where $g_{i_j} \in N$. Let H be the subgroup of N generated by g together with all of the g_{i_j}'s that appear in these products. Clearly, H is a finitely generated subgroup of N containing g and $1 - g \in A^n(H)$. Thus, $g \in D_n(H)$ and by assumption, $g \in \overline{\gamma}_n H$. Since $\gamma_n H \leq \gamma_n N$, it follows that $\overline{\gamma}_n H \leq \overline{\gamma}_n N$. Thus, $g \in \overline{\gamma}_n N$.

Reduction (2): We may assume that G is finitely generated torsion-free nilpotent.

Assume that (6.15) is true for every finitely generated torsion-free nilpotent group and let N be any finitely generated group. We assert that (6.15) holds for N. Set $K = \overline{\gamma}_n N$. By the Third Isomorphism Theorem,

$$N/K \cong \frac{N/\gamma_n N}{\tau(N/\gamma_n N)}.$$

Since $N/\gamma_n N$ is nilpotent, N/K is a finitely generated torsion-free nilpotent group of class at most $n-1$. Thus, $\overline{\gamma}_n(N/K) = 1$ because N/K is torsion-free. By assumption,

$D_n(N/K) \leq \overline{\gamma}_n(N/K)$ and thus, $D_n(N/K) = 1$. Moreover, $D_n(N/K) = D_n(N)/K$ by Lemma 6.10 because $K = \overline{\gamma}_n N \leq D_n(N)$. Therefore, $1 = D_n(N)/K$ and consequently, $D_n(N) = K = \overline{\gamma}_n N$ as asserted.

We have shown that if (6.15) is true for any finitely generated torsion-free nilpotent group, then it is true for all other groups. Thus, we need to establish that $D_n(G) \leq \overline{\gamma}_n G$ when G is finitely generated torsion-free nilpotent.

Reduction (3): It is enough to prove that if G is a finitely generated torsion-free nilpotent group, then $\overline{\gamma}_n(G) = 1$ implies $D_n(G) = 1$.

Assume that all finitely generated torsion-free nilpotent groups satisfy the assertion in Reduction (3) and let N be any finitely generated torsion-free nilpotent group. We claim that $D_n(N) \leq \overline{\gamma}_n(N)$ for every $n \in \mathbb{N}$.

Put $K = \overline{\gamma}_n(N)$. Then N/K is a finitely generated torsion-free nilpotent group of class at most $n - 1$. Since $K \leq D_n(N)$,

$$D_n(N/K) = D_n(N)/K$$

by Lemma 6.10. Since $\gamma_n (N/K) = 1$ and N/K is torsion-free, $\overline{\gamma}_n(N/K) = 1$. Thus, by assumption,

$$D_n(N/K) = D_n(N)/K = 1.$$

Hence, $D_n(N) \leq K = \overline{\gamma}_n(N)$ as promised.

For the remainder of the proof, we fix a finitely generated torsion-free nilpotent group G of class c. Note that $\overline{\gamma}_{c+1}G = 1$ and by Corollary 5.3, the series

$$G = \overline{\gamma}_1 G > \cdots > \overline{\gamma}_{c+1} G = 1$$

is central. We call this series the *isolated lower central series* of G. Now, each $\overline{\gamma}_i G/\overline{\gamma}_{i+1} G$ is free abelian of finite rank m_i for $1 \leq i \leq c$. Thus, we can refine this series to a poly-infinite cyclic and central series

$$G = G_0 > G_1 > \cdots > G_M = 1,$$

where $M = m_1 + m_2 + \cdots + m_c$. Let $G_{i-1} = gp(x_i, G_i)$ for $i = 1, 2, \ldots, M$. Then $\{x_1, x_2, \ldots, x_M\}$ is a Mal'cev basis for G.

Before proceeding with further reductions, we pause to construct a certain type of basis for the vector space FG.

Definition 6.3 A basis $\mathscr{B}$ of the vector space FG is called an *integral basis* if every element of G is a $\mathbb{Z}$-linear combination of the vectors in $\mathscr{B}$ and each vector in $\mathscr{B}$ is a $\mathbb{Z}$-linear combination of elements of G.

Note that if $u_i, u_j \in \mathscr{B}$, then we can always express the product $u_i u_j$ as a $\mathbb{Z}$-linear combination of the elements of $\mathscr{B}$.

We determine an integral basis for the group ring FH, where H is infinite cyclic.

Lemma 6.12 *Let $H = gp(x)$ be infinite cyclic and set $u = 1 - x$ in FH. Fix an integer $n \geq 0$. The set*

$$S = \{1, u, u^2, \ldots, u^n x^{-1}, u^n x^{-2}, \ldots\}$$

is an integral basis for FH.

Proof We prove by induction on n that every element of S is an integral linear combination of the elements of H. If $n = 0$, then for each $r > 0$,

$$u^r = (1 - x)^r = \sum_{s=0}^{r} (-1)^s \binom{r}{s} x^s.$$

Thus, each element of $\{1, u, u^2, \ldots, x^{-1}, x^{-2}, \ldots\}$ is an integral linear combination of the elements of H.

Suppose that $n > 0$ and assume that each element of the set

$$\{1, u, u^2, \ldots, u^{n-1} x^{-1}, u^{n-1} x^{-2}, \ldots\}$$

is an integral linear combination of the elements of H. We need to show that each element of

$$B = \{1, u, u^2, \ldots, u^n x^{-1}, u^n x^{-2}, \ldots\}$$

is also an integral linear combination of the elements of H. By induction, each element of the form u^i in B is an integral linear combination of the elements of H. Observe further that for $t > 0$,

$$x^{-t} = x^{-t+1} + x^{-t} - x^{-t+1} = x^{-(t-1)} + (1 - x)x^{-t}$$

and thus $u^n x^{-t} = u^{n-1} x^{-t} - u^{n-1} x^{-(t-1)}$. Applying induction again gives that each $u^n x^{-t}$ is also an integral linear combination of the elements of H. We omit the proof of linear independence. $\square$

Using Lemma 6.12, it is now apparent how to concoct an integral basis for FG. For $1 \leq i \leq M$, put $u_i = 1 - x_i$, where x_i is the ith element of the Mal'cev basis for G constructed above. By Lemma 6.12, the set

$$V = \{v_1 v_2 \cdots v_M \mid v_i = u_i^{r_i} \text{ or } v_i = u_i^n x_i^{-s_i} \ (r_i \geq 0, \ s_i > 0)\} \qquad (6.16)$$

is an integral basis for FG for every $n \geq 0$. We consider the specific integral basis V obtained by assuming, once and for all, that $n = c + 1 > 0$, where c is the class of G.

Lemma 6.13 $V \setminus \{1\}$ *is a basis for the vector space $A(G)$ over F.*

Proof An element $v \in V$ lies in $A(G)$ provided that not all of the r_i's equal zero whenever $v = u_1^{r_1} \cdots u_M^{r_M}$. Thus, a linear combination of elements of V is contained in $A(G)$ if and only if the identity element $1 = u_1^0 \cdots u_M^0$ in the linear combination has coefficient zero.   □

We associate a weight μ to each element of V. Define the *weight* of $u_i = 1 - x_i$ by setting $\mu(u_i) = k$ if and only if $x_i \in \overline{\gamma}_k G \setminus \overline{\gamma}_{k+1} G$. In this case, we abbreviate $\mu_i = \mu(u_i)$. Next, we define the weight of an arbitrary element of V. Suppose that $v = v_1 v_2 \cdots v_M$ is an element of V.

(i) If at least one of the factors $v_1, v_2, \ldots, v_M$ has the form $u_i^n x_i^{-s_i}$, then we declare that $\mu(v) \geq n$.

(ii) If $v = u_1^{r_1} u_2^{r_2} \cdots u_M^{r_M}$, then define $\mu(v) = \sum_{i=1}^{M} r_i \mu_i$.

Certain F-subspaces of FG which are spanned by elements of V of a specified weight will be of interest. For $0 \leq s \leq n$, let E_s be the F-subspace of FG spanned by $\{v \in V \mid \mu(v) \geq s\}$. If $s > n$, then we set $E_s = E_n$. Note that $E_0 = FG$. Moreover, since

$$1 = u_1^0 \cdots u_M^0$$

is the only element of V of weight 0, we have $E_1 = A(G)$ by Lemma 6.13. In what follows, for notational consistency, we write $A^0(G) = FG$.

Reduction (4): It suffices to prove that $A^s(G) \subseteq E_s$ for $0 \leq s \leq n$.

Observe that if $\overline{\gamma}_m G = 1$ for some m, then $\gamma_m G = 1$ since G is torsion-free. Thus, $m \geq n$ because G is of class $c = n - 1$. If we manage to prove that $D_n(G) = 1$, then we would also have $D_m(G) = 1$ because $D_m(G) \leq D_n(G)$. As a result, we can restate Reduction (4) as the following lemma:

Lemma 6.14 *Suppose that $A^s(G) \subseteq E_s$ for $0 \leq s \leq n$. Then $D_n(G) = 1$.*

Proof We begin by proving that $E_s \subseteq A^s(G)$ for $0 \leq s \leq n$. We must show that if $v \in V$ satisfies $\mu(v) \geq s$, then $v \in A^s(G)$. Let $u_i = 1 - x_i$ be of weight k, where $1 \leq i \leq M$ and $1 \leq k \leq n - 1$. Then $x_i \in \overline{\gamma}_k G \setminus \overline{\gamma}_{k+1} G$. Since $\overline{\gamma}_k G \leq D_k(G)$, we have $x_i \in D_k(G)$. And so, $u_i \in A^k(G)$.

Next, let $v = v_1 v_2 \cdots v_M$ be any element of V contained in E_s, so that $\mu(v) \geq s$. There are two possibilities for v:

- either $v = u_1^{r_1} \cdots u_M^{r_M}$ and $\mu(v) = \sum_{i=1}^{M} r_i \mu_i \geq s$, or
- at least one of the v_i equals $u_i^n x_i^{-s_i}$.

If $v = u_1^{r_1} \cdots u_M^{r_M}$, then $u_i \in A^{\mu_i}(G)$ implies that $u_i^{r_i} \in A^{r_i \mu_i}(G)$ for each $1 \leq i \leq M$. Hence,

$$v \in \prod_{i=1}^{M} A^{r_i \mu_i}(G) = A^{\sum_{i=1}^{M} r_i \mu_i}(G) = A^{\mu(v)}(G) \subseteq A^s(G).$$

Suppose on the other hand, that $u_i^n x_i^{-s_i}$ is a factor of v. In this case,

$$u_i^n = (1 - x_i)^n \in A^n(G).$$

Since $A^n(G)$ is an ideal of FG, it follows that $v \in A^n(G)$. Moreover, $A^n(G) \subseteq A^s(G)$ because $s \leq n$. In either case, $v \in A^s(G)$ and thus $E_s \subseteq A^s(G)$ as claimed. This, together with the hypothesis give $A^s(G) = E_s$ for $0 \leq s \leq n$. Hence, the elements of weight k span $A^k(G)$ modulo $A^{k+1}(G)$.

Let $1 \leq k \leq n - 1$ and assume that

$$u_j, \ u_{j+1}, \ \ldots, \ u_{j+l}$$

are those elements of $A^k(G)$ whose weight is exactly k. We assert that these elements are linearly independent modulo $A^{k+1}(G)$ and thus form a basis for $A^k(G)$ modulo $A^{k+1}(G)$. Let

$$a = \alpha_0 u_j + \cdots + \alpha_l u_{j+l} \in A^{k+1}(G)$$

for some $\alpha_0, \ \ldots, \ \alpha_l \in F$. Since V is an integral basis for FG, a basis for each $A^s(G) = E_s$ $(0 \leq s \leq n)$ consists of those $v \in V$ whose weight is at least s. Hence, a can also be written as a unique linear combination of v's of weight at least $(k+1)$. Thus, $\alpha_0 = \cdots = \alpha_l = 0$ as asserted.

Choose a nonidentity element $g \in G$. Since $\overline{\gamma}_n G = 1$, there exists a natural number $k < n$ such that $g \in \overline{\gamma}_k G \setminus \overline{\gamma}_{k+1} G$. Observe that the free abelian group $\overline{\gamma}_k G / \overline{\gamma}_{k+1} G$ is freely generated by

$$x_j \overline{\gamma}_{k+1} G, \ x_{j+1} \overline{\gamma}_{k+1} G, \ \ldots, \ x_{j+l} \overline{\gamma}_{k+1} G.$$

Hence, we may write g as

$$g = x_j^{r_0} x_{j+1}^{r_1} \cdots x_{j+l}^{r_l} z,$$

where $z \in \overline{\gamma}_{k+1} G$ and not all of the $r_0, \ r_1, \ \ldots, \ r_l$ are zero. Now, $1 - z \in A^{k+1}(G)$ since $\overline{\gamma}_{k+1} G \leq D_{k+1}(G)$. An application of the Binomial Theorem shows that for $m = 0, \ 1, \ \ldots, \ l$,

$$1 - x_{j+m}^{r_m} = 1 - \left(1 - u_{j+m}\right)^{r_m}$$

$$= r_m u_{j+m} + \text{terms contained in } A^{k+1}(G).$$

Hence,

$$1 - g = 1 - x_j^{r_0} x_{j+1}^{r_1} \cdots x_{j+l}^{r_l} z$$

$$= 1 - \left(1 - u_j\right)^{r_0} \left(1 - u_{j+1}\right)^{r_1} \cdots \left(1 - u_{j+l}\right)^{r_l} z$$

$$= \sum_{m=0}^{l} r_m u_{j+m} + \text{terms contained in } A^{k+1}(G).$$

Note that

$$\sum_{m=0}^{l} r_m u_{j+m} \notin A^{k+1}(G)$$

since the r_m are not all zero and the u_{j+m} are linearly independent modulo $A^{k+1}(G)$. Therefore, $1 - g \notin A^{k+1}(G)$ and consequently, $g \notin D_{k+1}(G)$. Since $k < n$, we have $D_{k+1}(G) \geq D_n(G)$ and thus $g \notin D_n(G)$. We conclude that $D_n(G) = 1$. $\square$

Reduction (5): It is enough to prove that $E_r E_s \subseteq E_{r+s}$ for $r, s \geq 0$.

For suppose that $E_r E_s \subseteq E_{r+s}$ for $r, s \geq 0$. We must prove that $A^t(G) \subseteq E_t$ whenever $0 \leq t \leq n$. Recall that $FG = A^0(G) = E_0$ and $A(G) = E_1$. Assume the result holds for $0 \leq t < n$. Then, by induction,

$$E_{t+1} \supseteq E_t E_1 \supseteq A^t(G)A(G) = A^{t+1}(G).$$

And so, we set out to prove that $E_r E_s \subseteq E_{r+s}$ by examining certain subspaces of E_s, for $s \geq 0$. Set $E_{s,1} = E_s$. For $1 < i \leq M$, let $E_{s,i}$ be the subspace of E_s spanned by those products $v = v_1 v_2 \cdots v_M$ in V such that $\mu(v) \geq s$ and

$$v_1 = v_2 = \cdots = v_{i-1} = 1.$$

If $s > n$, then we have $E_{s,i} = E_{n,i}$. Note that $E_{s,i} \subset FG_{i-1}$.

Reduction (6): It suffices to prove that $E_r\ _iE_{s,i} \subseteq E_{r+s,i}$ for $r, s \geq 0$ and $1 \leq i \leq M$.

For suppose that $E_r\ _iE_{s,i} \subseteq E_{r+s,i}$ for $r, s \geq 0$ and $1 \leq i \leq M$. Then, in particular, $E_r\ _1E_{s,1} \subseteq E_{r+s,1}$, or equivalently, $E_r E_s \subseteq E_{r+s}$.

Thus, our final task is to show that $E_r\ _iE_{s,i} \subseteq E_{r+s,i}$ for $r, s \geq 0$ and $1 \leq i \leq M$. We do this by induction on $M - i$. If $M - i = 0$, then the result follows. Suppose that $i < M$ and assume that $E_r\ _jE_{s,j} \subseteq E_{r+s,j}$ for $j > i$. In this situation, we prove the following:

Lemma 6.15 If $v = v_1 \cdots v_M \in E_p$, then $v(x_i - 1) \in E_{p+\mu_i, i}$.

Proof The bulk of the proof consists of establishing that

$$x_i^{-1} v_j x_i - v_j \in E_{\mu_i+\mu(v_j), i+1}$$

when $1 \leq i < j \leq M$, where v_j equals either $u_j^{r_j}$ for some $r_j \geq 0$ or $u_j^n x_j^{-s_j}$ for some $s_j > 0$ as before. For any $1 \leq i, j \leq M$, we have that

$$x_i^{-1} u_j x_i = 1 - x_j[x_j, x_i]$$
$$= u_j + z - u_j z$$

where $z = 1 - [x_j, x_i]$. By Corollary 5.3 and Theorem 2.14 (i),

$$\left[x_j, \; x_i\right] \in \left[\overline{\gamma}_{\mu_j} G, \; \overline{\gamma}_{\mu_i} G\right] \le \overline{\gamma}_{\mu_i+\mu_j} G.$$

And so, $z \in A\!\left(\overline{\gamma}_{\mu_i+\mu_j} G\right)$. It follows from Lemma 6.13 that z can be expressed as an F-linear combination of elements of V, each having weight at least $\mu_i + \mu_j$. Thus, $z \in E_{\mu_i+\mu_j, \; i+1}$.

Assume from this point on that $i < j$, so that $u_j \in E_{0, \; i+1}$. Since $E_{r, \; j} E_{s, \; j} \subseteq E_{r+s, \; j}$ when $i < j$ by assumption, we have

$$u_j z \in \left(E_{0, \; i+1}\right)\left(E_{\mu_i+\mu_j, \; i+1}\right) \subseteq E_{\mu_i+\mu_j, \; i+1}.$$

Thus,

$$x_i^{-1} u_j x_i - u_j = z - u_j z \in E_{\mu_i+\mu_j, \; i+1}.$$

We claim that

$$x_i^{-1} u_j^r x_i - u_j^r \in E_{\mu_i+r\mu_j, \; i+1} \tag{6.17}$$

whenever $r > 0$. Suppose that (6.17) is true for $r - 1$, so that

$$x_i^{-1} u_j^{r-1} x_i - u_j^{r-1} \in E_{\mu_i+(r-1)\mu_j, \; i+1}.$$

Then

$$
\begin{aligned}
x_i^{-1} u_j^r x_i &= \left(x_i^{-1} u_j^{r-1} x_i\right)\left(x_i^{-1} u_j x_i\right) \\
&= \left(u_j^{r-1} + s\right)\left(u_j + t\right) \quad \left(s \in E_{\mu_i+(r-1)\mu_j, \; i+1} \;\; \text{and} \;\; t \in E_{\mu_i+\mu_j, \; i+1}\right) \\
&= u_j^r + u_j^{r-1} t + s u_j + s t \\
&= u_j^r + w \quad \left(w \in E_{\mu_i+r\mu_j, \; i+1}\right),
\end{aligned}
$$

where the last equality follows from the fact that $u_j^{r-1} t$, $s u_j$, and st are all contained in $E_{\mu_i+r\mu_j, \; i+1}$ by the induction hypothesis, together with the observation that t also belongs to $E_{\mu_j, \; i+1}$. Thus, (6.17) is true as claimed. In particular,

$$x_i^{-1} u_j^n x_i - u_j^n \in E_{n, \; i+1}$$

because $n < \mu_i + n\mu_j$ and thus $E_{\mu_i+n\mu_j, \; i+1} = E_{n, \; i+1}$.

Since $\mu(u_j^n) = n\mu_j > n$, we have $u_j^n \in E_{n, \; i+1}$. And so, $x_i^{-1} u_j^n x_i \in E_{n, \; i+1}$. Furthermore,

$$x_i^{-1} x_j^{-s} x_i \in E_{0, \; i+1}$$

for any $s > 0$. Hence,

$$x_i^{-1} u_j^n x_j^{-s} x_i = \left(x_i^{-1} u_j^n x_i\right)\left(x_i^{-1} x_j^{-s} x_i\right) \in (E_{n,\,i+1})(E_{0,\,i+1}) \subseteq E_{n,\,i+1}$$

by induction. Since $u_j^n x_j^{-s} \in E_{n,\,i+1}$, we have

$$x_i^{-1} u_j^n x_j^{-s} x_i - u_j^n x_j^{-s} \in E_{n,\,i+1}.$$

This, together with (6.17), gives

$$x_i^{-1} v_j x_i - v_j \in E_{\mu_i + \mu(v_j),\,i+1} \tag{6.18}$$

for each $j > i$.

Now, $v = v_1 \cdots v_M \in E_p$ by hypothesis. Write $vx_i = v_1 \cdots v_M x_i$ as

$$vx_i = v_1 \cdots v_i (1 - u_i) \prod_{j=i+1}^{M} \left(x_i^{-1} v_j x_i\right).$$

An application of (6.18) gives $vx_i \equiv v \bmod E_{p+\mu_i,\,i}$.   □

We now finish the proof of Theorem 6.4. Suppose that $v = v_i \cdots v_M \in E_{p,\,i}$. By Lemma 6.15,

$$vu_i = v(1 - x_i) \in E_{p+\mu_i,\,i}.$$

We apply Lemma 6.15 repeatedly to conclude that for any $r > 0$,

$$vu_i^r = v(1 - x_i)^r \in E_{p+r\mu_i,\,i}.$$

In particular, $vu_i^n \in E_{p+n\mu_i,\,i} = E_{n,\,i}$. Repeating the proof of Lemma 6.15, replacing x_i by x_i^{-1}, gives

$$vu_i^n x_i^{-s} \in E_{n,\,i}.$$

Thus, if $v' = v_i' v_{i+1}' \cdots v_M' \in E_{q,\,i}$, then $vv_i' \in E_{p+\mu(v_i'),\,i}$. Therefore, by the induction hypothesis, $vv' \in E_{p+q,\,i}$ as required. We have shown that $E_{p,\,i} E_{q,\,i} \subseteq E_{p+q,\,i}$ for all i, completing the proof.

Corollary 6.4 *Let G be a torsion-free nilpotent group of class c. If F is any field of characteristic zero, then $D_{c+1}(F, G) = 1$. Consequently, G embeds in $1 + A_F(G)/A_F^{c+1}(G)$. Moreover, $D_{c+1}(\mathbb{Z}, G) = 1$.*

Proof Since $\gamma_{c+1} G = 1$ and G is torsion-free, $\overline{\gamma}_{c+1} G = 1$. Thus, $D_{c+1}(F, G) = 1$ by Theorem 6.4. The proof that G embeds in $1 + A_F(G)/A_F^{c+1}(G)$ is the same as the one given for Corollary 6.3. Since $D_i(\mathbb{Z}, G) \le D_i(F, G)$ for any $i \in \mathbb{N}$, we have $D_{c+1}(\mathbb{Z}, G) = 1$.   □

6.2.2 Faithful Representations

According to Example 2.17 in Chapter 1, the group $UT_n(R)$ consisting of all $n \times n$ upper unitriangular matrices with entries from a commutative ring with unity R is nilpotent of class less than n. A major result due to P. Hall shows that every finitely generated torsion-free nilpotent group can be represented by $UT_n(\mathbb{Z})$ for some n. In fact, the proof of this is implicit in the proof of Theorem 6.4. Our objective is to prove his theorem, along with a more general result regarding the representation of a polycyclic group in $GL_n(\mathbb{Z})$ for some n.

We begin with a brief discussion regarding representations. Suppose that R is a commutative ring with unity and let V be an R-module. We denote the group of all invertible R-module endomorphisms by $GL(V)$ and the group of non-singular $n \times n$ matrices over R under matrix multiplication by $GL_n(R)$.

Definition 6.4 Let G be any group. A *linear representation* of G (over R in V) is a group homomorphism $\varphi : G \to GL(V)$. If φ is one-to-one, then φ is called a *faithful* representation.

Suppose that $\varphi : G \to GL(V)$ is a linear representation of G. We explain how to turn V into an RG-module. Let $g \in G$, $v \in V$, and define a map

$$\psi : G \times V \to V \text{ by } \psi(g, v) = \varphi(g)(v).$$

It is easy to verify that ψ is a group action of G on V, which we will simply denote by gv from this point on. Thus, $\psi(g, v) = gv$. The following conditions hold for any g, g_1, $g_2 \in G$, v, $w \in V$, and r, $s \in R$:

  (i) $g(rv + sw) = r(gv) + s(gw)$;
  (ii) $(g_1 g_2)v = g_1(g_2 v)$;
  (iii) $1v = v$.

Hence, ψ is a *linear action* of G on V. Now, ψ can be extended linearly to an action Ψ of the group ring RG on V in a natural way,

$$\Psi\left(\sum_{g \in G} r_g g, \ v\right) = \sum_{g \in G} r_g(gv).$$

This action Ψ turns V into an RG-module.

Next, suppose that V is an RG-module for some group G. If $g \in G$, then the map

$$\varphi(g) : V \to V \text{ defined by } \varphi(g)(v) = gv$$

is a linear bijection and thus, an element of $GL(V)$. Furthermore, $\varphi : G \to GL(V)$ is a well-defined homomorphism. And so, φ is a linear representation of G. We summarize our discussion in a lemma.

Lemma 6.16 *If $\varphi : G \to GL(V)$ is a linear representation of G, then V can be viewed as an RG-module. Conversely, if V is an RG-module, then there exists a linear representation of G over R in V.*

Next, we turn to the specific case of free R-modules. Suppose that $V \cong R^n$ is a free R-module of rank n. If $\sigma : G \to GL(V)$ is a linear representation of G and $\mathscr{B}$ is an ordered basis for V, then we obtain a group homomorphism from G to $GL_n(R)$ defined by mapping $g \in G$ to the matrix of $\sigma(g)$ with respect to $\mathscr{B}$. Such a homomorphism is called a *matrix representation* of G (over R in V). If φ happens to be one-to-one, then the representation is *faithful*.

Lemma 6.16 carries over to matrix representations. More precisely, if G is a group and V is a free R-module, then V is an RG-module if and only if there exists a matrix representation of G over R in V relative to a basis of V.

We are interested in representing certain groups by groups of unitriangular matrices. This will amount to realizing a nilpotent action on a series.

Definition 6.5 Let G be a group with a fixed normal series

$$1 = G_0 \le G_1 \le \cdots \le G_r = G. \tag{6.19}$$

(i) An automorphism φ of G *stabilizes* (6.19) or *acts nilpotently* on (6.19) if

$$\varphi(xG_i) = xG_i \text{ for every } i = 0, \ldots, r-1 \text{ and } x \in G_{i+1}. \tag{6.20}$$

(ii) If A is a group of automorphisms of G and each element of A acts nilpotently on (6.19), then we say that A *stabilizes* (6.19) or *acts nilpotently* on (6.19). In this situation, A *acts nilpotently* on G.

Clearly, each G_i is invariant under φ (set $x = 1$ in (6.20)) and φ induces the trivial action on each factor group G_{i+1}/G_i.

Automorphism groups which act nilpotently are always nilpotent. This will be proven in the next chapter (see Theorem 7.9). It follows that if $A = Inn(G)$, then A acts nilpotently on G if and only if G is nilpotent. This is easy to see using Corollary 1.1 and Lemma 2.12.

Example 6.1 Let V be an r-dimensional vector space over a field F and choose a basis $\mathscr{B} = \{v_1, \ldots, v_r\}$ for V. Let V_0 denote the zero subspace of V and set $V_k = span\{v_1, \ldots, v_k\}$ for $k = 1, \ldots, r$. Suppose that $A \le GL(V)$ stabilizes the subgroup series

$$V_0 < V_1 < \cdots < V_r = V \tag{6.21}$$

in the additive abelian group V. Then A can be represented by a subgroup of $UT_r(F)$. To see this, let $g \in A$ and $M = \begin{bmatrix} m_{ij} \end{bmatrix}$, where M is the $r \times r$ matrix of g with respect to the basis $\mathscr{B}$. Thus, the entries of M satisfy the equations

$$g(v_j) = \sum_{i=1}^{r} m_{ij} v_i \quad (j = 1, \ldots, r).$$

Since g stabilizes (6.21), we have that $g(v_j) = v_j + \tilde{v}_{j-1}$ for some $\tilde{v}_{j-1} \in V_{j-1}$ for each $j = 1, \ldots, r$. One can deduce from this that M is upper unitriangular.

In what follows, $\mathbb{Z}^n$ will denote the n-fold direct sum of $\mathbb{Z}$.

Lemma 6.17 *Let A be a subgroup of $GL_n(\mathbb{Z})$ which acts nilpotently on $\mathbb{Z}^n$. Then $\mathbb{Z}^n$ has a basis with respect to which the action of A is represented by unitriangular matrices with entries in $\mathbb{Z}$.*

Proof Suppose that A acts nilpotently on a normal series

$$0 = G_0 \leq G_1 \leq \cdots \leq G_r = \mathbb{Z}^n.$$

We prove the lemma by induction on r. If $r = 1$, then the result is obvious. Assume that the lemma is true for $1 \leq k < r$ and let H denote the isolator of G_1 in $\mathbb{Z}^n$. It is not hard to see that A acts trivially on H. Since $\mathbb{Z}^n/H$ is torsion-free, $\mathbb{Z}^n$ splits over H and we obtain

$$\mathbb{Z}^n \cong H \oplus \mathbb{Z}^n/H.$$

Now, A induces a nilpotent action on the series

$$0 = H/H \leq (G_2 + H)/H \leq \cdots \leq (G_r + H)/H = \mathbb{Z}^n/H.$$

Since this series has length less than r, we conclude by induction that $\mathbb{Z}^n/H$ has a basis with respect to which action of A is represented by unitriangular matrices over $\mathbb{Z}$. We get a basis of $\mathbb{Z}^n$ of the required type by taking the basis for $\mathbb{Z}^n/H$ together with a basis for H. $\square$

We now prove the main theorem of P. Hall.

Theorem 6.5 *Every finitely generated torsion-free nilpotent group has a faithful representation in $UT_n(\mathbb{Z})$ for some n.*

We will refer to the proof of Theorem 6.4.

Proof Let G be a finitely generated torsion-free nilpotent group of class c with isolated lower central series

$$G = \overline{\gamma}_1 G > \cdots > \overline{\gamma}_{c+1} G = 1, \tag{6.22}$$

and let m_i be the rank of the free abelian group $\overline{\gamma}_i G / \overline{\gamma}_{i+1} G$ for $1 \leq i \leq c$. Suppose that $\{x_1, \ldots, x_M\}$ is the Mal'cev basis for G obtained by refining the series (6.22), so that

$$M = m_1 + \cdots + m_c.$$

Let F be a field of characteristic zero and put $A_F(G) = A$. Let d_i be the dimension of the F-vector space A^i/A^{i+1} and set $A^0 = FG$. Thus, $d_0 = dim(FG/A) = 1$.

Regard FG/A^{c+1} as an additive abelian group. We claim that G acts nilpotently on the series

$$FG/A^{c+1} > A/A^{c+1} > \cdots > A^{c+1}/A^{c+1} = 0 \qquad (6.23)$$

of FG/A^{c+1}. Clearly, G acts on FG by right multiplication. This induces an action of G on FG/A^{c+1}. It follows easily that A^i/A^{c+1} is G-invariant for $i = 0, 1, \ldots, c$. We need to show that the induced action is trivial on each factor group

$$\frac{A^i/A^{c+1}}{A^{i+1}/A^{c+1}} \cong A^i/A^{i+1} \qquad (i = 0, 1, \ldots, c).$$

In the proof of Lemma 6.14, we found that A^i/A^{i+1} has a basis consisting of vectors of the form $u + A^{i+1}$, where $u \in A^i$ is a vector in the integral basis for FG given in (6.16) and $\mu(u) = i$. Choose $1 \neq g \in G$ and $u' \in A^i \setminus A^{i+1}$, where u' is a basis element. Then $u'(g - 1) \in A^{i+1}$ since $g - 1 \in A$. Hence,

$$\left(u' + A^{i+1}\right)g = u'g + A^{i+1} = u' + A^{i+1},$$

and thus the action is trivial on each A^i/A^{i+1} as desired. Now, G embeds in FG/A^{c+1} by Corollary 6.4. It follows from Example 6.1 that G has a faithful representation in $UT_d(F)$ for $d = d_0 + d_1 + \cdots + d_c$. To get a faithful representation in $UT_d(\mathbb{Z})$, we use the fact that the basis (6.16) for FG is an integral basis. $\qquad \square$

Remark 6.3 Let G by a finitely generated torsion-free nilpotent group. According to Theorem 4.13, $G^{\mathbb{Q}}$ is a Mal'cev completion of G. Another way to obtain a Mal'cev completion of G is to make use of Theorem 6.5 and Corollary 5.2. Let

$$\psi : G \to UT_n(\mathbb{Z})$$

be an embedding for some $n > 0$. Clearly, G embeds in the nilpotent $\mathbb{Q}$-powered group $UT_n(\mathbb{Q})$. Thus, the isolator of $\psi(G)$ in $UT_n(\mathbb{Q})$ is a Mal'cev completion of G.

Example 6.2 Let G be a finitely generated torsion-free nilpotent group of class c with Mal'cev basis $\{u_1, \ldots, u_k\}$. Since G embeds into $UT_n(\mathbb{Z})$ for some $n > 0$ and $UT_n(\mathbb{Z})$ embeds in $UT_n(\mathbb{Q}[x])$, there exists an embedding

$$\Psi : G \to UT_n(\mathbb{Q}[x]).$$

View $UT_n(\mathbb{Q}[x])$ as a nilpotent $\mathbb{Q}[x]$-powered group by defining the $\mathbb{Q}[x]$-action on elements of $UT_n(\mathbb{Q}[x])$ in the same way as in Example 4.9, namely

$$(I_n + M)^\alpha = I_n + \alpha M + \cdots + \binom{\alpha}{n-1} M^{n-1},$$

where $\alpha \in \mathbb{Q}[x]$ and M is an $n \times n$ upper triangular matrix with 0's along the main diagonal. By Theorem 4.23, Ψ extends to a $\mathbb{Q}[x]$-homomorphism

$$\Phi : G^{\mathbb{Q}[x]} \to UT_n(\mathbb{Q}[x])$$

with respect to the basis $\{u_1, \ldots, u_k\}$ of G. In fact, Φ is a $\mathbb{Q}[x]$-monomorphism.

Every finitely generated nilpotent group is polycyclic by Theorem 4.4. Hence, certain polycyclic groups have faithful representations in $UT_n(\mathbb{Z})$ for some $n \in \mathbb{N}$. In general, every polycyclic group has a faithful representation in $GL_n(\mathbb{Z})$ for some $n \in \mathbb{N}$. This was proven by Auslander [1] using the theory of Lie groups. We provide a proof of a more general result due to Swan [17].

Our discussion is based on [8, 14] and [17]. First, we state a lemma whose proof can be found in [14] (see Lemma 10.12).

Lemma 6.18 *Let G be a finitely generated group and let I be a (two-sided) ideal of $\mathbb{Z}G$ such that $\mathbb{Z}G/I$ is finitely generated as an additive abelian group. Then I is a finitely generated ideal of $\mathbb{Z}G$. Furthermore, the additive abelian group $\mathbb{Z}G/I^k$ is finitely generated for every $k \in \mathbb{N}$.*

Theorem 6.6 (R. Swan) *Suppose that G is a group and N is a finitely generated torsion-free nilpotent normal subgroup of G such that G/N is finitely generated free abelian. There exists an integer d such that G has a faithful representation $\varrho : G \to GL_d(\mathbb{Z})$ for which $\varrho(N) \leq UT_d(\mathbb{Z})$.*

In what follows, we view $M_n(\mathbb{Z})$ as the ring of $n \times n$ matrices with integral entries, as well as an additive free abelian group of rank n^2.

Proof The proof is done by induction on the rank r of G/N. The case $r = 0$ (when $N = G$) has been done in Theorem 6.5.

Assume that the theorem holds when the rank is less than r. Since G/N is free abelian of rank r, there exists a normal subgroup H of G containing N such that H/N is free abelian of rank $r - 1$ and $(G/N)/(H/N)$ is infinite cyclic. By the Third Isomorphism Theorem, G/H is infinite cyclic. By the induction hypothesis, H has a faithful representation

$$\varrho : H \to GL_n(\mathbb{Z})$$

for some $n \in \mathbb{N}$ such that $\varrho(N) \leq UT_n(\mathbb{Z})$. This map can be naturally extended to a ring homomorphism

$$\overline{\varrho} : \mathbb{Z}H \to M_n(\mathbb{Z}).$$

Let $K = \ker \overline{\varrho}$. Observe that $\mathbb{Z}H/K$ is a finitely generated torsion-free additive abelian group since it is isomorphic to a subgroup of the additive group $M_n(\mathbb{Z})$. Let

L be the left ideal of $\mathbb{Z}H$ generated by the augmentation ideal of N. We claim that $L^n \leq K$. Since $N \trianglelefteq H$ and

$$(y - 1)h = h\left(h^{-1}yh - 1\right)$$

for all $y \in N$ and $h \in H$, L is a two-sided ideal. It follows that L^n is generated as a left ideal of $\mathbb{Z}H$ by all of the products of the form

$$(y_1 - 1) \cdots (y_n - 1) \qquad (y_1, \ldots, y_n \in N).$$

Let 0_n denote the $n \times n$ zero matrix. For any $y \in N$, $\varrho(y) - I_n$ is an $n \times n$ upper triangular matrix, all of whose main diagonal entries are zero. Thus,

$$\overline{\varrho}((y_1 - 1) \cdots (y_n - 1)) = (\varrho(y_1) - I_n) \cdots (\varrho(y_n) - I_n)$$
$$= 0_n$$

for any $y_1, \ldots, y_n \in N$. Therefore, $\overline{\varrho}(L^n) = 0_n$ and thus $L^n \leq K$ as claimed. Now,

$$(L + K)^n \subseteq L^n + K = K$$

because K is an ideal of $\mathbb{Z}H$ and $L^n \leq K$. Consider the set

$$T = \left\{a \in \mathbb{Z}H \mid sa \in (L + K)^n \text{ for some } s > 0\right\}.$$

It is clear that T is a two-sided ideal of $\mathbb{Z}H$ containing $(L+K)^n$ and $T/(L+K)^n$ is the torsion subgroup of the additive abelian group $\mathbb{Z}H/(L+K)^n$. We assert that $\mathbb{Z}H/T$ is a free $\mathbb{Z}$-module of finite rank. First, observe that $\mathbb{Z}H/(L+K)$ is finitely generated as an additive abelian group since it is a homomorphic image of the finitely generated abelian group $\mathbb{Z}H/K$. Clearly, H is finitely generated since it is an extension of H/N by N, both finitely generated. Hence, Lemma 6.18 guarantees that $\mathbb{Z}H/(L + K)^n$ is finitely generated as an abelian group. By the Third Isomorphism Theorem,

$$\mathbb{Z}H/T \cong \frac{\mathbb{Z}H/(L + K)^n}{T/(L + K)^n}$$

is an additive free abelian group of finite rank l, say. Put in another way, $\mathbb{Z}H/T$ is a free $\mathbb{Z}$-module of finite rank as asserted.

Our goal is to turn $\mathbb{Z}H/T$ into a $\mathbb{Z}G$-module. The first step is to let H act on $\mathbb{Z}H/T$ by right multiplication:

$$h \cdot \left(\sum_{i=1}^{k} n_i h_i + T\right) = \sum_{i=1}^{k} n_i h_i h + T \quad (n_i \in \mathbb{Z}, \ h_i \in H)$$

We assert that this action is faithful. Let

$$\bar{\sigma} : H \rightarrow Aut_{\mathbb{Z}}(\mathbb{Z}H/T)$$

be the homomorphism from H to the $\mathbb{Z}$-automorphisms of $\mathbb{Z}H/T$ induced by the action. We need to show that $\bar{\sigma}$ is a monomorphism. To do this, we begin by observing that H acts faithfully on $\mathbb{Z}H/K$ by right multiplication. To see this, let $h \in H$, and assume that

$$\sum_{i=1}^{k} r_i h_i h - \sum_{i=1}^{k} r_i h_i \in K = ker \, \bar{\varrho} \subseteq \mathbb{Z}H$$

for any $\sum_{i=1}^{k} r_i h_i \in \mathbb{Z}H$. In particular,

$$1 \cdot h - 1 = h - 1 \in K,$$

and thus

$$\bar{\varrho}(h - 1) = \varrho(h) - I_n = 0_n.$$

Since ϱ is faithful, $h = 1$ and thus H acts faithfully on $\mathbb{Z}H/K$ by right multiplication. Now, notice that $T \subseteq K$. To see this, suppose that $a \in T$. There exists $s > 0$ such that $sa \in (L + K)^n \subseteq K$. Hence,

$$\bar{\varrho}(sa) = s\bar{\varrho}(a) = 0_n.$$

Since $\bar{\varrho}(a) \in M_n(\mathbb{Z})$ and $M_n(\mathbb{Z})$ is an additively torsion-free abelian group, we have $\bar{\varrho}(a) = 0_n$. And so, $a \in K$. The fact that $T \subseteq K$ immediately gives that H acts faithfully on $\mathbb{Z}H/T$. Thus, $\bar{\sigma}$ is a monomorphism as claimed. We conclude that $\mathbb{Z}H/T$ is a faithful $\mathbb{Z}H$-module.

Next, we show that $N \leq H$ acts nilpotently on $\mathbb{Z}H/T$. Set $L^0 = \mathbb{Z}H$. Since

$$L^n \subseteq (L + K)^n \subseteq T,$$

there exists an integer $w_1 \geq 0$ such that $L^{w_1} \nsubseteq T$ and $L^{w_1+1} \subseteq T$. If $\beta \in N$ and $g \in L^{w_1} \setminus T$, then

$$g(\beta - 1) \in L^{w_1+1} \subseteq T.$$

This means that

$$(g + T)\beta = g\beta + T = g + T,$$

and thus N acts identically on $(L^{w_1} + T)/T$. In particular, N fixes some free generator $a_1 \in \mathbb{Z}H/T$. Let $T_1/T = gp(a_1)$. As before, $L^{w_2} \not\subseteq T \subseteq T_1$ and $L^{w_2+1} \subseteq T \subseteq T_1$ for some integer $w_2 \geq 0$. Hence, N acts as the identity on $(L^{w_2} + T_1)/T_1$. Thus, N fixes some free generator $a_2 \in \mathbb{Z}H/T_1$. Put $T_2/T_1 = gp(a_2)$ and continue in the same way until a series of the form

$$0 = T/T < T_1/T < T_2/T < \cdots < T_m/T < \mathbb{Z}H/T$$

has been reached. It follows that N stabilizes this series as claimed. By Lemma 6.17, a suitable $\mathbb{Z}$-basis can be chosen for $\mathbb{Z}H/T$ such that the induced monomorphism

$$\sigma : H \to GL_l(\mathbb{Z})$$

restricts to $UT_l(\mathbb{Z})$ on N.

We want to extend the action of H on $\mathbb{Z}H/T$ to an action of G on $\mathbb{Z}H/T$ and turn $\mathbb{Z}H/T$ into a $\mathbb{Z}G$-module. By hypothesis, G/H is infinite cyclic. Choose $x \in G$ such that $H \cap gp(x) \neq 1$ and $G = gp(x, H)$ and put $X = gp(x)$. We make $\mathbb{Z}H/T$ into a $\mathbb{Z}X$-module. Let $\alpha : \mathbb{Z}H \to \mathbb{Z}H$ be the ring automorphism defined by

$$\alpha\left(\sum_{i=1}^{k} n_i h_i\right) = \sum_{i=1}^{k} n_i \left(x^{-1} h_i x\right).$$

If $h \in H$, then $\alpha(h) = x^{-1}hx$. Therefore,

$$\alpha(h) - h = h\left(h^{-1}\alpha(h) - 1\right)$$
$$= h([h, x] - 1) \in L$$

because $[h, x] \in \gamma_2 G \leq N$. And so, $(\alpha - 1)\mathbb{Z}H \leq L$. One can now show that $\alpha(L + K) = L + K$ and thus $\alpha((L + K)^n) = (L + K)^n$ and $\alpha(T) = T$. Therefore, the mapping

$$\bar{\alpha} : \mathbb{Z}H/T \to \mathbb{Z}H/T \text{ defined by } \bar{\alpha}(s + T) = \alpha(s) + T$$

is an automorphism of $\mathbb{Z}H/T$. This means that if we let x act on $\mathbb{Z}H/T$ in the same way as $\bar{\alpha}$ by defining

$$x \cdot \left(\sum_{i=1}^{k} n_i h_i + T\right) = \sum_{i=1}^{k} n_i \left(x^{-1} h_i x\right) + T,$$

then $\mathbb{Z}H/T$ will become a $\mathbb{Z}X$-module. In order to turn $\mathbb{Z}H/T$ into a $\mathbb{Z}G$-module, we simultaneously extend the action of H on $\mathbb{Z}H/T$ and the action of X on $\mathbb{Z}H/T$ to obtain an action of G on $\mathbb{Z}H/T$. More precisely, if $g = hx^m$ for some $h \in H$ and $x^m \in X$, then the action of G on $\mathbb{Z}H/T$ is

$$g \cdot \left(\sum_{i=1}^{k} n_i h_i + T \right) = \sum_{i=1}^{k} n_i (x^{-m} h_i h x^m) + T.$$

This action need not be faithful. The final step is to construct the required faithful action. Let M be a free $\mathbb{Z}$-module of rank 2 and define an action of G on M by letting H act trivially on M and $X = gp(x)$ act on M according to the matrix representation induced by

$$x \mapsto \begin{pmatrix} 1 & 1 \\ 0 & 1 \end{pmatrix}$$

from X to $GL_2(\mathbb{Z})$. Consider the free $\mathbb{Z}$-module

$$D = \mathbb{Z}H/T \oplus M$$

of rank $l + 2$. Since H acts faithfully on $\mathbb{Z}H/T$ and trivially on M, one easily checks that G acts faithfully on D. This completes the proof.   □

Every polycyclic group has a subgroup of finite index of the type described in Theorem 6.6 (see [4]). Thus, every polycyclic group has a faithful representation in $GL_d(\mathbb{Z})$ for some d. We refer the reader to [15] for more on representations of polycyclic groups.

Remark 6.4 The method used to prove Theorem 6.6 can also be used to give a shorter proof of Theorem 6.5. Assume that $G = N$ in Theorem 6.6. As before, there exists a normal subgroup H of G such that G/H is infinite cyclic. By induction on the Hirsch length of G, we may assume that H has a faithful representation in $UT_l(\mathbb{Z})$ for some integer l. Proceeding as in the proof of Theorem 6.6, we obtain an action of G on $\mathbb{Z}H/T$, where $A^m(H) \subset T$ for some $m \in \mathbb{N}$. In order to complete the proof, we need to show that some power of $A(G)$ annihilates $\mathbb{Z}H/T$. This is just a consequence of Lemma 6.6. We refer the reader to Theorem 10.14 of [14] for a more detailed discussion.

6.3   The Lie Algebra of a Finitely Generated Torsion-Free Nilpotent Group

Lie algebras are essential in the study of Lie groups. Basically, one can translate a problem about Lie groups into a problem in linear algebra. In this section, we discuss the Lie algebra of a finitely generated torsion-free nilpotent group without imposing any topological structure on the group.

6.3.1 Lie Algebras

We begin with some generalities pertaining to Lie algebras which are relevant to our discussion.

Definition 6.6 A vector space V over a field k is called a *Lie algebra* over k if it comes equipped with a map

$$V \times V \to V \text{ defined by } (x, y) \mapsto [x, y]$$

which satisfies the following axioms for all x, y, $z \in V$ and a, $b \in k$:

(i) bilinearity: $[ax+by, z] = a[x, z]+b[y, z]$ and $[x, ay+bz] = a[x, y]+b[x, z]$;
(ii) skew symmetry: $[x, x] = 0$;
(iii) Jacobi identity: $[x, [y, z]] + [y, [z, x]] + [z, [x, y]] = 0$.

Evidently, a Lie algebra over k is an algebra over k which is simultaneously a Lie ring (see Definition 3.8). The properties in Lemma 3.3 hold for every Lie algebra. In particular, $[x, y] = -[y, x]$ for any x, $y \in V$ (anti-commutativity).

In order to avoid confusion between the commutator bracket of group elements and the Lie bracket of Lie algebra elements, we will write the Lie bracket as "$(,)$" from this point on.

Definition 6.7 Let V be a Lie algebra over k. A vector subspace W of V is a *Lie subalgebra* of V if $(w_1, w_2) \in W$ for all w_1, $w_2 \in W$. If $(w, v) \in W$ for every $w \in W$ and $v \in V$, then W is called an *ideal* of V.

Lemma 6.19 *Every ideal of a Lie algebra is two-sided.*

Proof Suppose that W is an ideal of a Lie algebra V. For any $v \in V$ and $w \in W$, $(v, w) \in W$ implies that $-(w, v) \in W$ by anti-commutativity. Since W is a vector subspace, $(w, v) \in W$. □

If V is a Lie algebra over k with Lie subalgebras U and W, then we use the notation

$$(U, W) = span\{(u, w) \mid u \in U, w \in W\}.$$

Lemma 6.20 *Let V be a Lie algebra. If U and W are ideals in V, then (U, W) is an ideal in V.*

Proof Let $u \in U$, $v \in V$, and $w \in W$. Since U and W are ideals, $(v, u) \in U$ and $(w, v) \in W$. By the Jacobi identity and anti-commutativity of the Lie bracket, we obtain

$$((u, w), v) = -(v, (u, w))$$

$$= (u, (w, v)) + (w, (v, u))$$

$$= (u, (w, v)) - ((v, u), w).$$

Therefore, $((u, w), v) \in (U, W)$, and thus (U, W) is an ideal in V. □

Definition 6.8 Let V be a Lie algebra over k. The *lower central series* of V is the sequence of Lie subalgebras

$$V = \gamma_1 V \supseteq \gamma_2 V \supseteq \cdots$$

recursively defined by $\gamma_{i+1} V = (\gamma_i V, \ V)$ for $i \in \mathbb{N}$.

By Lemma 6.20, each $\gamma_i V$ is an ideal of V. We say that V is *nilpotent* provided that $\gamma_n V = 0$ for some $n \in \mathbb{N}$. The *nilpotency class* or simply the *class* of a nilpotent Lie algebra V is the least k for which $\gamma_k V = 0$. If

$$\bigcap_{i=1}^{\infty} \gamma_i V = 0,$$

then V is termed *residually nilpotent*. One can endow such a Lie algebra with a topology by taking the set $\{\gamma_i V \mid i \geq 1\}$ as a neighborhood basis of $0 \in V$.

6.3.2 The A-Adic Topology on FG

Let G be a finitely generated torsion-free nilpotent group and let F be a field of characteristic zero. Let A denote the augmentation ideal of the group ring FG. Since A is residually nilpotent by Theorem 6.1, one can equip FG with the *A-adic topology* by taking $\{A^w \mid w \geq 1\}$ as a neighborhood basis of $0 \in FG$. The completion of FG and A in the A-adic topology is denoted by $\overline{FG}$ and $\overline{A}$ respectively. Thus,

$$\overline{FG} = \left\{ f_0 + \sum_{i=1}^{\infty} f_i a_i \ \middle| \ f_i \in F, \ a_i \in A^i \right\} \quad \text{and} \quad \overline{A} = \left\{ \sum_{i=1}^{\infty} f_i a_i \ \middle| \ f_i \in F, \ a_i \in A^i \right\}.$$

We will identify FG with its isomorphic image in $\overline{FG}$. It is clear that every element of $\overline{FG} \setminus \overline{A}$ is a unit in $\overline{FG}$. In particular, if $g \in G$, then its inverse in $\overline{FG}$ is

$$g^{-1} = (1 + (g - 1))^{-1} = \sum_{i=0}^{\infty} (-1)^i (g - 1)^i.$$

Furthermore, $\bigcap_{n=1}^{\infty} \overline{A}^n = 0$ and thus $\overline{A}$ is residually nilpotent. It follows that infinite sums of the form

$$\overline{a}_1 + \overline{a}_2 + \cdots + \overline{a}_i + \cdots$$

converge, where $\overline{a}_i \in \overline{A}^i$.

6.3.3   The Group $1 + \overline{A}$

Set $1 + \overline{A} = \{1 + a \mid a \in \overline{A}\}$ and let $1 + a$ and $1 + b$ be elements of $1 + \overline{A}$. Clearly,

$$(1 + a)(1 + b) = 1 + a + b + ab$$

and

$$(1 + a)^{-1} = 1 - a + a^2 - a^3 + \cdots.$$

Thus, $(1 + a)(1 + b)$ and $(1 + a)^{-1}$ are contained in $1 + \overline{A}$. Hence, $1 + \overline{A}$ is a group under ring multiplication in $\overline{FG}$. In fact, $1 + \overline{A}$ is a normal subgroup of the group of units of $\overline{FG}$ since $\overline{A}$ is an ideal in $\overline{FG}$. Since every $g \in G$ can be expressed in the form

$$g = 1 + (g - 1),$$

a similar computation shows that G is a subgroup of $1 + \overline{A}$.

Using our now standard notation, we let $\overline{A}^k$ be the ideal of $\overline{FG}$ generated by all the products of k elements in $\overline{A}$. For each $k \in \mathbb{N}$, set $\overline{D}_k = 1 + \overline{A}^k$. It is easy to verify that each $\overline{D}_k$ is a normal subgroup of $1 + \overline{A}$. We claim that the subgroups of the series

$$1 + \overline{A} = \overline{D}_1 > \overline{D}_2 > \overline{D}_3 > \cdots \tag{6.24}$$

satisfy the condition $\left[\overline{D}_i, \overline{D}_j\right] \leq \overline{D}_{i+j}$ for $i, j \in \mathbb{N}$. Let $1 + a_i \in \overline{D}_i$ and $1 + a_j \in \overline{D}_j$. Then

$$\left[1 + a_i,\ 1 + a_j\right] = \left(1 + a_i\right)^{-1}\left(1 + a_j\right)^{-1}\left(1 + a_i + a_j + a_i a_j\right)$$

$$= \left(1 + a_i\right)^{-1}\left(1 + a_j\right)^{-1}\left(1 + a_i + a_j + a_i a_j + a_j a_i - a_j a_i\right)$$

$$= \left(1 + a_i\right)^{-1}\left(1 + a_j\right)^{-1}\left[\left(1 + a_j\right)\left(1 + a_i\right) + a_i a_j - a_j a_i\right]$$

$$= 1 + \left(1 + a_i\right)^{-1}\left(1 + a_j\right)^{-1}\left(a_i a_j - a_j a_i\right)$$

is contained in $\overline{D}_{i+j}$ as claimed. In particular, $\left[\overline{D}_i,\ 1 + \overline{A}\right] \leq \overline{D}_{i+1}$ for $i \in \mathbb{N}$. This means $\gamma_n(1 + \overline{A}) \leq \overline{D}_n$ by Remark 2.6. Since $\overline{A}^n$ strictly contains $\overline{A}^{n+1}$, it follows that $\overline{D}_n = 1 + \overline{A}^n$ strictly contains $\overline{D}_{n+1} = 1 + \overline{A}^{n+1}$ for all $n \in \mathbb{N}$. Thus,

$$\bigcap_{n=1}^{\infty} \gamma_n(1 + \overline{A}) \leq \bigcap_{n=1}^{\infty} \overline{D}_n = 1,$$

and hence, $1 + \overline{A}$ is a residually nilpotent group.

6.3.4   The Maps Exp and Log

There is a way of corresponding elements of $\overline{A}$ to elements of $1 + \overline{A}$ and vice versa. Let $a \in \overline{A}$, and define

$$\exp a = 1 + a + \frac{a^2}{2!} + \cdots = \sum_{k=0}^{\infty} \frac{a^k}{k!}.$$

Clearly, this infinite sum converges in the A-adic topology to an element in $1 + \overline{A}$. Thus, exp is a mapping from $\overline{A}$ to $1 + \overline{A}$. Observe that exp a is a unit in $\overline{FG}$ for any $a \in \overline{A}$ since $1 + \overline{A}$ is a group.

Lemma 6.21   exp $a = 1$ *if and only if* $a = 0$.

Proof Clearly, exp $0 = 1$. Suppose that exp $a = 1$. Then

$$\exp a - 1 = a\left(1 + a + \frac{a}{2!} + \frac{a^2}{3!} + \cdots\right) = 0.$$

Since

$$1 + a + \frac{a}{2!} + \frac{a^2}{3!} + \cdots$$

is contained in $1 + \overline{A}$, and is therefore a unit in $\overline{FG}$, it follows that $a = 0$.   ☐

Lemma 6.22 *For any* $a \in \overline{A}$ *and* $p, \ q \in F$,

$$\exp(pa)\exp(qa) = \exp((p+q)a).$$

Proof Expand the product:

$$\exp(pa)\exp(qa) = \left(1 + pa + \frac{(pa)^2}{2!} + \cdots\right)\left(1 + qa + \frac{(qa)^2}{2!} + \cdots\right)$$

$$= \sum_{m=0}^{\infty} \sum_{k=0}^{m} \frac{(pa)^k}{k!} \cdot \frac{(qa)^{m-k}}{(m-k)!}$$

$$= \sum_{m=0}^{\infty} \frac{1}{m!} \sum_{k=0}^{m} \frac{m!}{k!(m-k)!} \cdot (pa)^k (qa)^{m-k}.$$

Since $(pa)(qa) = (qa)(pa)$, we have

$$\sum_{k=0}^{m} \frac{m!}{k!(m-k)!} \cdot (pa)^k (qa)^{m-k} = (pa + qa)^m$$

by the Binomial Theorem. Thus,

$$\exp(pa)\exp(qa) = \sum_{m=0}^{\infty} \frac{1}{m!}(pa + qa)^m. \tag{6.25}$$

Notice that the right-hand side of (6.25) is just $\exp((p + q)a)$. $\qquad\qquad\square$

Corollary 6.5 *For any $a \in \bar{A}$ and $n \in \mathbb{Z}$,*

$$(\exp a)^{-1} = \exp(-a) \ \ and \ \ (\exp a)^n = \exp(na).$$

Proof Apply Lemmas 6.21 and 6.22. $\qquad\qquad\qquad\qquad\qquad\qquad\square$

Next, we define a map which will serve as an inverse for $\exp$. Let $b \in 1 + \bar{A}$, so that $b - 1 \in \bar{A}$. Define

$$\log b = (b - 1) - \frac{(b-1)^2}{2} + \frac{(b-1)^3}{3} - \cdots = \sum_{k=0}^{\infty} \frac{(-1)^k (b-1)^{k+1}}{k+1}.$$

This series converges in the A-adic topology to an element in $\bar{A}$. Thus, $\log$ is a mapping from $1 + \bar{A}$ to $\bar{A}$. The usual relationships between the maps $\exp$ and $\log$ hold, namely

$$\log(\exp a) = a \ \ and \ \ \exp(\log b) = b.$$

Hence, $\exp$ and $\log$ are mutual inverses. Furthermore,

$$n \log b = \log \left(b^n \right)$$

for any $b \in 1 + \bar{A}$ and $n \in \mathbb{Z}$ by Corollary 6.5.

6.3.5 The Lie Algebra $\bar{\Lambda}$

We associate with the ring $\bar{A}$ a certain Lie algebra over F by equipping $\bar{A}$ with the commutation operation defined by

$$(a, b) = ab - ba \ \ for \ any \ a, b \in \bar{A}.$$

We write this Lie algebra as $\bar{\Lambda} = \mathscr{L}\left(\bar{A}\right)$. If $a_1, \ldots, a_n \in \bar{A}$, then we recursively define the *left-normed Lie product* by

$$(a_1, \ldots, a_{n-1}, a_n) = ((a_1, \ldots, a_{n-1}), a_n).$$

If $\overline{W}$ is any ideal of $\overline{A}$, then we shall write $\mathscr{L}(\overline{W})$ to denote the Lie subalgebra of $\overline{A}$ obtained by providing $\overline{W}$ with the commutation operation. One can show that $\mathscr{L}(\overline{W})$ is an ideal of the Lie algebra $\overline{A}$. In particular, we have a descending series of ideals

$$\overline{A} \supset \mathscr{L}(\overline{A}^2) \supset \mathscr{L}(\overline{A}^3) \supset \cdots.$$

Next, consider the lower central series of $\overline{A}$. We claim that $\gamma_k \overline{A} \subseteq \mathscr{L}(\overline{A}^k)$ for $k \in \mathbb{N}$. This will be shown by induction on k. By definition, $\gamma_1 \overline{A} = \mathscr{L}(\overline{A})$. Assume that $\gamma_m \overline{A} \subseteq \mathscr{L}(\overline{A}^m)$ for $m > 1$. If $a \in \gamma_m \overline{A}$ and $b \in \overline{A}$, then $a \in \mathscr{L}(\overline{A}^m)$. Therefore,

$$(a, b) = ab - ba \in \mathscr{L}(\overline{A}^{m+1}),$$

and thus $\gamma_{m+1} \overline{A} \subseteq \mathscr{L}(\overline{A}^{m+1})$ as claimed. Since $\overline{A}$ is residually nilpotent, it follows that

$$\bigcap_{k=1}^{\infty} \gamma_k \overline{A} = 0.$$

Thus, the Lie algebra $\overline{A}$ is residually nilpotent. Endow $\overline{A}$ with a topology by taking $\{\gamma_i \overline{A} \mid i \geq 1\}$ as a neighborhood basis of 0. We assert that $\overline{A}$ is complete in this topology.

Lemma 6.23 *If $a_i \in \gamma_i \overline{A}$ for $i \in \mathbb{N}$, then the infinite series*

$$a_1 + a_2 + \cdots + a_i + \cdots$$

is an element of $\overline{A}$.

Proof Since $\gamma_i \overline{A} \subseteq \mathscr{L}(\overline{A}^i)$ for $i \in \mathbb{N}$, each a_i can be regarded as an element of $\overline{A}^i$, the underlying ring of the Lie algebra $\mathscr{L}(\overline{A}^i)$. Thus, the sum

$$a_1 + a_2 + \cdots + a_i + \cdots$$

converges to a unique element of $\overline{A}$, which is also an element of $\overline{A}$. $\qquad\square$

6.3.6 The Baker-Campbell-Hausdorff Formula

There is a fundamental connection between the group $1 + \overline{A}$ and the Lie algebra $\overline{A}$. Let g and h be elements of $1 + \overline{A}$. There exist elements x, $y \in \overline{A}$ such that $x = \log g$ and $y = \log h$. Hence, $g = \exp x$ and $h = \exp y$. We may express the product

$$gh = (\exp x)(\exp y)$$

in the form $\exp z$, so that

$$z = \log((\exp x)(\exp y)).$$

The *Baker-Campbell-Hausdorff formula* gives an explicit formulation of z in terms of x, y, and Lie products involving x and y :

$$z = x + y + \frac{1}{2}(x, y) + \frac{1}{12}(x, y, y) + \frac{1}{12}(y, x, x) + \frac{1}{24}(y, x, x, y) + \cdots . \qquad (6.26)$$

The series in (6.26) is of the kind described in Lemma 6.23 and therefore converges. Hence, $z \in \overline{A}$. Furthermore, all of the coefficients appearing in the series turn out to be rational (see [6]).

Consider the equality $[g, h] = \exp w$. One can show that $w \in \overline{A}$ is an infinite sum of commutators in x and y :

$$w = \log [g, h] = (x, y) + \frac{1}{2}(x, y, x) + \frac{1}{2}(x, y, y) + \cdots .$$

It follows inductively that if $g_i = \exp x_i$ for $1 \leq i \leq n$ and $[g_1, g_2, \ldots, g_n] = \exp w_n$, then

$$w_n = \log [g_1, g_2, \ldots, g_n] = (x_1, x_2, \ldots, x_n) + \sum_{m=n+1}^{\infty} l_m, \qquad (6.27)$$

where each l_m is a rational linear combination of simple commutators of the form $(x_{i_1}, x_{i_2}, \ldots, x_{i_m})$, and $i_1, i_2, \ldots, i_m$ is a permutation of the integers $1, 2, \ldots, n$ with repetitions allowed, but each of the integers $1, 2, \ldots, n$ occurring at least once.

6.3.7 The Lie Algebra of a Nilpotent Group

We now arrive at the main theorem of this section. Let G be a finitely generated torsion-free nilpotent group of Hirsch length r and let $\{u_1, u_2, \ldots, u_r\}$ be a Mal'cev basis for G. Suppose that F is a field of characteristic zero. Recall that G can be regarded as a subgroup of $1 + \overline{A}$. Define the set

$$\log G = \{\log g \mid g \in G\} \subset \overline{A}.$$

Let $\mathscr{L}_F(G)$ be the F-subspace of $\overline{A}$ spanned by the elements of $\log G$.

Theorem 6.7 (S. A. Jennings) $\mathscr{L}_F(G)$ *is a nilpotent Lie subalgebra of $\overline{A}$ and the set*

$$\{\log u_1, \ldots, \log u_r\}$$

is a basis for its underlying vector space. Thus, the dimension of the vector space $\mathscr{L}_F(G)$ over F equals the Hirsch length of G.

We refer to $\mathscr{L}_F(G)$ as the *Lie algebra* of G over F. Later in this section, we use this Lie algebra to offer an alternate construction of the Mal'cev completion of G. In addition, we prove in Theorem 6.9 that $Aut(G)$ is isomorphic to the group of those $\mathbb{Q}$-linear automorphisms of the underlying vector space of $\mathscr{L}_F(G)$ that stabilize the set $\log G$. This is noteworthy since $\mathscr{L}_F(G)$ has finite dimension equal to the Hirsch length of G. It is in this sense that one may say that $\mathscr{L}_F(G)$ "linearizes" G. For a more detailed discussion of this and other applications, see [2] and [15].

The proof of Theorem 6.7 essentially follows from the next two lemmas.

Lemma 6.24 *Suppose that G is of class c. If $m > c$ and $\{g_1, g_2, \ldots, g_m\} \subset G$, then*

$$(\log g_1, \log g_2, \ldots, \log g_m) = 0.$$

Proof Put $x_i = \log g_i$. If $m > c$, then it follows from (6.27) that

$$1 = [g_1, g_2, \ldots, g_m] = \exp\left((x_1, x_2, \ldots, x_m) + \sum_{p=m+1}^{\infty} l_p\right),$$

where each l_p is a rational linear combination of simple commutators of the form $\left(x_{i_1}, x_{i_2}, \ldots, x_{i_p}\right)$, $i_1, i_2, \ldots, i_p$ is a permutation of integers $1, 2, \ldots, m$ with repetitions allowed and each of the integers $1, 2, \ldots, m$ occurrs at least once. By Lemma 6.21,

$$(x_1, x_2, \ldots, x_m) = -\sum_{p=m+1}^{\infty} l_p \qquad (6.28)$$

in the Lie algebra $\overline{A}$. Since each commutator $\left(x_{i_1}, x_{i_2}, \ldots, x_{i_p}\right)$ lies in $\overline{A}^p$, (6.28) implies that

$$(x_1, x_2, \ldots, x_m) \in \overline{A}^{m+1},$$

and this holds for every $m > c$. In particular, since $p > c$, every commutator $\left(x_{i_1}, x_{i_2}, \ldots, x_{i_p}\right)$ lies in $\overline{A}^{p+1}$ and thus (6.28) implies that

$$(x_1, x_2, \ldots, x_m) \in \overline{A}^{m+2}.$$

Continuing in this way leads to

$$(x_1, x_2, \ldots, x_m) \in \overline{A}^{m+N}$$

for all N. Since $\overline{A}$ is residually nilpotent, $(x_1, x_2, \ldots, x_m) = 0$.   □

It follows from Lemma 6.24 that the Lie algebra generated by $\log G$ is nilpotent of class at most c.

Lemma 6.25 *Every element of the set* $\log G$ *can be written uniquely as a $\mathbb{Q}$-linear combination of the elements* $\log u_1, \ldots, \log u_r$.

Proof Let $g \in G$. Since $\{u_1, \ldots, u_r\}$ is a Mal'cev basis for G, we can express g uniquely as

$$g = u_1^{\alpha_1} \cdots u_r^{\alpha_r} \quad (\alpha_1, \ldots, \alpha_r \in \mathbb{Z}).$$

Put $v_i = \log u_i$ for $i = 1, \ldots, r$ so that $u_i = \exp v_i$. By a straightforward application of (6.26), we get

$$
\begin{aligned}
g &= u_1^{\alpha_1} \cdots u_r^{\alpha_r} \\
&= \exp(\alpha_1 v_1) \cdots \exp(\alpha_r v_r) \\
&= \exp\left(\alpha_1 v_1 + \cdots + \alpha_r v_r + \sum_{\varrho=1}^{s} \beta_\varrho z_\varrho\right),
\end{aligned}
\tag{6.29}
$$

where each β_ϱ is rational and each z_ϱ is a commutator of the form $\left(v_{i_1}, \ldots, v_{i_t}\right)$ with $t \le c$. We emphasize that only finitely many z_ϱ occur as a consequence of Lemma 6.24. Thus,

$$\log g = \alpha_1 v_1 + \cdots + \alpha_r v_r + \sum_{\varrho=1}^{s} \beta_\varrho z_\varrho. \tag{6.30}$$

It suffices to prove that

$$\left(v_{i_1}, \ldots, v_{i_t}\right) = \sum_{j=1}^{r} \gamma_j v_j \quad (\gamma_j \in \mathbb{Q}) \tag{6.31}$$

for all $t \geq 2$. As we did after (6.5), define the weight of v_i to be 2^i and the weight of $\left(v_{i_1}, \ldots, v_{i_t}\right)$ to be

$$W(t) = 2^{i_1} + \cdots + 2^{i_t}.$$

We observe that $W_0 = c2^r$ is an upper bound on the weights of nonvanishing commutators of the form $(v_{i_1}, \ldots, v_{i_t})$. To see this, suppose that $W(t) > W_0$. Then

$$W_0 = c2^r < W(t) = 2^{i_1} + \cdots + 2^{i_t} \leq t2^r.$$

Hence, $c < t$, so that $(v_{i_1}, \ldots, v_{i_t}) = 0$ by Lemma 6.24. Thus, (6.31) is trivially true for commutators $\left(v_{i_1}, \ldots, v_{i_t}\right)$ of weight greater than W_0. Assume that (6.31) is true for all commutators of weight more than $W(t)$. By (6.27) and Lemma 6.24, we have

$$\left[u_{i_1}, u_{i_2}, \ldots, u_{i_t}\right] = \exp\left((v_{i_1}, v_{i_2}, \ldots, v_{i_t}) + \sum_{\varrho=1}^{s'} \beta'_\varrho z'_\varrho\right), \tag{6.32}$$

where $\beta'_\varrho \in \mathbb{Q}$ and each z'_ϱ is of the form $\left(v_{j_1}, v_{j_2}, \ldots, v_{j_l}\right)$ of weight $W(l) > W(t)$ since the $i_1, \ldots, i_t$ are contained among $j_1, \ldots, j_l$. Thus,

$$\sum_{\varrho=1}^{s'} \beta'_\varrho z'_\varrho = \sum_{j=1}^{r} \sigma_j v_j$$

for some $\sigma_1, \ldots, \sigma_r \in \mathbb{Q}$ by assumption. Consequently, (6.32) can be expressed as

$$\left[u_{i_1}, u_{i_2}, \ldots, u_{i_t}\right] = \exp\left((v_{i_1}, v_{i_2}, \ldots, v_{i_t}) + \sum_{j=1}^{r} \sigma_j v_j\right). \tag{6.33}$$

By Remark 4.4, there exist integers $\lambda_k, \ldots, \lambda_r$ such that

$$\left[u_{i_1}, u_{i_2}, \ldots, u_{i_t}\right] = u_k^{\lambda_k} u_{k+1}^{\lambda_{k+1}} \cdots u_r^{\lambda_r}, \tag{6.34}$$

where $k \geq t - 1 + max\{i_1, \ldots, i_t\}$. Put $m = max\{i_1, \ldots, i_t\}$ and take any v_s with $s \geq t - 1 + m$. The weight of v_s is 2^s and

$$W = W(t) = 2^{i_1} + \cdots + 2^{i_t} < 2^m t \leq 2^{t-1} 2^m \leq 2^s \tag{6.35}$$

because $i_1 \neq i_2$. Just as in (6.29), we can re-express (6.34) in the form

$$\left[u_{i_1}, u_{i_2}, \ldots, u_{i_t}\right] = \exp\left(\lambda_k v_k + \cdots + \lambda_r v_r + \sum_{\varrho=1}^{s''} \beta''_\varrho z''_\varrho\right), \tag{6.36}$$

where $\beta_\varrho'' \in \mathbb{Q}$ and each z_ϱ'' is a commutator of the form $\left(v_{k_1}, \ldots, v_{k_r}\right)$ with $k_d \geq k$ for $d = 1, 2, \ldots, r$. Since $k \geq t-1+m$, the weight of each z_ϱ'' exceeds W by (6.35).

By assumption, we can write each z_ϱ'' as a $\mathbb{Q}$-linear combination of $v_1, \ldots, v_r$. Therefore, (6.36) may be rewritten as

$$\left[u_{i_1}, u_{i_2}, \ldots, u_{i_t}\right] = \exp(\delta_1 v_1 + \cdots + \delta_r v_r) \tag{6.37}$$

for some $\delta_1, \ldots, \delta_r \in \mathbb{Q}$. Setting the right-hand sides of (6.33) and (6.37) equal to each other gives

$$\exp\left(\left(v_{i_1}, v_{i_2}, \ldots, v_{i_t}\right) + \sum_{j=1}^{r} \sigma_j v_j\right) = \exp(\delta_1 v_1 + \cdots + \delta_r v_r).$$

And so,

$$\left(v_{i_1}, v_{i_2}, \ldots, v_{i_t}\right) + \sum_{j=1}^{r} \sigma_j v_j = \delta_1 v_1 + \cdots + \delta_r v_r.$$

Therefore, (6.31) holds for any commutator of weight W if it holds for all commutators of weight exceeding W; that is, (6.31) holds in general.   □

The proof of Theorem 6.7 follows from the next two observations, both arising in the proof of Lemma 6.25.

- The set $\{\log u_1, \ldots, \log u_r\}$ is linearly independent over F since $\{u_1, \ldots, u_r\}$ is a Mal'cev basis for G.
- Every commutator of the form $\left(\log u_i, \log u_j\right)$ is a $\mathbb{Q}$-linear combination of the elements $\log u_1, \ldots, \log u_r$. Hence, $\left(\log u_i, \log u_j\right) \in \mathscr{L}_F(G)$.

6.3.8   The Group $\left(\overline{\Lambda}, *\right)$

The Lie algebra $\overline{\Lambda}$ can be equipped with a binary operation which turns it into a group. Let $g = \exp x$ and $h = \exp y$, where $x, y \in \overline{\Lambda}$. Using (6.26), we define an operation "$*$" on $\overline{\Lambda}$ as

$$x*y = x+y+\frac{1}{2}(x, y)+\frac{1}{12}(x, y, y)+\frac{1}{12}(y, x, x)+\frac{1}{24}(y, x, x, y)+\cdots. \tag{6.38}$$

As before, we have

$$\exp(x * y) = (\exp x)(\exp y) \quad \text{and} \quad x * y = \log\left((\exp x)(\exp y)\right).$$

Replacing x by $\log g$ and y by $\log h$ in the above gives

$$\log g * \log h = \log(gh).$$

Lemma 6.26 $\left(\overline{\Lambda}, *\right)$ *is a group.*

Proof Let w, x, and y be elements of $\overline{\Lambda}$. By our earlier discussion, we know that $x * y \in \overline{\Lambda}$. Thus, $\overline{\Lambda}$ is closed under $*$. Since

$$x * 0 = 0 * x = x \quad \text{and} \quad x * -x = -x * x = 0,$$

the identity element is 0 and each element of $\overline{\Lambda}$ has an inverse. We need to establish associativity. Consider the Magnus power series algebra $\mathbb{Q}[[a, b, c]]$ over the ring of rational numbers $\mathbb{Q}$ in the variables a, b, and c. Let W be the $\mathbb{Q}$-subalgebra of $\mathbb{Q}[[a, b, c]]$ consisting of those elements whose terms are of degree greater than zero and put $U = 1 + W$. We have seen in Section 3.2 that U is a subgroup of the group of units of $\mathbb{Q}[[a, b, c]]$. For any $w \in W$, we have $\exp w \in U$, where

$$\exp w = 1 + w + \frac{w^2}{2!} + \cdots = \sum_{k=0}^{\infty} \frac{w^k}{k!}.$$

Similarly, $\log u \in W$ whenever $u \in U$, where

$$\log u = (u - 1) - \frac{(u-1)^2}{2} + \frac{(u-1)^3}{3} - \cdots = \sum_{k=0}^{\infty} \frac{(-1)^k (u-1)^{k+1}}{k+1}.$$

As usual, $\exp$ and $\log$ are mutual inverse maps between W and U. For any elements w_1, $w_2 \in W$, define $w_1 * w_2 \in W$ by

$$\exp(w_1 * w_2) = (\exp w_1)(\exp w_2)$$

as before. Since $\{a, b, c\} \subset W$ and multiplication in $\mathbb{Q}[[a, b, c]]$ is associative, the equality

$$(\exp a)((\exp b)(\exp c)) = ((\exp a)(\exp b))(\exp c)$$

holds in U. Hence,

$$(\exp a)(\exp(b * c)) = (\exp(a * b))(\exp c),$$

and thus

$$\exp(a * (b * c)) = \exp((a * b) * c).$$

Using the fact that $\exp$ and $\log$ are mutual inverses, we obtain

$$a * (b * c) = (a * b) * c.$$

This will allow us to show that "$*$" is associative on $\overline{\Lambda}$. To see this, choose elements w, x, and y in $\overline{\Lambda}$. Put $S = \{a, b, c\}$ and define a set map

$$\varphi : S \to \overline{\Lambda} \quad \text{by} \quad a \mapsto w, \ b \mapsto x, \ c \mapsto y.$$

Clearly, S generates a free associative $\mathbb{Q}$-subalgebra of $\mathbb{Q}[[a, b, c]]$. It follows from Corollary 3.3 that the commutation Lie subalgebra M generated by S is free, freely generated by S. Hence, φ can be extended to a Lie algebra homomorphism from M to $\overline{\Lambda}$. This homomorphism can be further extended to a Lie algebra homomorphism from the completion of M in $\mathbb{Q}[[a, b, c]]$ to $\overline{\Lambda}$. Consequently,

$$w * (x * y) = (w * x) * y$$

as desired.   □

Suppose that G is a finitely generated torsion-free nilpotent group of class c with a Mal'cev basis $\{u_1, u_2, \ldots, u_r\}$. Let L denote the Lie algebra of G over $\mathbb{Q}$. By Theorem 6.7, L is a nilpotent Lie subalgebra of $\overline{\Lambda}$ of class c. Thus, the sum in the expansion (6.38) is always finite for any $x, y \in L$. It follows that $(L, *)$ is a subgroup of $(\overline{\Lambda}, *)$. In fact, $(L, *)$ is a nilpotent $\mathscr{D}$-group with $\mathbb{Q}$-exponentiation defined by

$$x^n = nx \quad (x \in L, n \in \mathbb{Q}).$$

Let x and y be any elements of $\log G$. Since $\exp x$ and $\exp y$ are contained in G,

$$x * y = \log ((\exp x)(\exp y)) \in \log G.$$

Hence, $\log G$ is closed under the operation $*$. Therefore, the bijection

$$\exp : \log G \to G$$

is a semigroup isomorphism between $(\log G, *)$ and G. More generally, the map $\exp$ gives an isomorphism between $(L, *)$ and a group $\widehat{G}$ such that

$$G \leq \widehat{G} \leq 1 + \overline{\Lambda}.$$

We immediately observe that $\widehat{G} = \exp(L, *)$ is a nilpotent $\mathscr{D}$-group since it is isomorphic to $(L, *)$. In fact, $\widehat{G}$ is a Mal'cev completion of G. One way to see this is to invoke the following result implicit in the work of Mal'cev [10]:

Theorem 6.8 *If $\hat{g} \in \widehat{G}$, then there exist unique rational numbers $\alpha_1, \ldots, \alpha_r$ such that $\hat{g} = \exp(\alpha_1 \log u_1 * \cdots * \alpha_r \log u_r)$.*

Since $\widehat{G}$ is a $\mathscr{D}$-group, rational exponentiation in $\widehat{G}$ is well defined. One can easily prove that

$$\alpha \log g = \log (g^\alpha) \quad (g \in G, \alpha \in \mathbb{Q}).$$

This, together with Theorem 6.8, give

$$\hat{g} = \exp \left(\alpha_1 \log \ u_1 * \cdots * \alpha_r \log \ u_r \right)$$
$$= \exp \left(\log \left(u_1^{\alpha_1} \right) * \cdots * \log \left(u_r^{\alpha_r} \right) \right)$$
$$= \exp \left(\log \left(u_1^{\alpha_1} \cdots u_r^{\alpha_r} \right) \right)$$
$$= u_1^{\alpha_1} \cdots u_r^{\alpha_r}.$$

Therefore, any element of $\widehat{G}$ can be written uniquely as

$$u_1^{\alpha_1} \cdots u_r^{\alpha_r} \quad (\alpha_1, \ \ldots, \ \alpha_r \in \mathbb{Q}).$$

These correspond precisely to the elements of $G^{\mathbb{Q}}$, the $\mathbb{Q}$-completion of G. By Theorem 4.13, $G^{\mathbb{Q}}$ is a Mal'cev completion of G. In light of this, the proof of Theorem 4.9 can be repeated in the case of $\widehat{G}$ and thus multiplication and exponentiation in $\widehat{G}$ are prescribed by the same polynomials as in $G^{\mathbb{Q}}$. It follows that the obvious mapping between $\widehat{G}$ and $G^{\mathbb{Q}}$ is an isomorphism. As a result, $\widehat{G}$ is a Mal'cev completion of G, as promised.

Remark 6.5 There are other ways to construct the Lie algebra and Mal'cev completion of a finitely generated torsion-free nilpotent group G. For example, one can take advantage of Theorem 6.5 and embed G into $UT_n(\mathbb{Q})$. Using the fact that the set S of $n \times n$ upper triangular matrices whose diagonal entries are zero can be made into a Lie algebra over $\mathbb{Q}$, one can associate the embedded image of G with such a Lie algebra using the exp and log maps. This idea allows one to study properties of G by examining its associated Lie subalgebra of S. Calculations in such a Lie subalgebra are quite manageable. This approach can be found in the work of Stewart [16] and Segal [15].

6.3.9   A Theorem on Automorphism Groups

The automorphism group of a finitely generated torsion-free nilpotent group can be represented as a group of linear transformations of a finite dimensional vector space over $\mathbb{Q}$. This is the point of the next theorem.

Theorem 6.9 *If G is a finitely generated torsion-free nilpotent group, then $Aut(G)$ is isomorphic to the group of automorphisms of $\mathscr{L}_{\mathbb{Q}}(G)$ which stabilize $\log G$.*

Proof Let $\varrho \in Aut(G)$. Then ϱ induces an automorphism $\overline{\varrho}$ of $\overline{\mathbb{Q}G}$ satisfying

$$\overline{\varrho} \left(\gamma_0 + \sum_{i=1}^{\infty} \gamma_i a_i \right) = \gamma_0 + \sum_{i=1}^{\infty} \gamma_i \overline{\varrho}(a_i),$$

where $\gamma_i \in \mathbb{Q}$, $a_i \in A^i(G)$, and the action of $\overline{\varrho}$ on $A^i(G)$ is induced by the action of ϱ on G. We show that $\overline{\varrho}$ stabilizes $\log G$. Indeed, for any $g \in G$,

$$\overline{\varrho}(\log g) = \overline{\varrho}\left(\sum_{i=1}^{\infty} \frac{(-1)^{i+1}}{i} (g-1)^i \right)$$

$$= \sum_{i=1}^{\infty} \frac{(-1)^{i+1}}{i} \overline{\varrho}\big((g-1)^i \big)$$

$$= \sum_{i=1}^{\infty} \frac{(-1)^{i+1}}{i} (\varrho(g) - 1)^i$$

$$= \log \varrho(g).$$

It readily follows that every automorphism of G gives rise to an automorphism of the Lie algebra $\mathscr{L}_{\mathbb{Q}}(G)$ which stabilizes $\log G$.

Next, suppose that $\varphi \in Aut\big(\mathscr{L}_{\mathbb{Q}}(G) \big)$ stabilizes $\log G$. Define a map

$$\eta : G \to G \ \text{by} \ \eta(g) = h \ \text{if} \ \varphi(\log g) = \log h.$$

We claim that $\eta \in Aut(G)$. Let g_1, g_2, and h be elements of G such that $\eta(g_1g_2) = h$. Then $\varphi(\log(g_1g_2)) = \log h$. However, $\log(g_1g_2) = \log(g_1) * \log(g_2)$. Hence,

$$\log h = \varphi(\log(g_1g_2))$$

$$= \varphi(\log(g_1) * \log(g_2))$$

$$= \varphi(\log(g_1)) * \varphi(\log(g_2)).$$

Thus, if $\varphi(\log(g_1)) = \log(h_1)$ and $\varphi(\log(g_2)) = \log(h_2)$, then

$$\log h = \log(h_1) * \log(h_2) = \log(h_1h_2).$$

Since $\log$ is injective, $h = h_1h_2$ and consequently, $\eta(g_1g_2) = \eta(g_1)\eta(g_2)$. Now, if $\eta(g_1) = \eta(g_2) = k$ for some $k \in G$, then $\varphi(\log(g_1)) = \varphi(\log(g_2)) = \log k$. Thus, $\log(g_1) = \log(g_2)$ because φ is injective. Since $\log$ is injective, $g_1 = g_2$ and thus η is a monomorphism. Lastly, suppose that $l \in G$. Since φ is bijective and φ^{-1} also stabilizes $\log G$, there exists $g \in G$ such that $\varphi(\log g) = \log l$. Hence, $\eta(g) = l$, and thus η is surjective. □

Applications of Theorem 6.9 and related results can be found in [2] and [13].

6.3.10   The Mal'cev Correspondence

Let G be a finitely generated torsion-free nilpotent group. We have seen that one can define a group operation "$*$" given by (6.38) and $\mathbb{Q}$-exponentiation by $x^n = nx$ for $x \in L$ and $n \in \mathbb{Q}$ on the underlying set of the nilpotent Lie algebra L of G. The group $(L, *)$ turns out to be a nilpotent $\mathscr{D}$-group.

This idea can be generalized for arbitrary nilpotent Lie algebras over $\mathbb{Q}$. In fact, one can also define Lie algebra addition, Lie commutation, and scalar multiplication on the underlying set of a nilpotent $\mathscr{D}$-group and obtain a nilpotent Lie algebra over $\mathbb{Q}$. This leads to the well-known *Mal'cev Correspondence*, a one-to-one correspondence between nilpotent $\mathscr{D}$-groups and nilpotent Lie algebras over $\mathbb{Q}$.

Our goal is to give an overview of how one defines the operations mentioned above. We refer the reader to [9] for an excellent discussion of the topic, along with some of its applications.

Define the formal power series for "exp" and "log" in the usual way,

$$\exp x = 1 + x + \frac{x^2}{2!} + \cdots \quad \text{and}$$

$$\log y = (y - 1) - \frac{(y - 1)^2}{2} + \frac{(y - 1)^3}{3} - \cdots .$$

- Let L be any nilpotent Lie algebra over $\mathbb{Q}$. Repeating what was done earlier, we may turn L into a nilpotent $\mathscr{D}$-group by defining the group multiplication "$*$" on the underlying set of L by

$$x * y = \log((\exp x)(\exp y))$$

$$= x + y + \frac{1}{2}(x, y) + \frac{1}{12}(x, y, y) + \frac{1}{12}(y, x, x) + \frac{1}{24}(y, x, x, y) + \cdots$$

and $\mathbb{Q}$-exponentiation on L by

$$x^n = nx \quad (n \in \mathbb{Q})$$

for $x, y \in L$. Note that the sum has finitely many terms since L is nilpotent.
- Suppose that G is a nilpotent $\mathscr{D}$-group. Three operations can be defined on the underlying set of G which turns it into a nilpotent Lie algebra over $\mathbb{Q}$. For any $g, h \in G$, the *inverse Baker-Campbell-Hausdorff formula* gives the expansion of

$$h_1(g, h) = \exp(\log g + \log h) \ \text{ and}$$

$$h_2(g, h) = \exp\{(\log g)(\log h) - (\log h)(\log g)\}$$

in terms of group commutators:

$$h_1(g, h) = gh[g, h]^{-\frac{1}{2}}[g, h, g]^{-\frac{1}{12}}[g, h, h]^{-\frac{1}{12}} \cdots$$

$$h_2(g, h) = [g, h][g, h, g]^{-\frac{1}{2}}[g, h, h]^{-\frac{1}{2}} \cdots.$$

Note that $h_1(g, h)$ and $h_2(g, h)$ are finite products since G is nilpotent. One can equip G with the structure of a Lie algebra over $\mathbb{Q}$ by defining Lie algebra addition, Lie commutation, and scalar multiplication by

$$g + h = h_1(g, h), \quad (g, h) = h_2(g, h), \quad \text{and} \quad ng = g^n \quad (n \in \mathbb{Q}) \tag{6.39}$$

respectively.

References

1. L. Auslander, On a problem of Philip Hall. Ann. Math. (2) **86**, 112–116 (1967). MR0218454
2. G. Baumslag, *Lecture Notes on Nilpotent Groups*. Regional Conference Series in Mathematics, No. 2 (American Mathematical Society, Providence, RI, 1971). MR0283082
3. E. Formanek, A short proof of a theorem of Jennings. Proc. Am. Math. Soc. **26**, 405–407 (1970). MR0272895
4. P. Hall, *The Edmonton Notes on Nilpotent Groups*. Queen Mary College Mathematics Notes. Mathematics Department (Queen Mary College, London, 1969). MR0283083
5. B. Hartley, The residual nilpotence of wreath products. Proc. Lond. Math. Soc. (3) **20**, 365–392 (1970). MR0258966
6. N. Jacobson, *Lie Algebras* (Dover Publications, New York, 1979). MR0559927
7. S.A. Jennings, The group ring of a class of infinite nilpotent groups. Can. J. Math. **7**, 169–187 (1955). MR0068540
8. M.I. Kargapolov, J.I. Merzljakov, *Fundamentals of the Theory of Groups*. Graduate Texts in Mathematics, vol. 62, (Springer, New York, 1979). MR0551207. Translated from the second Russian edition by Robert G. Burns
9. E.I. Khukhro, *p-Automorphisms of Finite p-Groups*. London Mathematical Society Lecture Note Series, vol. 246 (Cambridge University Press, Cambridge, 1998). MR1615819
10. A.I. Mal'cev, On a class of homogeneous spaces. Izv. Akad. Nauk. SSSR. Ser. Mat. **13**, 9–32 (1949). MR0028842
11. I.B.S. Passi, *Group Rings and Their Augmentation Ideals*. Lecture Notes in Mathematics, vol. 715 (Springer, Berlin, 1979). MR0537126
12. D.S. Passman, *The Algebraic Structure of Group Rings*. Pure and Applied Mathematics (Wiley-Interscience, New York, 1977). MR0470211
13. P.F. Pickel, Finitely generated nilpotent groups with isomorphic finite quotients. Trans. Am. Math. Soc. **160**, 327–341 (1971). MR0291287
14. D.J.S. Robinson, *Finiteness Conditions and Generalized Soluble Groups. Part 1* (Springer, New York, 1972). MR0332989
15. D. Segal, *Polycyclic Groups*. Cambridge Tracts in Mathematics, vol. 82 (Cambridge University Press, Cambridge, 1983). MR0713786
16. I.N. Stewart, An algebraic treatment of Mal'cev's theorems concerning nilpotent Lie groups and their Lie algebras. Composit. Math. **22**, 289–312 (1970). MR0288158
17. R.G. Swan, Representations of polycyclic groups. Proc. Am. Math. Soc. **18**, 573–574 (1967). MR0213442

Chapter 7
Additional Topics

This chapter contains a collection of miscellaneous topics. Section 7.1 pertains to M. Dehn's algorithmic problems for finitely generated nilpotent groups. In Section 7.2, we prove that finitely generated nilpotent groups are Hopfian. Section 7.3 contains useful facts about groups of upper unitriangular matrices over a commutative ring with unity R. In Section 7.4, we study certain groups of automorphisms that are themselves nilpotent. In particular, we prove that if G is a nilpotent group of class c, then the group of those automorphisms of G that induce the identity on the abelianization of G is nilpotent of class $c - 1$. Section 7.5 ends the chapter with an overview of the Frattini subgroup $\Phi(G)$ and Fitting subgroup $Fit(G)$ of a group G. Among other results, we prove that if G is a finite group, then $\Phi(G)$ is nilpotent, $Fit(G/\Phi(G)) = Fit(G)/\Phi(G)$, and $\Phi(G) \trianglelefteq Fit(G)$.

7.1 Decision Problems

In 1911, M. Dehn raised three decision problems about finitely presented groups. In what follows, let G be a group given by a finite presentation $G = \langle X \mid R \rangle$. An arbitrary (not necessarily reduced) word in the generators X is termed an *X-word*. We now state the problems.

The Word Problem: *Is there an algorithm which determines whether or not an X-word is the identity in G?*

The Conjugacy Problem: *Is there an algorithm which determines whether or not any pair of X-words g and h of G are conjugate in G? In other words, does an X-word k exist in G such that $g = k^{-1}hk$ in G?*

The Isomorphism Problem: *Let H be another finitely presented group with presentation*

$$H = \langle Y \mid S \rangle.$$

© Springer International Publishing AG 2017
A.E. Clement et al., *The Theory of Nilpotent Groups*,
DOI 10.1007/978-3-319-66213-8_7

Is there an algorithm which determines whether or not G and H are isomorphic?

If the answer to any such problem is "yes," then we say that the problem is *solvable*. In this section, we prove that the word problem and the conjugacy problem for finitely generated nilpotent groups are solvable. In fact, we prove the more general result that every finitely presented residually finite group has a solvable word problem.

The isomorphism problem was solved in the positive by Grunewald and Segal [4]. This is by far the most complicated of all of the decision problems for finitely generated nilpotent groups. The algorithms associated with it are quite lengthy and take up about 30 pages in the cited paper.

7.1.1   The Word Problem

We present the solution of the word problem given in [1] which uses residual finiteness. We begin with a key theorem.

Theorem 7.1 (J. C. C. McKinsey) *Every finitely presented residually finite group has a solvable word problem.*

Proof Let G be a finitely presented residually finite group and assume that G is given by an explicit finite presentation. Let w be a given word in the generators. We begin by describing two separate effective procedures.

- The first procedure simply enumerates all consequences of the defining relators. If w appears in this enumeration, then the procedure stops.
- The second procedure begins by enumerating all finite groups, say by constructing their multiplication tables. For each finite group F, the procedure then constructs the (finitely many) homomorphisms θ from G to F. This is done by assigning an element of F to each generator of G, then checking that each defining relator of G maps to the identity element in F. For each such homomorphism θ, the procedure then computes $\theta(w)$ in F. If there exists a finite group F and a homomorphism $\theta : G \to F$ such that $\theta(w) \neq 1$ in F, then the procedure stops.

Now, if $w = 1$ in G, then w will turn up as a consequence of the defining relators and the first procedure will stop. On the other hand, if $w \neq 1$, then the residual finiteness of G guarantees that $w \notin N$ for some $N \lhd G$ with $F = G/N$ finite. Thus, the image of w in F will be a nonidentity element and the second procedure will stop. We conclude that if the first procedure stops, then $w = 1$ in G, whereas if the second one stops, then $w \neq 1$ in G.   □

Polycyclic groups are always finitely presentable. This is an immediate result of the next theorem due to P. Hall.

Theorem 7.2 *If G is a group with $N \unlhd G$, and both N and G/N are finitely presented, then G is finitely presented.*

Proof Let $1_{G/N}$ denote the identity element of G/N. Suppose that

$$N = \langle x_1, \ldots, x_n \mid r_1(x_1, \ldots, x_n) = 1, \ldots, r_m(x_1, \ldots, x_n) = 1 \rangle$$

and

$$G/N = \langle g_1 N, \ldots, g_l N \mid s_1(g_1 N, \ldots, g_l N) = 1_{G/N}, \ldots, s_k(g_1 N, \ldots, g_l N) = 1_{G/N} \rangle.$$

Clearly,

$$G = gp(x_1, \ldots, x_n, g_1, \ldots, g_l),$$

and thus G is finitely generated. The relations in these generators are given by

$$r_i(x_1, \ldots, x_n) = 1 \quad (i = 1, \ldots, m),$$
$$s_j(g_1, \ldots, g_l) = t_j(x_1, \ldots, x_n) \quad (j = 1, \ldots, k),$$
$$g_j x_i g_j^{-1} = u_{ij}(x_1, \ldots, x_n) \quad (i = 1, \ldots, n \text{ and } j = 1, \ldots, l), \text{ and}$$
$$g_j^{-1} x_i g_j = v_{ij}(x_1, \ldots, x_n) \quad (i = 1, \ldots, n \text{ and } j = 1, \ldots, l).$$

Define $\overline{G}$ to be the group presented by generators $\overline{x}_1, \ldots, \overline{x}_n, \overline{g}_1, \ldots, \overline{g}_l$ and subject to the above relations in these generators. We claim that $G \cong \overline{G}$.

By Von Dyck's Lemma (see 2.2.1 in [11]), there exists a surjective homomorphism $\varphi : \overline{G} \to G$ determined by

$$\overline{x}_i \mapsto x_i \quad \text{and} \quad \overline{g}_j \mapsto g_j \quad (i = 1, \ldots, n \text{ and } j = 1, \ldots, l).$$

Let $K = ker\ \varphi$. We need to show that $K = 1$. Put $\overline{N} = gp(\overline{x}_1, \ldots, \overline{x}_n) < \overline{G}$. The restriction of φ to $\overline{N}$ determines an isomorphism with N because all of the relations in the elements x_i are consequences of the relations $r_i(x_1, \ldots, x_n) = 1$ for $i = 1, \ldots, m$. Consequently, $K \cap \overline{N} = 1$. Since

$$\overline{g}_j \overline{x}_i \overline{g}_j^{-1} \in \overline{N} \quad \text{and} \quad \overline{g}_j^{-1} \overline{x}_i \overline{g}_j \in \overline{N}$$

for $i = 1, \ldots, n$ and $j = 1, \ldots, l$, we have that $\overline{N} \lhd \overline{G}$. Furthermore, φ induces an injective map

$$\overline{\varphi} : \overline{G}/\overline{N} \to G/N \quad \text{such that} \quad \overline{g}_i \overline{N} \mapsto g_i N$$

because $\varphi(\overline{N}) = N$. Now, all relations in the elements $g_i N$ are consequences of the relations

$$s_j(g_1 N, \ldots, g_l N) = 1_{G/N} \quad (j = 1, \ldots, k).$$

Hence, $\overline{\varphi}$ is an isomorphism and thus has trivial kernel. However, $\ker \overline{\varphi} = K\overline{N}/\overline{N}$. And so, $K \leq \overline{N}$. Since $K \cap \overline{N} = 1$, we conclude that $K = 1$.   □

Corollary 7.1 *Every polycyclic group is finitely presentable.*

By Theorems 4.4 and 7.1, together with Corollaries 5.21 and 7.1, we now have:

Theorem 7.3 *Polycyclic groups have solvable word problem. In particular, finitely generated nilpotent groups have solvable word problem.*

7.1.2 The Conjugacy Problem

In order to prove that the conjugacy problem for finitely generated nilpotent groups is solvable, we need a theorem of N. Blackburn [2].

Definition 7.1 A group G is called *conjugacy separable* if, whenever two elements g and h of G are conjugate, then the images of g and h in every finite homomorphic image of G are conjugate.

Theorem 7.4 (N. Blackburn) *Every finitely generated nilpotent group is conjugacy separable.*

We introduce some notation: if u and v are conjugate elements of a group, then we write $u \sim v$; otherwise, we write $u \nsim v$.

Proof The proof, which is adopted from [1], is done by induction on the Hirsch length r of G. If $r = 0$, then G is finite and the result is immediate.

Suppose that $r > 0$ and assume the theorem is true for every finitely generated nilpotent group of Hirsch length less than r. Let g and h be elements of G such that $gN \sim hN$ for every $N \lhd G$ with G/N finite. We claim that $g \sim h$. Assume on the contrary, that $g \nsim h$. Since $r > 0$, G must be infinite. By Lemma 2.27, there exists of an element $a \in Z(G)$ of infinite order. Let

$$H_i = gp\left(a^{i!}\right) \quad (i = 1, 2, \ldots).$$

By Theorem 4.7, the Hirsch length of each G/H_i is $r - 1$. Suppose that $gH_i \nsim hH_i$ for some i. By induction and the Third Isomorphism Theorem, there exists a normal subgroup N_i/H_i of G/H_i such that

$$(G/H_i) / (N_i/H_i) \cong G/N_i$$

is finite and the images of g and h under the natural homomorphism $G \to G/N_i$ are not conjugate in G/N_i. As this contradicts our earlier assumption that the images of g and h are conjugate in every finite quotient of G, it must be the case that $gH_i \sim hH_i$ for all $i \in \mathbb{N}$. In particular, $gH_1 \sim hH_1$ in G/H_1. Since $H_1 = gp(a)$, we can find a nonzero integer m and an element $k \in G$ such that $h^k = ga^m$. It immediately follows that for all $i \in N$,

$$gH_i \sim ga^m H_i.$$

Thus, for each $i \in \mathbb{N}$, we can find $d_i \in G$ and $s_i \in \mathbb{Z}$ such that

$$g^{d_i} = ga^m \left(a^{i!}\right)^{s_i}. \tag{7.1}$$

Put $L = gp(d_1, d_2, \ldots, g, a) \leq G$. Since $a \in Z(G)$, it is clear that $a \in Z(L)$. Moreover, it follows from (7.1) that

$$[d_i, g] \in gp(a)$$

for all $i \in \mathbb{N}$. Since $gp(a) \leq Z(L)$, we apply the commutator calculus to conclude that $[x, g] \in gp(a)$ for all $x \in L$. As a result, we obtain a homomorphism

$$\varphi : L \to gp(a) \text{ defined by } \varphi(x) = [x, g].$$

It follows immediately that $L/ker \, \varphi$ is cyclic. Hence, there exists $b \in L$ such that

$$L = gp(ker \, \varphi, b). \tag{7.2}$$

We henceforth assume, on replacing b by b^{-1} if necessary, that

$$[b, g] = a^\alpha \quad (\alpha \geq 0). \tag{7.3}$$

We now argue that α cannot be zero. If it were the case that $\alpha = 0$, then $[b, g] = 1$. Since $ker \, \varphi = C_L(g)$, this would imply that $g \in Z(L)$. Thus, by (7.1),

$$1 = a^m \left(a^{i!}\right)^{s_i}$$

for all $i \in \mathbb{N}$. Since a has infinite order, it must be that $m + i! s_i = 0$ for all $i \in \mathbb{N}$. This is impossible since $m \neq 0$ and the sequence $\{i! \mid i \in \mathbb{N}\}$ is strictly increasing. We conclude that $\alpha > 0$ in (7.3).

Next, we find all of the conjugates of g in L. By (7.2), such a conjugate has the form g^{xb^n}, where $x \in ker \, \varphi$ and $n \in \mathbb{Z}$. Since $[g, b] \in Z(L)$ and $ker \, \varphi = C_L(g)$, it must be the case that

$$g^{xb^n} = g^{b^n} = g \, [g, b^n] = g[g, b]^n = ga^{-n\alpha}.$$

It follows that the conjugates of g in L are:

$$g, \, ga^{\pm\alpha}, \, ga^{\pm 2\alpha}, \, \ldots. \tag{7.4}$$

Since g and ga^m are not conjugate in G, they cannot be conjugate in L. Thus, α does not divide m. However, (7.1) yields

$$g^{d_\alpha} = ga^m \left(a^{\alpha!}\right)^{s_\alpha}$$

because $\alpha > 0$. This means that $ga^m \left(a^{\alpha!}\right)$ is one of the conjugates listed in (7.4), so $m + s_\alpha \alpha! = \lambda \alpha$ for some integer λ. Hence, α divides m, a contradiction. This completes the proof.   □

Corollary 7.2 *The conjugacy problem for finitely generated nilpotent groups is solvable.*

Proof Let G be a finitely generated nilpotent group. We can assume that G is given by an explicit finite presentation in light of Theorem 4.4 and Corollary 7.1. Let g and h be words in the generators of G. As we did in the proof of Theorem 7.1, we describe two separate procedures.

- The first procedure begins by enumerating the consequences of the finitely many relators. If there exists a word k in the generators of G such that $h^{-1}g^k$ appears in this enumeration, then the procedure stops.
- As before, the second procedure begins by listing all finite groups (up to isomorphism) by means of their multiplication tables. For every finite group F, we compute (the finitely many) homomorphisms from G to F. Once again, this is done by assigning an element of F to each generator of G and then verifying that the defining relators are mapped to 1. Finally, for every homomorphism $\varphi : G \to F$, we compute $\varphi(g)$ and $\varphi(h)$ in F. If there exists a finite group F and a homomorphism $\varphi : G \to F$ in which $\varphi(g)$ and $\varphi(h)$ are not conjugate, then the procedure stops.

Now, if g and h are conjugate in G, then there is a word k in the generators of G such that $h^{-1}g^k = 1$ in G. Hence, the word $h^{-1}g^k$ is a consequence of the defining relations and the first procedure will stop.

Next, assume that g and h are not conjugate in G. By Theorem 7.4, there exists a finite group F and a homomorphism $\varphi : G \to F$ such that $\varphi(g)$ and $\varphi(h)$ are not conjugate in F, so the second procedure stops.

Therefore, if the first procedure stops, g and h are conjugate in G; whereas if the second procedure stops, then g and h are not conjugate in G. This completes the solution to the conjugacy problem.   □

7.2   The Hopfian Property

In the 1930s, H. Hopf asked whether a finitely generated group could be isomorphic to a proper quotient of itself. Groups which are *not* isomorphic to proper quotients of themselves are known as *Hopfian*. In this section, we prove that every finitely generated nilpotent group is Hopfian. An example of a non-Hopfian group will be presented as well.

Definition 7.2 A group G is *Hopfian* if $G/N \cong G$ for some $N \trianglelefteq G$ implies that $N = 1$. Equivalently, every epimorphism of G is an isomorphism of G.

Recall that a group G satisfies Max if all of its subgroups are finitely generated (see Definition 2.14). By Theorem 2.16, G satisfies Max if and only if every ascending series of subgroups stabilizes.

Proposition 7.1 *If G satisfies Max, then G is Hopfian.*

Proof The proof is done by contradiction. Suppose that G satisfies Max and assume that there is an epimorphism $\Phi : G \to G$ with nontrivial kernel. For $j \geq 1$, set

$$\Phi^{\circ j} = \underbrace{\Phi \circ \cdots \circ \Phi}_{j} .$$

Clearly, $\Phi^{\circ j} : G \to G$ is an epimorphism for each $j \geq 1$. Let $K_j = ker\left(\Phi^{\circ j}\right)$. Observe that $\Phi^{\circ j}(K_j) = 1$ and $\Phi^{\circ j}(K_{j+1}) = ker\, \Phi$. We claim that

$$K_1 < K_2 < \cdots$$

is an infinite properly ascending chain of subgroups of G. Indeed, since $ker\, \Phi$ is nontrivial, there exists $k_{j+1} \in K_{j+1}$ such that

$$1 \neq \Phi^{\circ j}\left(k_{j+1}\right) \in ker\, \Phi.$$

Thus, $k_{j+1} \notin K_j$. This contradicts the fact that G satisfies Max. Therefore, the kernel of Φ must be trivial and thus Φ is a monomorphism. Consequently, Φ is an isomorphism. And so, G is Hopfian. □

Corollary 7.3 *Every finitely generated nilpotent group is Hopfian.*

Proof Finitely generated nilpotent groups satisfy Max by Theorem 2.18. Apply Proposition 7.1. □

Remark 7.1 More generally, A. I. Mal'cev proved that every finitely generated residually finite group is Hopfian.

An example of a non-Hopfian group is the *Baumslag-Solitar* group $BS(2, 3)$, where

$$BS(2, 3) = \left\langle a, b \mid b^{-1}a^2b = a^3 \right\rangle.$$

Consider the map

$$\theta : BS(2, 3) \to BS(2, 3) \text{ induced by } a \mapsto a^2 \text{ and } b \mapsto b.$$

The fact that θ is indeed a well-defined homomorphism follows from Von Dyck's Lemma (see 2.2.1 in [11]). Furthermore, b is clearly in the image of θ, and $a \in im\, \theta$ since

$$\theta\left(a^{-1}b^{-1}ab\right) = \theta\left(a^{-1}\right)\theta\left(b^{-1}\right)\theta\left(a\right)\theta\left(b\right)$$
$$= a^{-2}b^{-1}a^2b = a^{-2}a^3 = a.$$

Thus, θ is an epimorphism. We claim that θ has a nontrivial kernel. First, observe that the element $\left[b^{-1}ab,\ a\right]$ can be written as such:

$$\left[b^{-1}ab,\ a\right] = \left(b^{-1}ab\right)^{-1}a^{-1}b^{-1}aba$$
$$= b^{-1}a^{-1}ba^{-1}b^{-1}aba.$$

Hence, $\left[b^{-1}ab,\ a\right] \neq 1$ by Britton's Lemma (see Chapter IV in [9]). The element $\left[b^{-1}ab,\ a\right]$ belongs to the kernel of θ because

$$\theta\left(\left[b^{-1}ab,\ a\right]\right) = \left[\theta\left(b^{-1}ab\right),\ \theta(a)\right]$$
$$= \left[b^{-1}a^2b,\ a\right] = \left[a^3,\ a\right] = 1.$$

Since θ is an epimorphism of $BS(2,\ 3)$ onto itself,

$$BS(2,\ 3)/ker\ \theta \cong BS(2,\ 3)$$

by the First Isomorphism Theorem; that is, $BS(2,\ 3)$ is isomorphic to a proper quotient of itself.

7.3   The (Upper) Unitriangular Groups

In this section, R will always be a commutative ring with unity. An important collection of nilpotent groups are the *(upper) unitriangular groups of degree n over R*, denoted by $UT_n(R)$. These groups were first encountered in Example 2.17 of Section 1.3. The purpose of this section is to acquaint the reader with some of their fundamental properties.

Recall from Example 2.17 that if S is the set of $n \times n$ upper triangular matrix over R whose main diagonal entries are all 0, then

$$UT_n(R) = \{I_n + M \mid M \in S\}$$

is a nilpotent group of class less than n. For any $1 \leq m \leq n$, let $UT_n^m(R)$ be the normal subgroup of $UT_n(R)$ consisting of those matrices whose $m-1$ superdiagonals have 0's in their entries. For example,

$$UT_4^2(R) = \left\{ \begin{pmatrix} 1 & 0 & r & s \\ 0 & 1 & 0 & t \\ 0 & 0 & 1 & 0 \\ 0 & 0 & 0 & 1 \end{pmatrix} \middle| \; r,\, s,\, t \in R \right\} \quad \text{and} \quad UT_4^3(R) = \left\{ \begin{pmatrix} 1 & 0 & 0 & r \\ 0 & 1 & 0 & 0 \\ 0 & 0 & 1 & 0 \\ 0 & 0 & 0 & 1 \end{pmatrix} \middle| \; r \in R \right\}.$$

In particular, $UT_n^1(R) = UT_n(R)$ and $UT_n^n(R) = I_n$. Note that we have the inclusions

$$UT_n(R) > UT_n^2(R) > \cdots > UT_n^{n-1}(R) > I_n.$$

Many of the properties of $UT_n(R)$ can be derived using certain matrices called transvections. Let E_{ij} denote the $n \times n$ matrix with 1 in the (i,j) entry and 0's elsewhere. It is easy to see that

$$E_{ij} \cdot E_{kl} = \begin{cases} E_{il} & \text{if } j = k, \\ 0 & \text{otherwise.} \end{cases} \tag{7.5}$$

Let $r \in R$, and set

$$t_{i,j}(r) = I_n + rE_{ij} \quad (1 \le i,\, j \le n \text{ and } i \ne j).$$

If $r \ne 0$, then $t_{i,j}(r)$ is called a *transvection*. We abbreviate $t_{i,j}(1) = t_{i,j}$.

Lemma 7.1 *Let $r,\, s \in R$, and assume that $i \ne j$ and $k \ne l$. Then*

(i) $t_{i,j}(r) \cdot t_{i,j}(s) = t_{i,j}(r + s)$.

(ii) $\left(t_{i,j}(r) \right)^{-1} = t_{i,j}(-r)$.

(iii) $\left[t_{i,j}(r),\, t_{k,l}(s) \right] = \begin{cases} t_{i,l}(rs) & \text{if } j = k,\, i \ne l, \\ t_{k,j}(-rs) & \text{if } j \ne k,\, i = l, \\ I_n & \text{if } j \ne k,\, i \ne l. \end{cases}$

In case $R = \mathbb{Z}$, we also have

(iv) $t_{i,j}(r) = t_{i,j}(1)^r = t_{i,j}^r$.

Proof Observe that

$$\begin{aligned} t_{i,j}(r) \cdot t_{i,j}(s) &= \left(I_n + rE_{ij} \right)\left(I_n + sE_{ij} \right) \\ &= I_n + (r + s)E_{ij} + rsE_{ij}^2 \\ &= I_n + (r + s)E_{ij} \\ &= t_{i,j}(r + s). \end{aligned}$$

This proves (i). In particular,

$$t_{i,j}(r) \cdot t_{i,j}(-r) = t_{i,j}(r + (-r)) = I_n.$$

This gives (ii). To obtain (iii), we use the fact that

$$\left[t_{i,j}(r),\ t_{k,l}(s)\right] = t_{i,j}(-r) \cdot t_{k,l}(-s) \cdot t_{i,j}(r) \cdot t_{k,l}(-s)$$

$$= \left(I_n - rE_{ij}\right)\left(I_n - sE_{kl}\right)\left(I_n + rE_{ij}\right)\left(I_n + sE_{kl}\right)$$

by (ii). For instance, if $j = k$ but $i \neq l$, then a straightforward computation, together with (7.5), give

$$\left[t_{i,j}(r),\ t_{k,l}(s)\right] = I_n + rsE_{il} = t_{i,l}(rs).$$

The rest of (iii) follows similarly.

We prove (iv). The result is immediate for $r = 0$ since $t_{i,j}(0) = I_n$. Let $r > 0$. Using (i), we obtain

$$t_{i,j}(r) = t_{i,j}\left(\sum_{n=1}^{r} 1\right) = \prod_{n=1}^{r} t_{i,j}(1) = t_{i,j}^{r}. \qquad (7.6)$$

If $r < 0$, then (7.6), together with (ii), give

$$t_{i,j}(r) = \left(t_{i,j}(-r)\right)^{-1} = t_{i,j}^{r}$$

since $-r > 0$. □

Next, we find a convenient collection of transvections which generate $UT_n^m(R)$. First, we make a simple observation. Let $A = (a_{kl})$ be an $n \times n$ matrix with entries in R, so that

$$A = \sum_{1 \leq k,\ l \leq n} a_{kl}E_{kl}.$$

For $i \neq j$, we have

$$At_{i,j}(r) = A\left(I_n + rE_{ij}\right) = A + rAE_{ij}$$

$$= A + r\left(\sum_{1 \leq k,\ l \leq n} a_{kl}E_{kl}\right)E_{ij}$$

$$= A + r\sum_{k=1}^{n} a_{ki}E_{kj}$$

by (7.5). Thus, the product $At_{i,j}(r)$ is the matrix obtained by adding r times the ith column of A to the jth column of A.

Theorem 7.5 *The set of transvections* $\{t_{i,j}(r) \mid j - i \geq m,\ r \in R\}$ *generates* $UT_n^m(R)$. *If* $R = \mathbb{Z}$, *then the set* $\{t_{i,j} \mid j - i \geq m\}$ *generates* $UT_n^m(\mathbb{Z})$.

Proof Suppose that $A = (a_{ij}) \in UT_n^m(R)$. By the observation above, A can be reduced to the identity matrix by post-multiplying A by a suitable sequence of transvections. More specifically,

$$A\big(t_{1,\,m+1}(-a_{1,\,m+1}) \cdots t_{1,\,n}(-a_{1,\,n})\big)\big(t_{2,\,m+2}(-a_{2,\,m+2}) \cdots t_{2,\,n}(-a_{2,\,n})\big)$$
$$\cdots t_{n-m,\,n}(-a_{n-m,\,n}) = I_n.$$

After taking inverses of both sides and applying Lemma 7.1 (ii), the matrix A equals the product of transvections

$$t_{n-m,\,n}(a_{n-m,\,n}) \cdots \big(t_{2,\,n}(a_{2,\,n}) \cdots t_{2,\,m+2}(a_{2,\,m+2})\big)\big(t_{1,\,n}(a_{1,\,n}) \cdots t_{1,\,m+1}(a_{1,\,m+1})\big).$$

Thus, $\{t_{i,\,j}(r) \mid j - i \geq m,\ r \in R\}$ is a generating set for $UT_n^m(R)$. The case $R = \mathbb{Z}$ follows immediately from Lemma 7.1 (iv). $\qquad\square$

Setting $m = 1$ in Theorem 7.5 and applying Lemma 7.1 give:

Corollary 7.4 *The set* $\{t_{k-1,\,k}(r) \mid 2 \leq k \leq n,\ r \in R\}$ *generates* $UT_n(R)$, *and the set* $\{t_{k-1,\,k} \mid 2 \leq k \leq n\}$ *generates* $UT_n(\mathbb{Z})$.

Observe that the set $\{t_{k-1,\,k}(r) \mid 2 \leq k \leq n,\ r \in R\}$, together with the identities in Lemma 7.1, give a presentation for $UT_n(R)$.

The lower and upper central series of $UT_n(R)$ coincide. This is the point of the following theorem.

Theorem 7.6 *The series*

$$UT_n(R) > UT_n^2(R) > \cdots > UT_n^{n-1}(R) > 1 \tag{7.7}$$

is both the lower and the upper central series for $UT_n(R)$. *Hence,*

- *The nilpotency class of* $UT_n(R)$ *is exactly* $n - 1$,
- $UT_n^i(R) = \gamma_i UT_n(R) = \zeta_{n-i} UT_n(R)$ *for* $i = 1,\ 2,\ \ldots,\ n$ *and*
- $UT_n^j(R)/UT_n^{j+1}(R)$ *is abelian for* $j = 1,\ 2,\ \ldots,\ n - 1$.

Proof Put $UT_n(R) = U$ and $UT_n^i(R) = U_i$ for $i = 1,\ 2,\ \ldots,\ n$. We claim that $[U_i,\ U] = U_{i+1}$ for each $i = 1,\ 2,\ \ldots,\ n - 1$.

1. Let $[g,\ h] \in [U_m,\ U]$, where $g \in U_m$ and $h \in U$. By Theorem 7.5 and Lemma 7.1, g is a product of transvections of the form $t_{i,\,j}(r)$ with $j - i \geq m,\ r \in R$, and h is a product of transvections of the form $t_{k,\,l}(s)$ with $l - k \geq 1$ and $s \in R$. We prove the claim by induction on the number of transvections occurring in g and h combined.

   - For the basis of induction, assume that $g = t_{i,\,j}(r)$ and $h = t_{k,\,l}(s)$. Suppose that $j = k$ but $i \neq l$. By Lemma 7.1 (iii),

$$[g,\ h] = t_{i,\,l}(rs).$$

Since in this case $l - i = (j - i) + (l - k) \geq m + 1$, we have $[g, h] \in U_{m+1}$. The other cases are handled in a similar way.

- Suppose that $g = t_{i, j}(r)\bar{g}$ and $h = t_{k, l}(s)$, where $\bar{g} \in U_m$. By Lemma 1.4 (v),

$$[g, h] = \left[t_{i, j}(r)\bar{g},\ t_{k, l}(s)\right] = \left[t_{i, j}(r),\ t_{k, l}(s)\right]^{\bar{g}}\left[\bar{g},\ t_{k, l}(s)\right].$$

Now, $\left[\bar{g},\ t_{k, l}(s)\right] \in U_{m+1}$ and $\left[t_{i, j}(r), t_{k, l}(s)\right] \in U_{m+1}$ by induction. Hence,

$$\left[t_{i, j}(r),\ t_{k, l}(s)\right]^{\bar{g}} \in U_{m+1},$$

and thus $[g, h] \in U_{m+1}$.

- Suppose that $g = t_{i, j}(r)\bar{g}$ and $h = t_{k, l}(s)\bar{h}$, where $\bar{g} \in U_m$ and $\bar{h} \in U$. By Lemma 1.4 (vi),

$$[g, h] = \left[t_{i, j}(r)\bar{g},\ t_{k, l}(s)\bar{h}\right] = \left[t_{i, j}(r)\bar{g},\ \bar{h}\right]\left[t_{i, j}(r)\bar{g},\ t_{k, l}(s)\right]^{\bar{h}}.$$

Since $\left[t_{i, j}(r)\bar{g},\ \bar{h}\right] \in U_{m+1}$ by induction and $\left[t_{i, j}(r)\bar{g},\ t_{k, l}(s)\right] \in U_{m+1}$ by the previous case, we have $[g, h] \in U_{m+1}$.

Therefore, $[U_m,\ U] \subseteq U_{m+1}$.

2. In order to establish the reverse inclusion $U_{m+1} \subseteq [U_m,\ U]$, it suffices to show that $[U_m,\ U]$ contains every transvection $t_{i, j}(r)$ with $j - i \geq m + 1$ and $r \in R$. Consider the transvections $t_{i, i+m}(r) \in U_m$ and $t_{i+m, j} = t_{i+m, j}(1) \in U$. By Lemma 7.1 (iii),

$$t_{i, j}(r) = \left[t_{i, i+m}(r),\ t_{i+m, j}\right] \in [U_m,\ U].$$

Thus, $U_{m+1} \subseteq [U_m,\ U]$.

This proves the claim that $[U_i,\ U] = U_{i+1}$ for $i = 1, \ldots, n - 1$. Therefore, $\gamma_i U = U_i$, and thus (7.7) is the lower central series for $UT_n(R)$. In order to prove that (7.7) is the upper central series for $UT_n(R)$, one can use induction on $n - i$ to show that $\zeta_{n-i}UT_n(R) = \gamma_i UT_n(R)$. We omit the details. $\square$

One can easily find a set of generators for the factor groups of (7.7) by using Theorem 7.5.

Theorem 7.7 *For $m = 1, 2, \ldots, n - 1$, each factor group $UT_n^m(R)/UT_n^{m+1}(R)$ is generated, modulo $UT_n^{m+1}(R)$, by the set $\{t_{i, m+i}(r) \mid 1 \leq i \leq n - m, r \in R\}$. If $R = \mathbb{Z}$, then $UT_n^m(\mathbb{Z})/UT_n^{m+1}(\mathbb{Z})$ is generated, modulo $UT_n^{m+1}(\mathbb{Z})$, by the set $\{t_{i, m+i} \mid 1 \leq i \leq n - m\}$.*

We show next that $UT_n(\mathbb{Z})$ is torsion-free. In fact, we prove this for any ring whose characteristic is zero.

Theorem 7.8 *If R has characteristic zero, then $UT_n(R)$ is torsion-free. In particular, $UT_n(\mathbb{Z})$ is torsion-free.*

Proof By Theorem 7.6, the center of $UT_n(R)$ equals the subgroup $UT_n^{n-1}(R)$ consisting of those matrices with an arbitrary entry of R in the uppermost right corner and 0's elsewhere. It follows that $Z(UT_n(R))$ is isomorphic to the additive group of R. Since R has characteristic zero, $Z(UT_n(R))$ is torsion-free. The result follows from Corollary 2.21.   □

Remark 7.2 If the characteristic of R is different than zero, then $UT_n(R)$ need not be torsion-free. For example, let R be the polynomial ring in one variable over a finite field of characteristic a prime p and let $2 \leq k \leq n$. By Corollary 7.4, a typical generator for $UT_n(R)$ is a transvection of the form $t_{k-1,\,k}(r)$, where $r \in R$. Since R has characteristic p, we apply Lemma 7.1 (i) repeatedly to obtain

$$(t_{k-1,\,k}(r))^p = t_{k-1,\,k}(rp) = I.$$

Thus, every generator of $UT_n(R)$ is a torsion element.

There is a natural Mal'cev basis for $UT_n(\mathbb{Z})$. By Corollary 7.4 and Theorem 7.8, $UT_n(\mathbb{Z})$ is finitely generated and torsion-free. Each of the factor groups $UT_n^m(\mathbb{Z})/UT_n^{m+1}(\mathbb{Z})$ is torsion-free and finitely generated, modulo $UT_n^{m+1}(\mathbb{Z})$, by transvections of the form $t_{i,\,m+i}$, where $1 \leq i \leq n - m$. This follows directly from Corollary 2.20 and Theorem 7.7. Taking this into consideration, we have:

Lemma 7.2 *The set of transvections*

$$\left\{ t_{1,\,2},\ t_{2,\,3},\ \ldots,\ t_{n-1,\,n},\ t_{1,\,3},\ t_{2,\,4},\ \ldots,\ t_{n-2,\,n},\ \ldots,\ t_{1,\,n} \right\}$$

is a Mal'cev basis for $UT_n(\mathbb{Z})$.

7.4 Nilpotent Groups of Automorphisms

In this section, we prove that certain subgroups of the automorphism group of a given group are nilpotent. We begin by defining the holomorph of a group.

Let G be a group and define the set

$$\widetilde{G} = \{\varphi g \mid \varphi \in Aut(G),\ g \in G\}.$$

We can regard $\widetilde{G}$ as the Cartesian product $Aut(G) \times G$. This set becomes a group under the operation

$$(\varphi g)(\varphi' g') = \varphi\varphi' g^{\varphi'} g',$$

where $g^{\varphi'} = \varphi'(g) \in G$. Thus, we have a semi-direct product of G by $Aut(G)$. This group is called the *holomorph* of G, written as $Hol(G)$.

7.4.1   The Stability Group

Definition 7.3 Let G be any group and let

$$G = G_0 \geq G_1 \geq \cdots \geq G_r = 1 \qquad (7.8)$$

be a series of subgroups of G. The *stability group* of G relative to the series (7.8) is the group A of all automorphisms φ of G such that

$$\varphi(xG_i) = xG_i \text{ for every } i = 0, \ldots, r-1 \text{ and } x \in G_{i+1}. \qquad (7.9)$$

The group A and the automorphism $\varphi \in A$ are said to *stabilize* the series (7.8).

Putting $x = 1$ in (7.9) shows that each G_i is invariant under φ. In [7], L. Kalužnin proved that the stability group of a group is always solvable of "solvability length" at most m. If the series happens to be a normal series, then the stability group is, in fact, nilpotent.

Theorem 7.9 *Let*

$$G = G_0 \geq G_1 \geq \cdots \geq G_r = 1 \qquad (7.10)$$

be a normal series of a group G. If A is the stability group of G relative to (7.10), then A and $[G, A]$ are nilpotent of class less than r. Here, $[G, A]$ is viewed as a subgroup of $\mathrm{Hol}(G)$.

In this situation, the stability group acts nilpotently on G according to Definition 6.5. The proof relies on a property of commutator subgroups.

Theorem 7.10 *Let H and K be subgroups of a group G and suppose that*

$$H = H_0 \geq H_1 \geq \cdots$$

is a descending series of normal subgroups of H such that $[H_i, K] \leq H_{i+1}$ for each integer $i \geq 0$. Put $K = K_1$, and for $j > 1$, define

$$K_j = \{x \in K \mid [H_i, x] \leq H_{i+j} \text{ for all } i \geq 0\}.$$

Then $[K_j, K_l] \leq K_{j+l}$ for all j, $l \geq 1$, and $[H_i, \gamma_j K] \leq H_{i+j}$ for all $i \geq 0$ and $j \geq 1$.

Proof We show first that $[K_j, K_l] \leq K_{j+l}$. By definition, $[H_i, K_j] \leq H_{i+j}$ and $[H_{i+j}, K_l] \leq H_{i+j+l}$. Hence,

$$[H_i, K_j, K_l] \leq [H_{i+j}, K_l] \leq H_{i+j+l}$$

by Proposition 1.1 (iii). Likewise,

$$[K_l, H_i, K_j] = [H_i, K_l, K_j] \leq H_{i+j+l}$$

by Proposition 1.1 (i) and (iii). By hypothesis, H_{i+j+l} is normal in H. Moreover, for $x \in K$,

$$x^{-1} H_{i+j+l} x = H_{i+j+l} [H_{i+j+l}, x].$$

Since $[H_{i+j+l}, x] \leq H_{i+j+l+1}$, we conclude that

$$x^{-1} H_{i+j+l} x \leq H_{i+j+l}.$$

This shows that H_{i+j+l} is normal in the subgroup of G generated by H and K. Thus,

$$[K_j, K_l, H_i] = [H_i, [K_j, K_l]] \leq H_{i+j+l}$$

by Proposition 1.1 (i) and Lemma 2.18. Therefore, by definition, $[K_j, K_l] \leq K_{j+l}$. In particular, $[K_j, K] \leq K_{j+1}$. Hence, $\gamma_j K \leq K_j$ and thus

$$[H_i, \gamma_j K] \leq [H_i, K_j] \leq H_{i+j}$$

by Proposition 1.1 (iii). This completes the proof. $\square$

Remark 7.3 Note that Theorem 2.14 (i) can be obtained from Theorem 7.10 by setting $H_i = \gamma_{i+1} G$ and $K = G$.

We now prove Theorem 7.9. To begin with, we assert that $[G_i, A] \leq G_{i+1}$ for $i = 0, 1, \ldots, r - 1$. If $x \in G_i$ and $\alpha \in A$, then

$$[x, \alpha] = x^{-1} \alpha^{-1} x \alpha = x^{-1} x^\alpha \in G_{i+1}$$

since α induces the identity on G_i/G_{i+1}. Put $H_i = G_i$ $(i = 0, 1, \ldots, r)$ and $K = A$ in Theorem 7.10 and regard all of these as subgroups of the $Hol(G)$. Then

$$[G, \gamma_r A] = [G_0, \gamma_r A] \leq G_r = 1,$$

and thus $[G, \gamma_r A] = 1$. Now, let $\alpha \in \gamma_r A$ and $x \in G$. Then $[x, \alpha] = x^{-1} x^\alpha = 1$, which shows that $\alpha = 1$. Consequently, $\gamma_r A = 1$ and A is nilpotent of class less than r as asserted.

Next, we prove that $[G, A]$ is nilpotent of class less than r. Since $G_i \trianglelefteq G$ for $i = 0, 1, \ldots, r$, we have $[G_i, G] \leq G_i$. Hence,

$$[G_{i-1}, G, A] \leq [G_{i-1}, A] \leq G_i.$$

Furthermore,

$$[A, G_{i-1}, G] = [G_{i-1}, A, G] \leq [G_i, G] \leq G_i$$

by Proposition 1.1 (i). By assumption, $x^\alpha \in G_i$ for $x \in G_i$ and $\alpha \in A$. Thus, G_i is normal in the subgroup of $Hol(G)$ generated by A and G. Therefore,

$$[G, A, G_{i-1}] = [G_{i-1}, [G, A]] \le G_i$$

by Proposition 1.1 (i). By letting $K = [G, A]$ and $G_i = H_i$ in Theorem 7.10, we have

$$[G_1, \gamma_{r-1}[G, A]] \le G_r = 1.$$

However, $[G, A] = [G_0, A] \le G_1$. Therefore,

$$[[G, A], \gamma_{r-1}[G, A]] = 1,$$

and thus $\gamma_r[G, A] = 1$. This completes the proof of Theorem 7.9.

Remark 7.4 By repeating what was done in Example 6.1 with a free R-module of finite rank over a ring with unity, one can conclude from Theorem 7.9 that $UT_n(R)$ is a nilpotent group of class less than n. This was established in Chapter 1 using a different method (see Example 2.17).

The next theorem is a more general result due to P. Hall [5].

Theorem 7.11 *The stability group A relative to any series of subgroups of length $m \ge 1$ of a group G is nilpotent of class at most $m(m-1)/2$.*

We begin by proving Theorem 7.12, from which Theorem 7.11 will follow. Regard both $Aut(G)$ and G as subgroups of $Hol(G)$ and note that the stability group A relative to the series (7.8) also becomes a subgroup of $Hol(G)$. Thus, A is characterized by the property

$$[G_{i-1}, A] \le G_i \quad (i = 1, \cdots, m)$$

since $\alpha(g) = g^\alpha = \alpha^{-1}g\alpha$ in $Hol(G)$ for every $\alpha \in Aut(G)$. We have already encountered this property in the proof of Theorem 7.9.

Theorem 7.12 *Let H and K be two subgroups of a group G. If*

$$[H, \underbrace{K, \cdots, K}_{m}] = 1$$

for some $m \in \mathbb{N}$, then $[\gamma_{n+1}K, H] = 1$, where $n = m(m-1)/2$.

To obtain Theorem 7.11 from Theorem 7.12, observe that $[G_{i-1}, A] \le G_i$ implies that

$$[G, \underbrace{A, \cdots, A}_{m}] \le [G_1, \underbrace{A, \cdots, A}_{m-1}] \le \cdots \le [G_{m-1}, A] \le G_m = 1,$$

so that

$$[G, \underbrace{A, \cdots, A}_{m}] = 1.$$

Theorem 7.12 now implies that $[\gamma_{n+1}A, G] = 1$, where $n = m(m-1)/2$. However, this means that $\alpha(g) = g^{\alpha} = g$ for every $\alpha \in \gamma_{n+1}A$ and $g \in G$. Therefore, $\gamma_{n+1}A$ contains only the identity isomorphism and thus A is nilpotent of class at most n.

The proof of Theorem 7.12 relies on the next lemma.

Lemma 7.3 *Let H, K, and L be subgroups of a group G such that $G = gp(H, K)$. If $[H, K, L] = 1$, then $[L, H, K] = [K, L, H] \trianglelefteq G$.*

Proof Let C be the centralizer of $[H, K]$ in G. By Corollary 1.4, $[H, K] \trianglelefteq gp(H, K)$ and thus $[H, K] \trianglelefteq G$ by assumption. Since $L \leq C$ by hypothesis, a direct computation shows that $[L, H] \leq C$.

Let $x \in H$, $y \in K$, and $t \in [L, H]$. Set $z = [x, y^{-1}]$. Then $t \in C$ and $z \in [H, K]$, so that t and z commute. Since $y^x = zy$, Lemma 1.4 (vi) gives

$$[t, y^x] = [t, zy] = [t, y][t, z]^y = [t, y].$$

This means that $[L, H, K^x] = [L, H, K]$. Since $[L, H] \trianglelefteq gp(L, H)$ by Corollary 1.4 and $x \in H$, we also have $[L, H] = [L, H]^x$. Hence,

$$[L, H, K]^x = [[L, H]^x, K^x] = [L, H, K^x] = [L, H, K],$$

which means that H normalizes $[L, H, K]$. Since $[L, H, K] \trianglelefteq gp([L, H], K)$ by Corollary 1.4, K also normalizes $[L, H, K]$. Therefore, $[L, H, K]$ is normal in G.

By Proposition 1.1 (i), $[K, H, L] = [H, K, L] = 1$. Thus, we interchange the roles of H and K and conclude that $[L, K, H] = [K, L, H]$ is also normal in G. It follows that $[H, K, L] = 1$ and both $[L, H, K]$ and $[K, L, H]$ are normal in G. By Lemma 2.18, the subgroups $[L, H, K]$ and $[K, L, H]$ are contained in one another. The result now follows. $\square$

We now prove Theorem 7.12 by induction on m. If $m = 1$, then $[H, K] = 1$ and $n = 0$. Thus, $[K, H] = [H, K] = 1$ by Proposition 1.1 (i). This gives the basis of induction.

Let $m > 1$, and put $n = m(m-1)/2$. Let $H_1 = [H, K]$, and notice that

$$[H_1, \underbrace{K, \cdots, K}_{m-1}] = 1$$

since, by hypothesis, $[H, \underbrace{K, \cdots, K}_{m}] = 1$. Given that $[H_1, \underbrace{K, \cdots, K}_{m-1}] = 1$, we may assume inductively that

$$[\gamma_{l+1}K, H_1] = 1,$$

where $l = (m-1)(m-2)/2$. A direct calculation shows that $l = n - m + 1$, where $n = m(m-1)/2$. Let $r \geq l + 1$, and observe that

$$[\gamma_r K, H_1] \leq [\gamma_{l+1} K, \ H_1] = 1.$$

By Proposition 1.1 (i),

$$[H, \ K, \ \gamma_r K] = [H_1, \ \gamma_r K] = [\gamma_r K, \ H_1] = 1.$$

By Lemma 7.3 and Proposition 1.1 (i),

$$[\gamma_r K, \ H, \ K] = [K, \ \gamma_r K, \ H] = [\gamma_r K, \ K, \ H] = [\gamma_{r+1} K, \ H]$$

for $r > l$. Hence,

$$[\gamma_{l+1} K, H, \underbrace{K, \cdots, K}_{m-1}] = [\gamma_{l+2} K, H, \underbrace{K, \cdots, K}_{m-2}] = \cdots = [\gamma_n K, H, K] = [\gamma_{n+1} K, H].$$

Since $\gamma_{l+1} K \leq K$, it follows from Proposition 1.1 (i) and (iii) that

$$[\gamma_{l+1} K, \ H] \leq [K, \ H] = [H, \ K].$$

Thus,

$$[\gamma_{n+1} K, \ H] = [\gamma_{l+1} K, \ H, \ \underbrace{K, \ \cdots, \ K}_{m-1}] \leq [H, \ \underbrace{K, \ \cdots, \ K}_{m}].$$

However, $[H, \ \underbrace{K, \ \cdots, \ K}_{m}] = 1$. This completes the proof of Theorem 7.12.

7.4.2 The IA-Group of a Nilpotent Group

If G is a nilpotent group of class c, then we conclude from Corollary 1.1 and Lemma 2.12 that $Inn(G)$ is nilpotent of class $c - 1$. We turn our attention to a certain nilpotent subgroup of $Aut(G)$ that contains $Inn(G)$ and whose class is also $c - 1$. The material that appears in this section is based on [6] and Section 1.2 of [13].

Definition 7.4 An *IA-automorphism* of a group G is an automorphism of G that induces the identity on the abelianization of G.

Thus, α is an *IA*-automorphism of G if and only if α belongs to the kernel of the natural homomorphism

$$Aut(G) \rightarrow Aut(G/\gamma_2 G).$$

It is easy to show that the set of all IA-automorphisms of G is a subgroup of $Aut(G)$. This subgroup is called the IA-*group* of G and is denoted by $IA(G)$. Hence,

$$IA(G) = \{\alpha \in Aut(G) \mid g^{-1}\alpha(g) \in \gamma_2 G \text{ for all } g \in G\}.$$

Clearly, $IA(G) = 1$ whenever G is abelian. Furthermore, $IA(G)$ contains $Inn(G)$. For if $\varphi_h \in Inn(G)$, then

$$\varphi_h(g) = g^h = g[g, h]$$

for every $g \in G$. Since $[g, h] \in \gamma_2 G$, $\varphi_h \in IA(G)$.

Our ultimate goal in this section is to prove that the IA-group of a finitely generated torsion-free nilpotent group of class c is finitely generated torsion-free nilpotent of class $c - 1$. This is essentially a result due to P. Hall [6]. We begin by proving a lemma which is the main ingredient in establishing that the IA-group of a nilpotent group of class c is nilpotent of class $c - 1$.

Lemma 7.4 *Let H and K be subgroups of a group G such that $[H, K] \leq \gamma_2 H$. Then $\left[\gamma_i H, \gamma_j K\right] \leq \gamma_{i+j} H$ for all $i, j \in \mathbb{N}$.*

Proof We do "double induction." Let $j = 1$. We prove that $[\gamma_i H, K] \leq \gamma_{i+1} H$ by induction on i. If $i = 1$, then the result holds by hypothesis. Assume that $[\gamma_{i-1} H, K] \leq \gamma_i H$ for $i > 1$. By the induction hypothesis and Proposition 1.1 (iii),

$$[\gamma_{i-1} H, K, H] \leq [\gamma_i H, H] = \gamma_{i+1} H.$$

Furthermore,

$$[K, H, \gamma_{i-1} H] \leq [\gamma_2 H, \gamma_{i-1} H] \leq \gamma_{i+1} H$$

by hypothesis, Proposition 1.1 (i) and (iii), and Theorem 2.14 (i). Using Lemma 2.18, we have

$$[\gamma_i H, K] = [H, \gamma_{i-1} H, K] \leq \gamma_{i+1} H.$$

The basis of induction for j now follows.

The induction hypothesis for j is $\left[\gamma_i H, \gamma_{j-1} K\right] \leq \gamma_{i+j-1} H$ for all i. By Proposition 1.1 (i),

$$\left[\gamma_i H, \gamma_j K\right] = \left[\gamma_i H, \left[K, \gamma_{j-1} K\right]\right] = \left[K, \gamma_{j-1} K, \gamma_i H\right].$$

By Proposition 1.1 (iii) and the induction hypothesis on j, we have

$$\left[\gamma_i H, K, \gamma_{j-1} K\right] \leq \left[\gamma_{i+1} H, \gamma_{j-1} K\right] \leq \gamma_{i+j} H$$

and

$$\left[\gamma_{j-1}K,\ \gamma_iH,\ K\right] \le \left[\gamma_{i+j-1}H,\ K\right] \le \gamma_{i+j}H.$$

Invoking Lemma 2.18 once again, we conclude that $\left[\gamma_iH,\ \gamma_jK\right] \le \gamma_{i+j}H$. □

Theorem 7.13 *For all $j \in \mathbb{N}$ and any group G, the elements of $\gamma_j IA(G)$ induce the identity on each $\gamma_i G/\gamma_{i+j}G$. If G is nilpotent of class c, then $IA(G)$ is nilpotent of class $c - 1$.*

In particular, $IA(G)$ is abelian when G is nilpotent of class 2.

Proof To prove the first assertion, notice that if we choose $x \in G$ and $\psi \in IA(G)$, then

$$[x,\ \psi] = x^{-1}x^{\psi} \in \gamma_2 G$$

in $Hol(G)$. Hence, $[G,\ IA(G)] \le \gamma_2 G$, and by Lemma 7.4,

$$\left[\gamma_i G,\ \gamma_j IA(G)\right] \le \gamma_{i+j}G.$$

This means that the elements of $\gamma_j IA(G)$ induce the identity on $\gamma_i G/\gamma_{i+j}G$ as asserted.

Next, suppose that G is nilpotent of class c. We claim that $IA(G)$ is nilpotent of class $c - 1$. Let $\alpha \in \gamma_c IA(G)$. By the previous result, α induces the identity on $\gamma_1 G/\gamma_{c+1}G$. Since $\gamma_1 G/\gamma_{c+1}G \cong G$, α is the identity automorphism and thus $\gamma_c IA(G) = 1$. By Corollary 2.3, $IA(G)$ is nilpotent of class less than c. However,

$$IA(G) \ge Inn(G) \cong G/Z(G),$$

and $G/Z(G)$ is of class $c - 1$ by Lemma 2.12. Thus, $IA(G)$ is of class $c - 1$. □

Remark 7.5

(i) Theorem 7.13 implies that if G is a nilpotent group, then $IA(G)$ is the stability group of G relative to its lower central series.
(ii) While Theorem 7.11 already implies that $IA(G)$ is nilpotent whenever G is nilpotent, Theorem 7.13 guarantees that the class of $IA(G)$ is exactly one less than the class of G.

If G is a torsion-free nilpotent group, then so is $G/Z(G)$ by Corollary 2.22. Thus, $Inn(G)$ is also torsion-free nilpotent. The fact that every inner automorphism is an IA-automorphism suggests that $IA(G)$ may be torsion-free as well. This is indeed the case.

Lemma 7.5 *If G is a torsion-free nilpotent group, then $IA(G)$ is torsion-free.*

First, we prove an auxiliary result. Let $\overline{\gamma}_c G$ denote the isolator of $\gamma_c G$ in G.

Lemma 7.6 *If G is a torsion-free nilpotent group of class c, then $\overline{\gamma}_c G$ is a central subgroup of G and $G/\overline{\gamma}_c G$ is torsion-free.*

Proof We first prove that $\overline{\gamma}_c G$ is central in G. If $g \in \overline{\gamma}_c G$, then there exists $m \in \mathbb{N}$ such that $g^m \in \gamma_c G$. Since G is of class c, $\gamma_c G \leq Z(G)$, and thus $g^m \in Z(G)$. By Theorem 2.7 and Lemma 5.15, $g \in Z(G)$ as desired. To establish that $G/\overline{\gamma}_c G$ is torsion-free, observe that $\tau(G/\gamma_c G) = \overline{\gamma}_c G/\gamma_c G$ and thus

$$G/\overline{\gamma}_c G \cong (G/\gamma_c G)/\tau(G/\gamma_c G)$$

by the Third Isomorphism Theorem. The result immediately follows from Corollary 2.15. □

We now prove Lemma 7.5 by induction on the class c of G. If $c = 2$, then $IA(G)$ is abelian by Theorem 7.13. Let $\varphi \in IA(G)$ and $x \in G$. Then $\varphi(x) = xd$, where $d \in \gamma_2 G$. Suppose that $\varphi^m = 1$, where $m > 0$. Since G is of class 2, φ acts as the identity on $\gamma_2 G$ by Theorem 7.13. Thus,

$$\varphi^m(x) = xd^m = x,$$

and consequently, $d^m = 1$. Since $\gamma_2 G$ is torsion-free, $d = 1$. This completes the basis of induction.

Assume that the IA-group of a torsion-free nilpotent group of class less than c is always torsion-free. By Lemma 7.6, $G/\overline{\gamma}_c G$ is torsion-free nilpotent and clearly of class less than c. The induction hypothesis gives that $IA(G/\overline{\gamma}_c G)$ is torsion-free. To prove that $IA(G)$ is torsion-free, let $\varphi \in IA(G)$ and assume that $\varphi^m = 1$ where $m > 0$. For $g \in G$, let $[g]$ denote the equivalence class of g in $G/\overline{\gamma}_c G$. Consider the natural homomorphism

$$\pi : IA(G) \to IA\big(G/\overline{\gamma}_c G\big) \text{ defined by } \chi \mapsto \widehat{\chi}, \text{ where } \widehat{\chi}([g]) = [\chi(g)].$$

Since φ^m is the identity and $IA\big(G/\overline{\gamma}_c G\big)$ is torsion-free, $\widehat{\varphi}$ is the identity on $G/\overline{\gamma}_c G$. Let $x \in G$ and write $\varphi(x) = xd$, where $d \in \gamma_2 G$. Then

$$\widehat{\varphi}([x]) = [\varphi(x)] = [xd] = [x].$$

Hence, $d \in \overline{\gamma}_c G$. Since G is of class c, φ acts trivially on $\gamma_c G$ by Theorem 7.13. We claim that φ also acts trivially on $\overline{\gamma}_c G$. To see this, let $y \in \overline{\gamma}_c G$. There exists $m \in \mathbb{N}$ such that $y^m \in \gamma_c G$. Moreover, there exists $y_1 \in \gamma_2 G$ such that $\varphi(y) = yy_1$. Hence,

$$y^m = \varphi(y^m) = \varphi(y)^m = (yy_1)^m = y^m y_1^m$$

since y is central by Lemma 7.6. And so, $y_1^m = 1$. Since G is torsion-free, $y_1 = 1$, and thus φ acts trivially on $\overline{\gamma}_c G$ as claimed. This, together with $\varphi(x) = xd$ and $\varphi^m = 1$, gives

$$\varphi^m(x) = xd^m = x.$$

Therefore, $d^m = 1$, and consequently, $d = 1$. This completes the proof of Lemma 7.5.

Lemma 7.7 *If G is a finitely generated nilpotent group, then IA(G) is finitely generated.*

Proof The proof is done by induction on the class c of G. If $c = 2$, then $IA(G)$ is abelian by Theorem 7.13. Since G is finitely generated, so is $\gamma_2 G$ by Theorem 2.18. A typical member of a generating set for $IA(G)$ can be constructed as follows: let X be a finite set of generators for G and Y a finite set of generators for $\gamma_2 G$. For every $x \in X$ and $y \in Y$, construct the IA-automorphism that sends x to xy and each remaining generator of G to itself. It is clear that the set of all such IA-automorphisms generates $IA(G)$. Since X and Y are finite, $IA(G)$ is finitely generated.

Assume that the IA-group of any finitely generated nilpotent group of class less than c is finitely generated. Consider the natural homomorphism

$$\pi : IA(G) \to IA(G/\gamma_c G) \text{ defined by } \chi \mapsto \widehat{\chi}, \text{ where } \widehat{\chi}([g]) = [\chi(g)]$$

and $[g]$ now denotes the equivalence class of g in $G/\gamma_c G$. The kernel of π is the subgroup

$$I_c = \left\{ \alpha \in IA(G) \;\middle|\; g^{-1}\alpha(g) \in \gamma_c G \text{ for all } g \in G \right\}$$

of $IA(G)$, and it is finitely generated. This can be established by an analogous construction as before, where Y is taken to be a finite generating set for $\gamma_c G$. By the induction hypothesis, $IA(G/\gamma_c G)$ is finitely generated. Thus, the image of π, being a subgroup of a finitely generated nilpotent group, is also finitely generated by Theorem 2.18, as well as isomorphic to $IA(G)/I_c$. Since $IA(G)$ is an extension of a finitely generated group by another, it must be finitely generated. □

Theorem 7.13, together with Lemmas 7.5 and 7.7, gives:

Theorem 7.14 *If G is a finitely generated torsion-free nilpotent group of class c, then IA(G) is finitely generated, torsion-free nilpotent of class c − 1.*

Using basic sequences and the commutator calculus, M. Zyman [13] has shown that if G is a finitely generated nilpotent group such that $\gamma_2 G$ is abelian, then $IA(G_p) \cong \big(IA(G)\big)_p$, where p is any prime and G_p and $\big(IA(G)\big)_p$ denote the p-localizations of G and $IA(G)$ respectively. A group G for which $\gamma_2 G$ is abelian is termed *metabelian*.

7.5   The Frattini and Fitting Subgroups

Two subgroups which provide useful information about the structure of a group are the Frattini and Fitting subgroups. In this section, we give a brief overview of the properties of these subgroups and describe their connection to nilpotent groups.

7.5.1   The Frattini Subgroup

Definition 7.5 Let G be any group. The *Frattini subgroup* of G, denoted by $\Phi(G)$, is the intersection of all of the maximal subgroups of G.

By convention, $\Phi(G) = G$ if G has no maximal subgroups. Thus, $\Phi(\mathbb{Q}) = \mathbb{Q}$ and $\Phi(\mathbb{Z}_{p^\infty}) = \mathbb{Z}_{p^\infty}$. Clearly, the Frattini subgroup of a group is characteristic since every group automorphism maps maximal subgroups to maximal subgroups.

Example 7.1 We give the Frattini subgroup of certain groups.

1. For each prime p, the subgroup $gp(p)$ of $\mathbb{Z}$ is maximal. Thus, $\Phi(\mathbb{Z}) = \{0\}$.
2. If p is a prime and $G = gp(g)$ is a cyclic group of order p^2, then $\Phi(G) = gp(g^p)$.
3. $\Phi(S_3) = \{e\}$. To see this, notice that the distinct subgroups

$$H = gp((1\ \ 2))\ \text{ and }\ K = gp((1\ \ 2\ \ 3))$$

   of S_3 each have prime index in S_3. Thus, H and K are maximal subgroups. Therefore, $\Phi(S_3)$ is a subgroup of $H \cap K = \{e\}$.
4. This example appears in [8]. Let p be a prime and let U_{ip} be the subgroup of $UT_n(\mathbb{Z})$ consisting of all matrices $A = (a_{ij})$ whose superdiagonal entries $a_{i,\ i+1}$ are contained in $\langle p \rangle$ for $i = 1,\ 2,\ \ldots,\ n-1$. Then U_{ip} is maximal in $UT_n(\mathbb{Z})$. Since the intersection of all U_{ip} lies in $UT_n^2(\mathbb{Z})$, we have $\Phi(UT_n(\mathbb{Z})) \leq UT_n^2(\mathbb{Z})$.

Another way to define the Frattini subgroup of a group is in terms of its set of non-generators.

Definition 7.6 Let G be a group. An element $g \in G$ is called a *non-generator* of G if $G = gp(g, X)$ implies that $G = gp(X)$ whenever $X \subset G$.

Thus, the set of non-generators of a group are precisely the elements that can be excluded from any generating set.

Theorem 7.15 (G. Frattini) *If G is any group, then $\Phi(G)$ is the set of all non-generating elements of G.*

The proof uses the next lemma.

Lemma 7.8 *Let G be any group and suppose that $H < G$ with $g \in G \smallsetminus H$. There exists a subgroup $K < G$ which is maximal in G with respect to the properties $H \leq K$ and $g \notin K$.*

Proof Let $R = \{J < G \mid H \leq J \text{ and } g \notin J\}$. Clearly, $R \neq \emptyset$ since $H \in R$. Furthermore, R is partially ordered by inclusion and the union of any chain in R is again in R. By Zorn's lemma, R has a maximal element.   □

We now prove Theorem 7.15. Let $g \in \Phi(G)$. We prove by contradiction that g is a non-generating element. Assume that there exists $X \subset G$ such that $G = gp(g, X)$, but $G \neq gp(X)$. Clearly, $g \notin gp(X)$. By Lemma 7.8, there exists a subgroup M of G which is maximal in G with respect to the properties $gp(X) \leq M$ and $g \notin M$. If $M < H \leq G$ for some H, then $g \in H$. Consequently, $H = G$. Hence, M is maximal in G. This implies that $g \in \Phi(G) \leq M$, which is a contradiction.

Conversely, let g be a non-generator of G. We prove by contradiction that $g \in \Phi(G)$. Suppose on the contrary, that $g \notin \Phi(G)$. There exists a maximal subgroup M of G for which $g \notin M$. Hence, $M \neq gp(g, M)$, and thus $G = gp(g, M)$ by the maximality of M in G. Since g is a non-generator of G, we have $G = M$, a contradiction. This completes the proof of Theorem 7.15.

The next two corollaries are immediate.

Corollary 7.5 *If G is a finitely generated group and $G = H\Phi(G)$ for some $H \leq G$, then $G = H$.*

Corollary 7.6 *If G is any group and $\Phi(G)$ is finitely generated, then the only subgroup H of G such that $G = H\Phi(G)$ is $H = G$.*

Lemma 7.9 *Let G be any group and suppose that H is a finitely generated subgroup of G. If $N \triangleleft G$ and $N \leq \Phi(H)$, then $N \leq \Phi(G)$.*

Proof Assume on the contrary, that N is not a subgroup of $\Phi(G)$. There exists a maximal subgroup M of G that does not contain N as a subgroup and thus, satisfies $G = MN$. Hence,

$$H = H \cap G = H \cap (MN) = (H \cap M)N.$$

Since $N \leq \Phi(H)$, we have $H = (H \cap M)\Phi(H)$. By Corollary 7.5, $H = H \cap M$, and thus $H \leq M$. However, H contains N. And so, $N \leq M$, a contradiction.   □

Corollary 7.7 *If G is any group and H is a finitely generated normal subgroup of G, then $\Phi(H) \leq \Phi(G)$.*

Proof By Lemma 1.8, $\Phi(H) \trianglelefteq G$ because $\Phi(H)$ is characteristic in H and $H \trianglelefteq G$. Put $N = \Phi(H)$ in Lemma 7.9.   □

The next two lemmas deal with the Frattini subgroup of a direct product. We give the proofs which appear in [3].

Lemma 7.10 *If $G = H \times K$, then $\Phi(G) \leq \Phi(H) \times \Phi(K)$.*

Proof If M is a maximal subgroup of H, then $M \times K$ is a maximal subgroup of $H \times K$. Hence, $\Phi(G) \leq \Phi(H) \times K$. Similarly, $\Phi(G) \leq H \times \Phi(K)$. Therefore,

$$\Phi(G) \leq (\Phi(H) \times K) \cap (H \times \Phi(K)) = \Phi(H) \times \Phi(K),$$

as desired.   □

Lemma 7.11 *If G is a finitely generated group where* $G = H \times K$ *for some subgroups H and K of G, then* $\Phi(G) = \Phi(H) \times \Phi(K)$.

Proof Clearly, H and K are finitely generated and normal in G since G is finitely generated and $G = H \times K$. Thus, $\Phi(H) \leq \Phi(G)$ and $\Phi(K) \leq \Phi(G)$ by Corollary 7.7. It follows that $\Phi(G) \geq \Phi(H) \times \Phi(K)$. Lemma 7.10 takes care of the reverse inclusion. □

Lemma 7.12 *Let G be any group and* $N \leq G$. *If* $N \leq \Phi(G)$, *then N is normal in G and* $\Phi(G/N) = \Phi(G)/N$.

Proof Since $\Phi(G)$ is characteristic in G and $N \leq \Phi(G)$, N must be normal in G. Furthermore, $N \leq \Phi(G)$ implies that N is contained in every maximal subgroup of G. If $gN \in \Phi(G/N)$, then $gN \in M/N$ for every maximal subgroup M/N in G/N. Thus, for each maximal subgroup M in G, there exists $h_m \in M$ such that $gN = h_mN$; that is, $gh_m^{-1} \in N$. It follows that $gN \in \Phi(G)/N$. We reverse the above steps to conclude that $\Phi(G)/N \leq \Phi(G/N)$. □

Theorem 7.16 *Let G be a finite group. If H is a normal subgroup of G containing* $\Phi(G)$ *and* $H/\Phi(G)$ *is nilpotent, then H is nilpotent.*

In particular, if G is finite and $G/\Phi(G)$ is nilpotent, then G is nilpotent.

Proof In light of Theorem 2.13, it is enough to prove that the Sylow subgroups of H are normal. Let P be a Sylow p-subgroup of H. Since $\Phi(G)$ is contained in H by hypothesis, $\Phi(G) \trianglelefteq H$. By Lemma 2.16 (ii), $P\Phi(G)/\Phi(G)$ is a Sylow p-subgroup of $H/\Phi(G)$. Since $H/\Phi(G)$ is nilpotent, its Sylow p-subgroups are normal by Theorem 2.13. Thus, $P\Phi(G)/\Phi(G) \trianglelefteq H/\Phi(G)$. Furthermore, $P\Phi(G)/\Phi(G)$ is characteristic in $H/\Phi(G)$. This is due to the fact that normal Sylow p-subgroups are unique by Corollary 2.8 and every automorphism maps a subgroup of a given order into a subgroup of the same order. Since $H/\Phi(G) \trianglelefteq G/\Phi(G)$, it follows that $P\Phi(G)/\Phi(G) \trianglelefteq G/\Phi(G)$, and thus $P\Phi(G) \trianglelefteq G$.

Now, P is a Sylow p-subgroup of $P\Phi(G)$ because it is a Sylow p-subgroup of H. Since $P\Phi(G) \trianglelefteq G$, Lemma 2.17 gives $N_G(P)P\Phi(G) = G$. However, $P \leq N_G(P)$, and thus $N_G(P)\Phi(G) = G$. By Theorem 7.15, $N_G(P) = G$. Therefore, $P \trianglelefteq G$, and consequently, $P \trianglelefteq H$. □

Setting $H = \Phi(G)$ in Theorem 7.16 gives:

Corollary 7.8 (G. Frattini) *If G is a finite group, then* $\Phi(G)$ *is nilpotent.*

Remark 7.6 If G is infinite, then $\Phi(G)$ need not be nilpotent. Consider, for instance, the wreath product $G = \mathbb{Z}_{p^\infty} \wr \mathbb{Z}_{p^\infty}$. By Remark 2.8, G is an infinite non-nilpotent group. Since G has no maximal subgroups, it coincides with its Frattini subgroup as we noted after Definition 7.5. Thus, $\Phi(G)$ is not nilpotent.

Theorem 7.17 (W. Gaschütz) *If G is a finite group, then* $\gamma_2 G \cap Z(G) \leq \Phi(G)$.

Proof It is enough to prove that $\gamma_2 G \cap Z(G)$ is contained in each maximal subgroup of G. If M is a maximal subgroup of G, then either $Z(G) \leq M$ or $G = MZ(G)$. If it is the case that $Z(G) \leq M$, then it is clear that $\gamma_2 G \cap Z(G) \leq M$. If, on the other hand, $G = MZ(G)$, then $M \trianglelefteq G$ and G/M is abelian. By Lemma 1.6, $\gamma_2 G \leq M$. Hence, $\gamma_2 G \cap Z(G) \leq M$.   □

We now turn our attention to the Frattini subgroup of nilpotent groups. We prove an important result which states that the commutator subgroup of a nilpotent group G is a set of non-generating elements of G.

Theorem 7.18 (K. A. Hirsch) *Let G be a nilpotent group. If $H \leq G$ and $H\gamma_2 G = G$, then $H = G$.*

Proof The proof is done by induction on the class c of G. If $c = 1$, then $\gamma_2 G = 1$ and the result is immediate.

Suppose that the result holds for all nilpotent groups of class less than c, and consider the natural homomorphism $\pi : G \to G/\gamma_c G$. By Lemma 2.8, $G/\gamma_c G$ is nilpotent of class $c - 1$. If $H\gamma_2 G = G$, then $\pi(H)\pi(\gamma_2 G) = \pi(G)$ holds in $G/\gamma_c G$. By Lemma 2.5, $\pi(\gamma_2 G) = \gamma_2 \pi(G)$. Thus, $\pi(H) = \pi(G) = G/\gamma_c(G)$ by induction; that is, $H\gamma_c G = G$. Using the commutator calculus and the fact that $\gamma_c G \leq Z(G)$, we have

$$\gamma_2 G = [H\gamma_c G, \ H\gamma_c G]$$

$$= [H, \ H][H, \ \gamma_c G][\gamma_c G, \ H][\gamma_c G, \ \gamma_c G]$$

$$= [H, \ H] = \gamma_2 H.$$

Therefore, $\gamma_2 G = \gamma_2 H \leq H$, and thus $H = G$.   □

Corollary 7.9 *If G is a nilpotent group, then $\gamma_2 G \leq \Phi(G)$.*

Proof This is immediate from Theorems 7.15 and 7.18.   □

Remark 7.7 By Theorem 7.6, $\gamma_2 UT_n(\mathbb{Z}) = UT_n^2(\mathbb{Z})$. Thus, $\Phi(UT_n(\mathbb{Z})) = UT_n^2(\mathbb{Z})$ by Example 7.1 and Corollary 7.9.

Corollary 7.10 *Let G be a nilpotent group and $X \subseteq G$. If $\pi : G \to Ab(G)$ is the natural homomorphism, then $G = gp(X)$ if and only if $Ab(G) = gp(\pi(X))$.*

Proof Suppose that $Ab(G) = gp(\pi(X))$, and let $H = gp(X)$. Then $G = H\gamma_2 G$, and thus $G = H = gp(X)$ by Theorem 7.18. The converse is obvious.   □

Corollary 7.11 *Let G and H be nilpotent groups. If $\varphi : G \to H$ is a homomorphism, then φ is surjective if and only if $\overline{\varphi} : Ab(G) \to Ab(H)$ is surjective.*

Proof Assume that $\overline{\varphi}$ is surjective and let $h \in H$. There exists an element $g\gamma_2 G$ in $Ab(G)$ such that

$$\overline{\varphi}(g\gamma_2 G) = \varphi(g)\gamma_2 H = h\gamma_2 H.$$

Hence, $h \in \varphi(G)\gamma_2 H$, and thus $H = \varphi(G)\gamma_2 H$. By Theorem 7.18, $H = \varphi(G)$. The converse is clear.   □

For finite groups, the converse of Corollary 7.9 holds.

Theorem 7.19 (H. Wielandt) *If G is a finite group and $\gamma_2 G \leq \Phi(G)$, then G is nilpotent.*

Proof If $\gamma_2 G \leq \Phi(G)$, then $\gamma_2 G \leq M$ for every maximal subgroup M of G. By Lemma 1.6, each M is normal in G. Thus, G is nilpotent by Theorem 2.13.   □

For any prime p, the quotient of a finite p-group by its Frattini subgroup has a nice structure. It is always abelian and each of its elements has order p. This motivates the next definition.

Definition 7.7 Let p be a prime. A group G is called an *elementary abelian p-group* if it is abelian and every element of G has order p.

Clearly, every finite elementary abelian p-group is isomorphic to a direct product of n copies of $\mathbb{Z}_p$ for some $n < \infty$. Moreover, the Frattini subgroup of such a group is always trivial.

Lemma 7.13 *If G is a finite elementary abelian p-group, then $\Phi(G) = 1$.*

Proof Suppose that G is a direct product of n copies of $\mathbb{Z}_p$. The subgroup

$$G_i = \big\{ (g_1, \ldots, g_{i-1}, 1, g_{i+1}, \ldots, g_n) \mid g_j \in \mathbb{Z}_p \big\}$$

is maximal in G, and $\cap_{i=1}^{n} G_i = 1$. Hence, $\Phi(G) \leq 1$.   □

Lemma 7.14 *If G is a finite p-group and $H \leq G$, then $\Phi(G) \leq H$ if and only if $H \trianglelefteq G$ and G/H is an elementary abelian p-group.*

Proof Suppose that $H \trianglelefteq G$ and G/H is an elementary abelian p-group. Then $\Phi(G/H) = H$ by Lemma 7.13. Thus, $\Phi(G) \leq H$ by Lemma 7.12.

Conversely, suppose that $\Phi(G) \leq H$. By Corollary 7.9, $\gamma_2 G \leq \Phi(G)$. Thus, $H \trianglelefteq G$ and G/H is abelian by Lemma 1.6. Suppose that $g \in G$, and let M be any maximal proper subgroup of G. By Theorems 2.3 and 2.13 (iv), $M \triangleleft G$. Furthermore, $(gM)^p = M$ because $|G/M| = p$. Hence, $g^p \in M$, and thus g^p lies in all maximal proper subgroups of G. Therefore, $g^p \in \Phi(G)$ and G/H is an elementary abelian p-group since $\Phi(G) \leq H$.   □

Setting $H = \Phi(G)$ in Lemma 7.14 proves our earlier remark:

Lemma 7.15 *If G is a finite p-group, then $G/\Phi(G)$ is an elementary abelian p-group.*

Lemma 7.16 *If G is a finite p-group, then*

$$\Phi(G) = G^p \gamma_2 G = gp(a^p, [b, c] \mid a, b, c \in G).$$

Hence, $N = \Phi(G)$ is the smallest normal subgroup of a finite p-group G such that G/N is an elementary abelian p-group.

Proof Since G is a finite p-group, $G/\Phi(G)$ is elementary abelian by Lemma 7.15. It follows from this and Lemma 1.6 (ii) that $g^p \in \Phi(G)$ for all $g \in G$, and $\gamma_2 G \le \Phi(G)$. Therefore, $G^p \gamma_2 G \le \Phi(G)$.

We establish the reverse inclusion. By Lemma 1.6 (i), $G/G^p \gamma_2 G$ is abelian. Since $G^p \le G^p \gamma_2 G$, the factor group $G/G^p \gamma_2 G$ is a finite elementary abelian p-group. By the previous observation, $G^p \gamma_2 G \le \Phi(G)$. Thus,

$$\Phi(G/G^p \gamma_2 G) = \Phi(G)/G^p \gamma_2 G = G^p \gamma_2 G$$

by Lemmas 7.12 and 7.13. And so, $G^p \gamma_2 G \ge \Phi(G)$. $\qquad\qquad\square$

By Lemma 7.15, the quotient $G/\Phi(G)$ can be viewed as a vector space over $\mathbb{F}_p$, the finite field containing p elements. Furthermore, any set of group generators for $G/\Phi(G)$ is also a spanning set of $G/\Phi(G)$ as a vector space over $\mathbb{F}_p$.

Theorem 7.20 (Burnside's Basis Theorem) *Let G be a finite p-group. Consider $\overline{G} = G/\Phi(G)$ to be a vector space over $\mathbb{F}_p$, and suppose that $[G : \Phi(G)] = p^d$.*

(i) *The dimension of $\overline{G}$ over $\mathbb{F}_p$ is d.*
(ii) *If $G = gp(g_1, \ldots, g_k)$, then $k \ge d$. More generally, $G = gp(g_1, \ldots, g_k)$ if and only if $\overline{G} = span\{g_1\Phi(G), \ldots, g_k\Phi(G)\}$.*
(iii) *G can be generated by exactly d elements. Furthermore, the set $\{g_1, \ldots, g_d\}$ generates G if and only if the set $\{g_1\Phi(G), \ldots, g_d\Phi(G)\}$ is a basis for $\overline{G}$ over $\mathbb{F}_p$.*

Proof

(i) The dimension of a vector space over $\mathbb{F}_p$ is d if and only if it contains precisely p^d elements.
(ii) Since $G = gp(g_1, \ldots, g_k)$, the vectors $g_1\Phi(G), \ldots, g_k\Phi(G)$ span $\overline{G}$. By (i), the dimension of $\overline{G}$ over $\mathbb{F}_p$ is d. And so, $k \ge d$.
   If $\overline{G} = span\{g_1\Phi(G), \ldots, g_k\Phi(G)\}$, then $G = gp(g_1, \ldots, g_k, \Phi(G))$. By Theorem 7.15, $G = gp(g_1, \ldots, g_k)$.
(iii) If $G = gp(g_1, \ldots, g_d)$, then $\overline{G}$ is spanned by $g_1\Phi(G), \ldots, g_d\Phi(G)$ by (ii). Since $G/\Phi(G)$ has dimension d by (i), the vectors $g_1\Phi(G), \ldots, g_d\Phi(G)$ are linearly independent. Thus, $\{g_1\Phi(G), \ldots, g_d\Phi(G)\}$ is a basis for $\overline{G}$. The converse follows from (ii). $\qquad\square$

Lemma 7.17 *If G is a finite nilpotent group, then $G/\Phi(G)$ is a direct product of elementary abelian p-groups.*

Proof Let P_i denote the Sylow p_i-subgroup of G for $i = 1, \ldots, k$. Since G is finite and nilpotent, $G \cong P_1 \times \cdots \times P_k$ by Theorem 2.13. Thus,

$$\frac{G}{\Phi(G)} \cong \frac{P_1 \times \cdots \times P_k}{\Phi(P_1 \times \cdots \times P_k)} = \frac{P_1}{\Phi(P_1)} \times \cdots \times \frac{P_k}{\Phi(P_k)},$$

where the last equality follows from Lemma 7.11. Since each P_i is a finite p_i-group, each factor group $P_i/\Phi(P_i)$ is an elementary abelian p_i-group by Lemma 7.15. $\quad\square$

7.5.2 The Fitting Subgroup

Definition 7.8 The *Fitting subgroup* of a group G, denoted by $Fit(G)$, is the subgroup of G generated by all of the normal nilpotent subgroups of G.

Clearly, $Fit(G)$ is a characteristic subgroup of G, and thus $Fit(G) \trianglelefteq G$.

Lemma 7.18 *If G is a finite group, then $Fit(G)$ is the nilpotent radical of G.*

Proof The result follows at once from Theorem 2.11. $\square$

Remark 7.8 If G is an infinite group, then $Fit(G)$ is not necessarily nilpotent. For example, choose any prime p, and let A be a countably infinite elementary abelian p-group. One can show that the group $G = \mathbb{Z}_p \wr A$ is not nilpotent and $G = Fit(G)$. See pp. 3–4 in [10] for details.

Lemma 7.19 *If G is a nontrivial finite group, then the centralizer of $Fit(G)$ in G contains every minimal normal subgroup of G.*

Proof Set $F = Fit(G)$, and let N be a minimal normal subgroup of G. There are two cases to consider.

- If N is not a subgroup of F, then $F \cap N = 1$ because $F \cap N$ is a proper subgroup of N and $F \cap N \trianglelefteq G$. By Theorem 1.4, $[F, N] = 1$. And so, $N \leq C_G(F)$.
- If $N \leq F$, then there is a minimal normal subgroup M of F with $M \leq N$. By Theorem 2.29 and Lemma 7.18, we have $M \leq Z(F)$. Hence, $Z(F) \cap N \neq 1$. However, $Z(F) \trianglelefteq G$, and thus $Z(F) \cap N \trianglelefteq G$. Consequently, $Z(F) \cap N = N$ because N is a minimal normal subgroup of G. Therefore, $N \leq Z(F) \leq C_G(F)$. $\square$

Another description of $Fit(G)$ for a finite group G is in terms of its chief factors.

Theorem 7.21 *If G is a finite group, then $Fit(G)$ is the intersection of the centralizers of the chief factors of G.*

Proof We adopt the proof from [12]. Set $F = Fit(G)$, and let

$$1 = G_0 \leq G_1 \leq \cdots \leq G_n = G$$

be a chief series of G. Set

$$I = \bigcap_{i=0}^{n-1} C_G(G_{i+1}/G_i).$$

(The notation used above can be found in Definition 5.6.) It is clear that $I \trianglelefteq G$. Furthermore, $[G_{i+1}, I] \leq G_i$ for each $i = 0, 1, \ldots, n-1$. Consequently, I is nilpotent and thus, $I \leq F$.

To show that $F \leq I$, we prove that $F \leq C_G(G_{i+1}/G_i)$ for each $i = 0, 1, \ldots, n-1$. Consider the factor group $FG_i/G_i \leq G/G_i$. Clearly, FG_i/G_i is normal in G/G_i and $F \cap G_i \trianglelefteq F$. By the Second Isomorphism Theorem, $FG_i/G_i \cong F/(F \cap G_i)$.

Since F is nilpotent by Lemma 7.18, FG_i/G_i is a normal nilpotent subgroup of G/G_i by Corollary 2.5. Hence, $FG_i/G_i \leq Fit(G/G_i)$. Now, G_{i+1}/G_i is a minimal normal subgroup of G/G_i by Lemma 2.23. According to Lemma 7.19, it follows that $Fit(G/G_i)$ must centralize G_{i+1}/G_i. This means that

$$FG_i/G_i \leq C_{G/G_i}(G_{i+1}/G_i),$$

or equivalently, $F \leq C_G(G_{i+1}/G_i)$. □

We record a couple of properties of the Fitting subgroup of a finite group.

Proposition 7.2 *Let G be a finite group.*

(i) If $H \trianglelefteq G$, then $H \cap Fit(G) = Fit(H)$.
(ii) If $K \leq Z(G)$, then $Fit(G/K) = Fit(G)/K$.

Proof

(i) By Lemma 1.8, $Fit(H)$ is a normal subgroup of G because it is characteristic in H. Since $Fit(H)$ is also nilpotent by Lemma 7.18, $Fit(H) \leq H \cap Fit(G)$. Furthermore, $H \cap Fit(G) \trianglelefteq H$ because $Fit(G) \trianglelefteq G$. Since $H \cap Fit(G)$ is nilpotent, we have $H \cap Fit(G) \trianglelefteq Fit(H)$. Hence, $H \cap Fit(G) = Fit(H)$.

(ii) Set $M/K = Fit(G/K)$. We claim that $M = Fit(G)$. By Lemma 7.18, M/K is nilpotent. Thus, M is nilpotent by Theorem 2.6 because K is central in M. Moreover, $M \trianglelefteq G$ since M/K is characteristic, hence normal, in G/K. Thus, $M \leq Fit(G)$. Now, since $Fit(G)$ is nilpotent and $K \leq Fit(G)$, it must be the case that $Fit(G)/K$ is nilpotent. Since $Fit(G)/K$ is normal in G/K, we have

$$Fit(G)/K \leq Fit(G/K) = M/K.$$

Consequently, $Fit(G) \leq M$, and thus $M = Fit(G)$. □

There are natural connections between the Fitting and Frattini subgroups. We state two of them in our last theorem.

Theorem 7.22 (W. Gaschütz) *If G is a finite group, then $\Phi(G) \trianglelefteq Fit(G)$ and $Fit(G/\Phi(G)) = Fit(G)/\Phi(G)$.*

Proof The fact that $\Phi(G) \trianglelefteq Fit(G)$ is immediate from Corollary 7.8. We prove the second assertion. Clearly, $Fit(G)/\Phi(G) \leq Fit(G/\Phi(G))$ since $Fit(G)/\Phi(G)$ is a normal nilpotent subgroup of $G/\Phi(G)$. Suppose that $H/\Phi(G)$ is a normal nilpotent subgroup of $G/\Phi(G)$. Then H is a normal nilpotent subgroup of G by Theorem 7.16. Consequently, $H \leq Fit(G)$, and thus $Fit(G/\Phi(G)) \leq Fit(G)/\Phi(G)$. Therefore, $Fit(G/\Phi(G)) = Fit(G)/\Phi(G)$. □

References

1. G. Baumslag, *Lecture Notes on Nilpotent Groups*. Regional Conference Series in Mathematics, No. 2 (American Mathematical Society, Providence, RI, 1971). MR0283082
2. N. Blackburn, Conjugacy in nilpotent groups. Proc. Am. Math. Soc. **16**, 143–148 (1965). MR0172925
3. J.D. Dixon, *Problems in Group Theory* (Dover, New York, 1973)
4. F. Grunewald, D. Segal, Some general algorithms. II. Nilpotent groups. Ann. Math. (2) **112**(3), 585–617 (1980). MR0595207
5. P. Hall, Some sufficient conditions for a group to be nilpotent. Ill. J. Math. **2**, 787–801 (1958). MR0105441
6. P. Hall, *The Edmonton Notes on Nilpotent Groups*. Queen Mary College Mathematics Notes. Mathematics Department (Queen Mary College, London, 1969). MR0283083
7. L. Kalužnin, Über gewisse Beziehungen zwischen einer Gruppe und ihren Automorphismen. Berliner Mathematische Tagung, pp. 164–172 (1953)
8. M.I. Kargapolov, J.I. Merzljakov, *Fundamentals of the Theory of Groups*. Graduate Texts in Mathematics, vol. 62 (Springer, New York, 1979). MR0551207. Translated from the second Russian edition by Robert G. Burns
9. R.C. Lyndon, P.E. Schupp, *Combinatorial Group Theory* (Springer, Berlin, 1977). MR0577064
10. D.J.S. Robinson, *Finiteness Conditions and Generalized Soluble Groups. Part 2* (Springer, New York, 1972). MR0332990
11. D.J.S. Robinson, *A Course in the Theory of Groups*. Graduate Texts in Mathematics, vol. 80, 2nd edn. (Springer, New York, 1996). MR1357169
12. J.S. Rose, *A Course on Group Theory* (Dover Publications, New York, 1994). MR1298629. Reprint of the 1978 original [Dover, New York; MR0498810 (58 #16847)]
13. M. Zyman, *IA-Automorphisms and Localization of Nilpotent Groups* (ProQuest LLC, Ann Arbor, MI, 2007). MR2711065

Index

© Springer International Publishing AG 2017
A.E. Clement et al., *The Theory of Nilpotent Groups*,
DOI 10.1007/978-3-319-66213-8

Printed in the United States
By Bookmasters